OPTIMIZATION OF COMPUTER NETWORKS – MODELING AND ALGORITHMS

OPTIMIZATION OF COMPUTER NETWORKS – MODELING AND ALGORITHMS

A HANDS-ON APPROACH

Pablo Pavón Mariño

WILEY

Library of Congress Cataloging-in-Publication Data

Names: Marino, Pablo Pavon, author.
Title: Optimization of computer networks : modeling and algorithms : a
    hands-on approach / Pablo Pavon Marino.
Description: Chichester, West Sussex, United Kingdom : John Wiley & Sons,
    Inc., [2016] | Includes bibliographical references and index.
Identifiers: LCCN 2015044522 (print) | LCCN 2016000694 (ebook) | ISBN
    9781119013358 (cloth) | ISBN 9781119013334 (ePub) | ISBN 9781119013341
    (Adobe PDF)
Subjects: LCSH: Network performance (Telecommunication)–Mathematical models.
    | Computer networks–Mathematical models. | Computer algorithms.
Classification: LCC TK5102.83 .M37 2016 (print) | LCC TK5102.83 (ebook) | DDC
    004.601–dc23
LC record available at http://lccn.loc.gov/2015044522

A catalogue record for this book is available from the British Library.

1   2016

*To my sons, Pablo and Guille, and to my wife Victoria,
the smiles of my life.*

Contents

About the Author

Pablo Pavón Mariño is Associate Professor at the Universidad Politécnica de Cartagena (Spain) and Head of GIRTEL research group, MSc and Ph.D in Telecommunications, and MSc in Mathematics, with specialization in operations research. His research interests in the last 15 years are in optimization, planning, and performance evaluation of computer networks. He has more than a decade track as a lecturer in network optimization courses. He is author or co-author of more than 100 research papers in the field, published in top journals and international conferences, as well as several patents. He leads the Net2Plan open-source initiative, which includes the Net2Plan tool and its associated public repository of algorithms and network optimization resources (www.net2plan.com). Pablo Pavón has served as chair in international conferences like IEEE HPSR 2011, ICTON 2013 or ONDM 2016. He is Technical Editor of the *Optical Switching and Networking* journal, and has participated as Guest Editor in other journals such as *Computer Networks*, *Photonic Network Communications*, and *IEEE/OSA Journal of Optical Communications and Networking*.

Preface

Computer and communication networks have evolved into more and more complex structures of heterogeneous technologies with multiple interactions between different protocols and layers. From a didactic point of view, it is challenging to show newcomers how things are, and more importantly, why they are like that.

This is the task I faced in 2011 when starting the preparation of two basic courses in network theory fundamentals at the Universidad Politécnica de Cartagena in Spain for second and third year students doing telecommunications engineering degrees. My wish list included three musts: keep it simple, provide technology-agnostic fundamentals enriched with application examples, and make it practical.

Keep it Simple

There is added value in simplicity *per se*. In this case, it was also a constraint given the still incipient mathematical skills of the undergraduate students targeted. In plain words, there was no room to cover the different mathematical disciplines traditionally used in network courses, mainly queuing theory, control theory, and game theory for analysis of network protocols and the network as a distributed dynamic system, and optimization for static provisioning and dimensioning.

Within this *less is more* philosophy, I got convinced that optimization was the most convenient choice for a sort of didactic *theory to rule them all*[1]:

- Optimization is a popular approach in network provisioning and dimensioning problems, and fits well when the network is seen as a static system.
- The work initiated in the 1990s extended the application of optimization to capture the *macroscopic* dynamic behavior of network protocols. This methodology is complementary to queueing theory or a stochastic network characterization (e.g., using Markovian analysis): these are left for studying the fast timescale interactions, and their main average and system equilibrium results are then introduced and exploited in macroscopic network optimization models. Gradient projection algorithms have a prominent role in this framework for understanding network dynamics. Network protocols and their interactions among different layers are seen as gradient schemes that globally optimize a network problem. Then, stability

---

[1] This is the only quote from *The Lord of the Rings*, I promise.

in a fair equilibrium solution emerges easily from the convergence properties of gradient iterations under asynchronous distributed executions, subject to delays or losses in the signaling between the nodes, or noisy observations of the network.

Eventually the goal of simplicity led to a careful compilation of a relatively reduced mathematical optimization corpus (summarized in the book's appendices), and the quest of making the most of it to describe networks.

Technology-Agnostic Fundamentals

The book intends to develop a methodology for understanding and optimizing computer networks, applicable to any network technology. With this aim, the material is separated into two parts: problem modeling (Part I) and algorithm design (Part II).

- Part I identifies and models as constrained optimization formulations, the essential network design problems appearing in any network technology: routing the traffic, allocating capacities to the links, controlling the source rates, and deciding the network topology. Multiple real-life problem variants are included to illustrate the modeling process. When possible, Karush–Kuhn–Tucker (KKT) optimality conditions are used to give insight as to what the optimum network designs look like.
- Part II covers a set of mathematical techniques suitable for computer network problems. We concentrate on gradient-based algorithms for creating distributed schemes and network protocols and heuristics for offline algorithms suitable, for example, for capacity planning. Also, we show how the same technique can be applied to apparently different problems leading to different protocols. For instance, a dual decomposition approach can help to devise a decentralized transmission power allocation scheme in wireless networks or provide a cross-layer algorithm where congestion control and traffic routing cooperate to globally optimize network performance.

The book is full of examples and applications in IP, optical, and wireless networks to illustrate how the theory applies into real algorithms. We hope this prepares the reader to adapt this methodology to other existing technologies, or new technologies appearing.

Make it Practical

The hands-on philosophy of the book aims to permit students to perform practical optimization of networks in their homework, and the general reader to see how the ideas and mathematical approaches take real form in algorithms and models. Three practical skills are pursued:

- Formulate and obtain numerical solutions to the network problem instances, interfacing with numerical solvers.
- Implement and fine-tune the parameters of distributed network algorithms, and observe their performances and convergence under realistic scenarios, with asynchronous executions, random delays, or losses in signaling messages and subject to noisy observations of the network.

- Implement heuristic-based offline algorithms for network dimensioning and adjust their parameters to perform an efficient exploration of the solution space.

At the moment of designing the network optimization courses in 2011, no software tool was even close to match these requirements in the form I expected: easy to use, technology-agnostic, and open-source. This was the motivation to start JOM and Net2Plan open-source initiatives, the latter in collaboration up to release 0.3.1 with my Ph.D. student and colleague José Luis Izquierdo Zaragoza. JOM (Java Optimization Modeler, www.net2plan.com/jom) is a library to solve constrained optimization models written inside Java programs in a human-readable syntax, interfacing with several solvers (at this moment, GLPK and CPLEX for mixed integer linear programs, and IPOPT for differentiable programs). Net2Plan (www.net2plan.com/) is a network optimization tool that supports the fast-prototyping in Java of offline and online (dynamic) network algorithms.

Net2Plan and JOM are enabling tools for the reader interested in gaining practical skills in network optimization. All the models and algorithms in the book's text, examples, and selected exercises are included as Net2Plan algorithms, and are freely available for inspection and reuse in:

www.net2plan.com/ocn-book

The reader is encouraged to access the web page and follow the instructions there to use Net2Plan, and get the most of this book.

Both Net2Plan and JOM are stable software. As a resource for network optimization courses they are used today by several hundreds of students in my university and other institutions. Net2Plan has also become a powerful software tool for research and industry, and is present in a number of ongoing projects.

Reader Requisites and Intended Audience

Reader prerequisites are just the basic skills to handle functions of multiple variables, at the level of a first-year university course in calculus. The book appendices are then all that is needed to follow the results in the book. These appendices include a fair amount of examples and can be the base of introductory lectures.

As a textbook, this book can support courses in different forms. Several examples follow:

- Courses in the mathematical fundamentals of computer networks. In particular, those following the technology-agnostic view, accompanied by examples in different technologies. In this context, I use Part I (modeling) in a second year course for a four-year degree in Telecommunications Engineering.
- Courses in design of distributed network algorithms. Part II of the book can help to illustrate the network dynamics and the design, implementation, and test of network distributed algorithms.
- Courses in heuristic algorithms for network planning, including development of planning algorithms. Chapter 12 and Appendix C support a third year course in my University for a four-year degree in Telecommunications Engineering, specializing in networking.
- Ph.D. courses in network optimization can benefit from advanced material in the book, such as the chapters devoted to decomposition techniques and cross-layer algorithms.

- The book can be a secondary resource for different courses in computer networks and related degrees that focus on a particular technology (e.g., wireless networks, optical networks, IP networks), and rely on this book for network dimensioning, or protocol and algorithm design for these technologies.

The book can be useful for researchers and practitioners in network planning, or protocol design for multiple network technologies. For instance:

- This book, and in particular its hands-on approach, would be quite appealing for network specialists with a limited background in optimization, to address the network problems variants appearing in their technologies of interest, making benefit of a rigorous methodology that leads to successful models and algorithms.
- In addition, practitioners in operations research willing to specialize in computer networks, will appreciate the systematic approach to categorize network optimization problems, and the consistent methodology showed in the book to apply classical optimization results to communication networks.

Acknowledgments

I am heartily grateful to the many people who have contributed in some way in the preparation of this book. First, the many colleagues and students who helped me with their day-to-day discussions. Among them, I would like to give special thanks to José Luis Izquierdo Zaragoza and Victoria Bueno Delgado for their thoughtful feedback when reading the book drafts. I am thankful to the members of the GIRTEL group for the time spent together in this fascinating work. I want to express my gratitude to the Wiley editors for their confidence, help, and cooperation in the process.

Last, but foremost, I owe my family and my friends a huge debt of gratitude. Thank you, Victoria, for your love, for putting up with my mental absences this last year, and your flexibility and support. Big thanks also to Marta, María Jesús, Sonia, and Sara for helping us with the kids. I am grateful to my sisters and my parents for being there any time I needed. Finally, Pablo and Guille. Still too small to read this. You are everything to me.

1

Introduction

1.1 What is a Communication Network?

We are surrounded by communication networks, they are part of our life. If asked, we can easily enumerate examples of them: the fixed or mobile telephone network, the Internet, or someone's Ethernet home network would probably be the most popular answers. However, difficulties appear when we try to define the concept "communication network" more formally, without mentioning any specific network technology, looking for a definition applicable to any of them. We will start this book addressing precisely this basic question.

There are two basic elements on which networks are constructed: *telecommunication systems* and *switching systems*. A telecommunication system or "link" consists of a transmitter and one or more receivers connected through a medium that propagates the involved electromagnetic signals. Applying this definition, two telephones A and B directly connected through a bidirectional cable pair (Fig. 1.1) contain two telecommunication systems: (i) one composed of the transmitter at A, the medium A → B, and the receptor at B, and (ii) another system formed by the transmitter at B, the medium B → A and the receiver at A.

Telecommunication systems are the basis for assembling any communications service. However, no *service* can be reasonably built by pure aggregation of telecommunication systems. As an example, imagine we want to provide the telephone service to four users A, B, C, and D, using just "links". To do that, we would need two telephones and one cable *dedicated* for each different *user pair*. The result would be something like Fig. 1.2.

The previous example illustrates that, although possible, it is not *economically feasible* to provide a communication service using just links. Inefficiency in Fig. 1.2 appears because each link is *dedicated* exclusively to a particular user pair, and is thus idle when corresponding users are not talking each other. Improving the efficiency in our example requires adding new elements to the picture that permits a link to be *shared* among several communications. And that is precisely where switching systems come into play.

A switching system or "node" is a device that connects telecommunication systems (links) among them, so that the information from one link can be forwarded to other. We can represent a switching system with a node with N_{in} input ports and N_{out} output ports, as shown in

Optimization of Computer Networks – Modeling and Algorithms: A Hands-On Approach,
First Edition. Pablo Pavón Mariño.
© 2016 John Wiley & Sons, Ltd. Published 2016 by John Wiley & Sons, Ltd.
Companion Website: www.wiley.com/go/PavonMarinoSol16

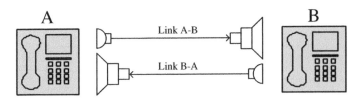

Figure 1.1 Two telecommunication systems

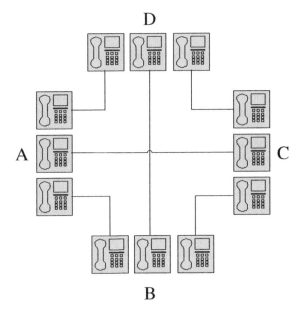

Figure 1.2 Communication service built with dedicated telecommunication systems

Fig. 1.3a. Each input port represents the receiver side of a link, each output port represents the transmitter side of other. The state of the switching system at a given moment is defined by the particular scheme in which input and output ports are internally connected. For instance, Fig. 1.3b represents a node configuration where information from input port 1 is forwarded to output port 2, while input port 3 is connected to output port 1.

The defining aspect to make a system like the one in Fig. 1.3 a switching system, is that it must be *reconfigurable*. That is, using mechanical, electronic, optical, or any other physical procedure, it should be possible to rearrange the internal connections between input and output ports, so that an outgoing link can be carrying at different moments the information received from different input links. Reconfigurability is the enabling feature to make output links become a *shared resource* among the input links.

Figure 1.4 helps us to illustrate how a combination of nodes an links supports a telephone service to seven users, using four nodes and seven (bidirectional) links. Naturally, a network like the one shown, requires extra elements to enable end-to-end communications. First,

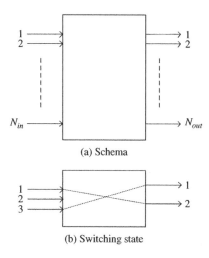

(a) Schema

(b) Switching state

Figure 1.3   Switching system (node)

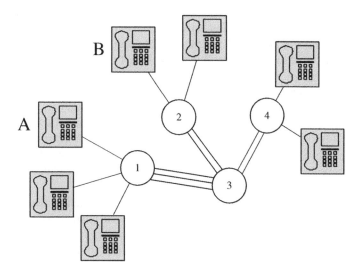

Figure 1.4   Communication network example

telephones become now a more sophisticated device, since a single telephone can be used in Fig. 1.4 to communicate with any other telephone. They are actually the *sources* and *destinations* of information. An addressing or numbering scheme is now required to identify each telephone individually, so that each user can decide the destination telephone to call to. Second, a decision on the sequence of links (route) to be traversed by each telephone call should be taken and signaled to the switching nodes that must reconfigure accordingly.

The previous network scheme should be enriched with the concept of *multiplexed links*. Previous examples have shown links as elements that can either carry one single telephone call, or otherwise be idle. In its turn, a multiplexed link is able to simultaneously carry several communications, so that the aggregated traffic can be de-aggregated again at the link receiver side. Many multiplexing technologies exist, according to the physical form in which the aggregation/deaggregation takes place, the most frequent being frequency multiplexing, time multiplexing, code multiplexing, or a combination of them. The particular multiplexing technique has no importance for the abstract network model we are pursuing. What is important, to capture the essence of multiplexing, is that links should be characterized by a *link capacity*, measured in arbitrary units (e.g., bits-per-second – bps – in digital links). In turn, sources should now be characterized as producers of traffic measured in the same units as the link capacities. Eventually, the link capacities become the shared resource, so that we will be able to compare the capacity of a link with the sum of the traffic of the sources traversing it.

Put together, we have identified four elements in a communication network: (i) information sources and destinations, (ii) links capable of propagating information between their end points, (iii) switching nodes capable of interconnecting links in a reconfigurable form, and (iv) addressing and signaling systems for controlling network operation.

1.2   Capturing the Random User Behavior

A characteristic aspect of communication networks is that they have to deal with the *random nature* of the user behavior. In Fig. 1.4 this means that we do not know beforehand who each user is willing to talk to at each moment. Random behavior is crucial for the dimensioning of resources such as link capacities in the network.

In Fig. 1.4, capacities are set so that two idle users will always be able to call each other, irrespective of what other users are doing. This is called a worst-case network dimensioning. However, this approach is clearly economically unfeasible for nontrivial scenarios. Just imagine that in Fig. 1.4 10,000 phones were connected to nodes 1 and 2. A worst-case dimensioning would need a capacity of 10,000 calls for links 1-3 and 2-3. These links would be clearly under-utilized, since the probability of the two sets of 10,000 users using the phone at the same time is small.

For this reason, communication network design has been historically based on probabilistic models that characterize the random user behavior. A statistical characterization of the performances observed by the users, permits dimensioning the network resources so that the service degradation (or the probability of this to happen) becomes small enough. The statistical metrics of interest can be quite variable. However, two main alternatives appear corresponding to the two main strategies network apply for link capacity sharing: *delay systems* and *loss systems*:

- **Delay systems**. Delay systems typically correspond to the form in which *packet switching networks* operate. In packet switching networks, traffic sources split the information into fragments called packets. Each packet is attached to a header, with sufficient control information to permit the packet reach its destination. Nodes process incoming packets one by one and forward them to their corresponding output link. Link capacities are dimensioned so that the *average* flow of packets that traverses a link does not exceed its capacity, potentially with some safety margins. However, the random nature of packet arrival times makes that packets could find their output links busy with other packets. For this reason, nodes incorporate memories for storing packets waiting for their turn to be transmitted. The storage time or queue time is an added delay to the end-to-end communication. Moreover, traffic fluctuations can fill up these memories and force the nodes to discard packets. In these systems, the delay and the packet loss probabilities are the performance metrics of interest.

- **Loss systems.** Loss systems can appear in networks where the traffic takes the form of end-to-end connections. These connections can be circuits, in so-called *circuit switching networks* (e.g., telephone calls in the telephone network, lightpaths in WDM optical networks), or virtual circuits over packet switching networks (e.g., virtual circuits in MPLS networks). Each connection reserves a given amount of capacity in each of the traversed links. The reserved bandwidth is kept throughout the communication. If a connection request does not find a sequence of links with enough capacity, it is discarded (*lost* or *blocked*). The probability that a new connection request is blocked, or *blocking probability*, is the main performance metric for dimensioning loss systems.

1.3 Queueing Theory and Optimization Theory

Queueing Theory has been traditionally the mathematical corpus supporting the design of computer or communication networks, capturing the non-deterministic user behavior and network occupation. This theory focuses on systems where a set of resources (e.g., links in a network) are shared among different users, so that the moment in which each user requests a resource (e.g., a new connection or a packet to transmit), and/or the time in which the resource is to be occupied (connection/packet duration), are known statistically. As such, Queueing Theory is a branch of Applied Probability.

Queueing Theory has been extremely successful and elegant for analyzing network subsystems such as the performances of the traffic traversing a link or a switching node. However, applying this strategy to a network view is extraordinary challenging. Actually, the probability models used in Queueing Theory become intractable when applied to non-trivial communication networks, unless strong simplifying assumptions are made. The essential disadvantage is that the exact statistical characterization of the traffic traversing several links and/or merging with other users traffic, can be mathematically intractable. For these situations, the combination of Queuing Theory and Network Optimization has shown to be a successful approach.

Optimization is a branch of Applied Mathematics, studying the maximization or minimization of functions, subject or not to a set of constraints. Network optimization is just the application of standard general optimization results to network design problems, exploiting any special structure they may have. Decision variables in these problems take the form of, for example, the capacity to assign to each link or the amount of a user traffic to route in each valid path.

The application of network optimization requires getting rid of the unreachable goal of a full statistical characterization of the traffic in the links. In its turn, in most of the cases (as happens in this book), the traffic in network design problems is modeled as a continuous *flow*, not identifying individual connections or packets in it. Flows are simply characterized by a real number: its *average intensity*, measured in bits-per-second, number of active connections, or any suitable unit. Characterizing a traffic flow just by its average yields a significant simplification: *the intensity of the traffic in a link aggregating the contributions of several flows, is just the sum of the intensities of the traversing flows.*

Previous simplification, opens the door to introducing into the optimization models expressions coming from Queuing Theory that estimate performance metrics like delay or blocking probability, as a function of *average* flow intensities. In general, these relations are closed formulas that assume as constants other parameters (e.g., variance or Hurst parameter of the traffic flows), and result in simple expressions. Thanks to that, these network performance estimations can be introduced into the optimization model as objective functions (e.g., minimize the average network delay), or design constraints (e.g., blocking probabilities should be below 1%), without significantly augmenting the problem complexity.

1.4   The Rationale and Organization of this Book

The casuistic of the different network design problems that can exist is unlimited, as the multitude of technologies, protocols and heterogeneous network conditions yield to an unnumbered amount of variants. Since it is impossible to study them all, the target of this book is providing a methodology that can be applied whatever network technology we focus on, and eventually results in (i) insights to understand the network behaviors and (ii) design keys to produce algorithms that solve network problems.

The mathematical corpus of our approach is network optimization, or the application of general optimization theory to network problems. In this strategy, we distinguish two main steps, elaborated into the two parts (Part I and Part II) in which this book is divided: problem modeling and algorithm design. This is followed by a small set of appendices covering basic optimization concepts used throughout the book.

1.4.1   Part I: Modeling

In this part, we pursue the *modeling* of network design problems appearing in communication networks, as *optimization problems*. This means translating them into the problem of finding the values of a vector *x* of *decision variables* that maximize or minimize an *objective function*, subject to some *constraints*. The reasoning of the approach is that, once a problem is modeled, we can benefit from optimization theory to characterize what their optimum solutions look like and how to reach them.

For didactic purposes, four problem types are addressed separately: (i) traffic routing, (ii) link capacity dimensioning, (iii) bandwidth sharing among network users (i.e., congestion control), and (iv) topology design. These problems appear in different forms in all network technologies. In this part of the book, multiple variants of each are described and analyzed, covering most of the aspects appearing in production networks. The source code for Net2Plan tool of the examples and selected exercises is available for the reader. Then, the he/she will be able to find numerical solutions to these problems in any network instance, using the solvers interfacing with Net2Plan.

Part I chapter organization is described below:

- *Chapter 2. Definitions and notation.* Key definitions and the notation that will be used along the book is presented. We elaborate on concepts like network links, nodes, traffic demands (unicast, anycast, multicast, broadcast), and routing (bifurcated, non-bifurcated, integral).
- *Chapter 3. Performance metrics.* This chapter describes different performance metrics that are commonly used within network optimization models: delay and blocking probabilities in packet-switched and circuit-switched networks, respectively, average number of hops, network congestion, network cost, network availability in failure scenarios, and fairness

among competing entities in the allocation of shared resources. We pursue, when possible, simplified estimations of these metrics that can be introduced in network design models. In such case, their convexity properties are analyzed, since this is a property that can be exploited when designing optimization algorithms.

- *Chapter 4. Traffic routing*. The routing problem consists of deciding the sequence of links that traverse each user's traffic. Three main modeling strategies are addressed: flow-path and flow-link formulations, for flow-based routing (e.g., suitable for MPLS, SDH, WDM networks), and destination-link formulation for destination-based routing (e.g., IP, Ethernet). Main routing problem variants are presented, including routing in anycast and multicast traffic, bifurcated/non-bifurcated routing, routing under protection or restoration schemes, and multi-hour routing.

- *Chapter 5. Link capacity dimensioning*. This chapter covers the modeling of the capacity assignment problem in two typical contexts: (i) the capacity assignment, as an offline problem to be solved every, for example, 6 months for updating the capacities in network core links, (ii) capacity assignment as a network control problem, for example, solved at a sub-second pace in wireless networks where each node optimizes its rate to adapt to varying medium and traffic conditions, affecting and being affected by the interferences from neighbor nodes. In offline capacity planning, one of the difficulties is handling the concavity of the link cost with respect to its capacity. We present a case study for which a closed formula is reached to characterize the optimum and give insight on different trade-offs appearing in network design, in particular, between complete and hub- and -spoke topologies. For the wireless case, we first model the medium-access control problems in random-access and in CSMA-based networks (like Wi-Fi). Then, we focus on a network where the link capacities are adjusted by tuning the transmission power in the antennas.

- *Chapter 6. Congestion control*. This chapter covers the modeling of the classical network congestion or bandwidth sharing problem, consisting of assigning the amount of traffic to be injected by each traffic demand without oversubscribing the links. Typically, network congestion control is implemented in a decentralized form, by rate adjustment schemes in the traffic sources. We present here a popular model based on the Network Utility Maximization (NUM) framework, where each demand is associated to an utility function. It is shown how fairness in congestion control is related to the concave shape of the utility function, so that by choosing different functions it is possible to enforce different fairness notions. We then concentrate on NUM modeling of the two main versions of TCP protocol (Reno and Vegas), showing how, despite its decentralized nature, it enforces a fair distribution of the link bandwidth among the competing connections. The role of NUM modeling to understand the behavior of active queue management (AQM) techniques is also illustrated.

- *Chapter 7. Topology design*. This chapter covers the modeling of topology design problems. First, we address the classical node location problem variants, where the position of the network nodes is optimized. Then, we include some topology design variants where the network links are a part of the problem output. Case studies and numerical examples are provided to illustrate some of the trade-offs in topology design.

1.4.2   Part II: Algorithms

This part targets the development of algorithms to solve network problems like the ones modeled in Part I. In this task, we emphasize the importance of the *algorithm context*, describing the scenario where the algorithm should be implemented. This comprises relevant information like the scale of the network targeted, the time and computing resources available, or the acceptable accuracy. Actually, *good* algorithms in a particular context, can be *bad* or simply inapplicable schemes in other contexts.

In this book, we clearly distinguish two families of algorithms: those targeted to be implemented by distributed protocols and those that are amenable to a centralized execution:

- *Distributed algorithms*. In multiple scenarios it is not possible to communicate to a centralized server the full network picture so that it can run the optimization algorithm. These problems should be solved in a *distributed* fashion by properly designed network protocols, where nodes cooperate and exchange signaling information such that they can iteratively and autonomously adjust its configuration. An example of distributed implementation is congestion control schemes, where traffic sources adjust their rates autonomously and asynchronously receiving a limited signaling information from their traversed links (or estimating it implicitly, e.g., as in TCP, monitoring its packet losses or round-trip-time delays).
- *Centralized algorithms*. A centralized algorithm execution is possible when a central computer can be fed with the full network knowledge required to solve the problem. Centralized algorithms are typical, for example, of those design problems executed offline by network planning departments. For instance, the upgrade plans for the link capacities or the placement of new links for the upcoming year, in order to cope with a forecasted traffic demand. Once the current network topology and traffic forecasts are collected, these algorithms can typically run for hours or even days until an acceptable solution is found.

This distinction between distributed and centralized executions is not gratuitous, since the mathematical approach is different for both. In this book, the theoretical corpus for the design of distributed algorithms is the application of the standard gradient or subgradient projection iterative methods to the primal or the dual version of the network problems. Convergence guarantees are strong in many situations, as long as the optimization problem is convex. In its turn, many offline planning problems to be solved by a centralized algorithm are non-convex and with $\mathcal{NP}$-hard complexity. For them, heuristic and approximation algorithms are needed.

In Part II of the book, we guide the reader in the design of distributed and centralized network algorithms. The source code for Net2Plan tool of the case studies, examples, and exercises is available. In particular, the reader will be able to easily repeat the experiments in the book, or extend them to other network instances, or different contexts in terms of signaling delays or losses.

The organization of the material is described below:

- *Chapter 8. Gradient algorithms in network design*. This chapter initiates the reader in the basic but necessary theory related to the standard gradient and subgradient schemes. Their convergence properties are also summarized when they are executed in a distributed form by separated entities that operate asynchronously, potentially using outdated information; for example, because of signaling delays or affected by noisy observations of the network. Techniques and hints for dimensioning the step size are described and later applied in multiple examples in the subsequent chapters.

- *Chapter* 9. *Primal gradient algorithms.* In this chapter we present methods based on the application of the gradient projection iteration to the primal (original) network problem. First, we present penalty techniques to deal with non-separable feasible sets, for which the projection operation is not easy. Then, we apply the primal methodology in several case studies providing distributed primal algorithms for adaptive routing, congestion control, persistence probability adjustment in MAC protocols and transmission power assignment in wireless networks. In each case, the convergence is empirically tested and connected to the theoretical convergence guidelines, under asynchronous executions, handling delayed and noisy observations, and applying convergence improving techniques described in Chapter 8.

- *Chapter* 10. *Dual gradient algorithms.* In this chapter we elaborate on the application of gradient projection iterations to the dual version of network problems. Convergence properties of dual methods are summarized, emphasizing the importance of the strict convexity/concavity of the objective function, and the role of problem regularization when this does not hold. The dual approach is exemplified in several case studies, providing distributed dual algorithms for adaptive routing, backpressure (center-free) routing, congestion control, and decentralized optimization of CSMA window sizes. Empirical convergence tests are provided under realistic asynchronous executions, with delayed and noisy observations.

- *Chapter* 11. *Decomposition techniques.* This chapter presents a framework for applying decomposition techniques to network problems, splitting them into simpler subproblems that can be solved independently, but coordinated by a master program. First, we describe the two baseline decomposition approaches: primal and dual decomposition, together with some reformulation techniques. Then, these strategies are applied in selected examples. We start putting the emphasis on problem decomposition as a theoretical support for so-called *cross-layer algorithms* that make protocols at different network layers cooperate to achieve a common goal. In this case, each network layer is a subproblem, which is solved by a different protocol (potentially itself a distributed protocol), and the master program defines the signaling to coordinate the layers. Three examples with empirical tests are provided: cross-layer congestion control and capacity allocation to traffic sources with different QoS requirements, cross-layer congestion control and backpressure routing, and cross-layer congestion control and power allocation in wireless networks.

  Afterward, we use decomposition techniques to coordinate different agents in a single-layer problem. We present a case study where multiple interconnected network carriers can cooperate in an asynchronous form to globally optimize the routing, for example in the Internet, without disclosing sensitive information like their internal topologies and traffics. Finally, we also include a case study for a $\mathcal{NP}$-hard joint capacity and routing design problem to be solved *offline*. In this case, the target of the decomposition is reducing the overall computational complexity and also permitting parallel executions of the subproblems at a cost of losing some convergence properties. Illustrative numerical tests are included for all the examples.

- *Chapter* 12. *Heuristic algorithms.* This chapter focuses on the design of heuristic algorithms for solving large instances of $\mathcal{NP}$-hard network planning problems in offline and centralized executions. The main heuristic techniques are described in a didactic form: local search, simulated annealing (SAN), tabu search (TS), greedy algorithms, GRASP, ant colony optimization, and evolutionary algorithms. Recommendations for parameter tuning are provided. Each scheme is illustrated by applying it to a common traffic engineering example: finding the OSPF link weights in IP networks that minimize congestion. Then, a realistic network planning full case study is described where a backbone optical network is optimized for three different protection and restoration schemes. This exemplifies an algorithm where a heuristic scheme is used to guide the search and control the complexity of small ILP formulations.

1.4.3 Basic Optimization Requisites: Appendices I, II, and III

The defining rationale of this book is the systematic application to network problems of selected optimization theory results. The fact that a relatively simple mathematical corpus can be exploited to provide insight on complex network interactions, and guide the design of successful network algorithms, is by all means a strength of this approach.

Still, queuing theory alone has been often the spine of network courses, leaving optimization basics for separated operations research subjects. The fruitful application of optimization to network design along the last decades has changed this inertia, and optimization theory is gaining more and more momentum in the computer networks curricula.

Appendices I, II, and III in this book are motivated by this. As a whole, their target is providing the reader not familiar with optimization theory, the necessary results repeatedly applied throughout the book. Despite being appendices, the material is organized such that it can be used as an introductory part in those computer network courses that prefer to integrate this theory. This is the approach I make in my lectures.

A short description of these appendices follows:

- *Appendix I. Convex sets. Convex functions.* Convexity is a keystone concept in optimization, which helps to later characterize the frontier between those problems that are *easy* to solve (solvable in polynomial time) from those that are not. This appendix defines and summarize the key properties of convex sets, and of convex and concave functions.
- *Appendix II. Mathematical optimization basics.* This appendix introduces optimization theory in a simple and didactic form, suitable for beginners. Optimization problems are first classified according to classical taxonomies, emphasizing the role of convex optimization. The dual of an optimization problem is defined and their main properties exposed. Then, we present and provide a graphical interpretation of Karush–Kunt–Tucker (KKT) optimality conditions for problems with a strong duality property, like the convex network problems appearing throughout the book. KKT conditions are a simple but powerful tool recurrently applied to characterize the optimum of these problems.
- *Appendix III. Complexity theory basics.* Complexity theory helps us to assess the computational complexity of problems and algorithms solving them. For problems, we define the concept of deterministic and non-deterministic machines that helps us to define P and $\mathcal{NP}$ problem complexity classes. For algorithms, we define algorithm complexity and emphasize the practical difference between polynomial and non-polynomial algorithms. Then, we define $\mathcal{NP}$-complete complexity class and discuss its importance in network design: $\mathcal{NP}$-complete problems cannot be solved to optimality for real network sizes. Then, we elaborate on problems for which it is possible to find algorithms with approximation guarantees ($\mathcal{APX}$ class and $\mathcal{PTAS}$ and $\mathcal{FPTAS}$ approximation schemes), and problems for which it is conjectured that no approximation algorithm exists ($\mathcal{NPO}$-complete problems).

1.4.4   Net2Plan Tool: Appendix IV

This appendix briefly presents Net2Plan, an open-source freeware Java-based software application for network optimization and planning, and its role as a supporting tool in the hands-on approach pursued. The book does not rely on, but is well supported by, Net2Plan for demonstrating competency and practice. The mathematical formulations and algorithms in the book are implemented as Net2Plan algorithms. They are available, indexed and documented, for the interested readers in the website www.net2plan.com/ocn-book. Then, the user will be able to see how the ideas and mathematical approaches take real form in algorithms, reuse, or execute them to obtain numerical solutions, and end up developing his/her own algorithms.

Part One

Modeling

2

Definitions and Notation

2.1 Notation for Sets, Vectors and Matrices

Throughout, we use the following notation. We use capital calligraphic letters like $\mathcal{N}, \mathcal{E}, \mathcal{X}$, to denote sets. We denote as $\mathcal{X} - \mathcal{Y}$ to the set of elements of $\mathcal{X}$ that are not in $\mathcal{Y}$. Given a set $\mathcal{X}$, the number of elements of $\mathcal{X}$ is noted as $|\mathcal{X}|$. The notation $f : \mathcal{X} \to \mathcal{Y}$ means that f is a function from the set $\mathcal{X}$ into the set $\mathcal{Y}$. The set of real numbers is denoted as $\mathbb{R}$, and $\mathbb{R}_+$ is the set of non-negative real numbers. Similarly, the sets of integer and non-negative integer numbers are $\mathbb{Z}$ and $\mathbb{Z}_+$, respectively. The Cartesian product of $\mathbb{R}, \mathbb{R}_+, \mathbb{Z}$ or $\mathbb{Z}_+$, multiplied n times by itself is denoted as $\mathbb{R}^n, \mathbb{R}_+^n, \mathbb{Z}^n$ and $\mathbb{Z}_+^n$, respectively.

Vectors and scalars are represented with lower case letters, for example x, u, h, while matrices are denoted using capital letters, for example H, R. The ith coordinate of a vector x is denoted as x_i. The (i, j) coordinate of matrix X is noted as X_{ij}. Unless stated otherwise, vectors have a column form. That is, a vector x of k coordinates is supposed to be a matrix of dimension $k \times 1$. The transpose of a vector x and a matrix X are denoted x^T and X^T, respectively, although we may omit the transpose sign to simplify the writing when it is evident from the context. Finally, we use $\mathbb{E}(x)$ to denote the expectation of the scalar or vectorial random variable x.

2.1.1 Norm Basics

A function $\|x\| : \mathbb{R}^n \to \mathbb{R}$ is called a norm, if: (i) $\|x\| \geq 0, \forall x$, (ii) $\|x\| = 0 \Leftrightarrow x = 0$, (iii) $\|tx\| = |t| \|x\|, \forall t \in \mathbb{R}, \forall x$, and (iv) $\|x + y\| \leq \|x\| + \|y\|, \forall x, y$.

A norm is a form of measuring the length of a vector. A well-known family of norms parametrized by a constant $p \geq 1$, called p-norms, is given by:

$$\|x\|_p = \left(\sum_i |x_i|^p \right)^{1/p} \tag{2.1}$$

Most common p-norms are the Euclidean norm $\left(p = 2, \|x\|_2 = \sqrt{\sum_i x_i^2} \right)$, the maximum or infinite norm $(p = \infty, \|x\|_\infty = \max\{|x_i|, i = 1, \ldots, n\})$, and the sum-absolute norm $(p = 1, \|x\|_1 = \sum_i |x_i|)$. Unless specified otherwise, $\|x\|$ will represent the Euclidean norm.

Optimization of Computer Networks – Modeling and Algorithms: A Hands-On Approach,
First Edition. Pablo Pavón Mariño.
© 2016 John Wiley & Sons, Ltd. Published 2016 by John Wiley & Sons, Ltd.
Companion Website: www.wiley.com/go/PavonMarinoSol16

We refer to dist $(x, y) = \|x - y\|$, as a measure of the distance between two vectors. In the Euclidean norm, the distance corresponds to the standard expression: dist $(x, y) = \sqrt{\sum_i (x_i - y_i)^2}$.

Let A be a $m \times n$ real matrix. A vectorial norm $\|x\|_p$ induces a matrix norm in matrices given by:

$$\|A\|_p = \sup_{\|x\|_p = 1} \|Ax\|_p = \sup_{x \in \mathbb{R}^n} \frac{\|Ax\|_p}{\|x\|_p}$$

In the case $p = 1$ and $p = \infty$, the matrix norms can be computed as:

$$\|A\|_1 = \max_j \sum_i |A_{ij}|, \quad \|A\|_\infty = \max_i \sum_j |A_{ij}|$$

If A is a square matrix, $\|A\|_2$ is the modulus of its largest eigenvalue, and is called the matrix spectral radius. The following inequality holds:

$$\|A\|_2 \le \sqrt{\|A\|_1 \|A\|_\infty} \tag{2.2}$$

Note that if A is symmetric, $\|A\|_1 = \|A\|_\infty$, and previous inequality reads $\|A\|_2 \le \|A\|_1 = \|A\|_\infty$.

Example 2.1 Let A be a symmetric matrix:

$$A = \begin{pmatrix} 1 & -2 \\ -2 & 4 \end{pmatrix}$$

The $p = 1$ and $p = \infty$ norms of A are given by:

$$\|A\|_1 = \|A\|_\infty = \max\{|1| + |-2|, |-2| + |4|\} = 6$$

The eigenvalues of A are 0 and 5, satisfying that $\|A\|_2 = 5 \le \|A\|_1 = \|A\|_\infty = 6$.

2.1.2 Set Basics

We define a ball B of radius $r > 0$ centered in x_0 as the set:

$$B(x_0, r) = \{x \in \mathbb{R}^n : \|x - x_0\| \le r\} \subset \mathbb{R}^n$$

A vector $x \in \mathcal{X} \subset \mathbb{R}^n$ is called an interior point of $\mathcal{X}$, if it is possible to find a ball $B(x, r)$, with $r > 0$ that is contained in $\mathcal{X}$. The set of all interior points of $\mathcal{X}$ is denoted as **int** $(\mathcal{X})$. The norm used for defining the ball is not important, an interior point with a norm, is interior with any other norm, a non interior point according to a norm, is not interior according to any other norm. This is true here since all norms defined in a finite dimension space $\mathbb{R}^n$ are *equivalent*: given any two norms $\|x\|_a$ and $\|x\|_b$, there exists positive constants α, β such that for all $x \in \mathbb{R}^n$:

$$\alpha \|x\|_a \le \|x\|_b \le \beta \|x\|_a$$

A set $\mathcal{X} \subset \mathbb{R}^n$ is said to be open, when all its points are interior points ($\mathcal{X} = \mathbf{int}\,(\mathcal{X})$). A set $\mathcal{X} \in \mathbb{R}^n$ is said to be closed when its complement $\mathbb{R}^n - \mathcal{X}$ is open. Both $\mathbb{R}^n$ and the empty set are considered open and closed sets. The boundary of a set $\mathcal{X}$ (denoted $\mathbf{bd}\,(\mathcal{X})$) is the set of points x for which every ball $B(x, r)$ of any radius, centered in x, intersects with $\mathcal{X}$ and with $\mathbb{R}^n - \mathcal{X}$. That is, arbitrarily close points in $\mathcal{X}$ exist, as well as arbitrarily close points not in $\mathcal{X}$. Finally, a set $\mathcal{X} \subset \mathbb{R}^n$ is compact if and only if it is closed and bounded. Being bounded means that it can be contained in a ball of finite radius.

Example 2.2 In $\mathbb{R}$, open intervals (a, b) are open sets, closed intervals $[a, b]$ are closed sets. The boundary of $[a, b]$ is the set of end points of the interval: $\mathbf{bd}([a, b]) = \{a, b\}$. The interval $[a, b)$ is not closed nor open and its boundary is also $\{a, b\}$. Set $[a, b]$ is closed and bounded and thus compact, but $[a, \infty)$ is not compact, since it is not bounded.

2.2   Network Topology

The network topology is represented as a graph $\mathcal{G}(\mathcal{N}, \mathcal{E})$, being $\mathcal{N}$ the set of network nodes, and $\mathcal{E}$ the set of links composing the network. Each link $e \in \mathcal{E}$ has a single initial node denoted as $a(e)$ and a single end node denoted as $b(e)$, both belonging to $\mathcal{N}$. Thus, only unidirectional links are considered. Given a node n, we denote as $\delta^+(n)$ to the set of *outgoings* links from n (that is $\delta^+(n) = \{e \in \mathcal{E} : a(e) = n\}$). Also, we denote $\delta^-(n)$ as the set of incoming links to node n: $\delta^-(n) = \{e \in \mathcal{E} : b(e) = n\}$[1]. Figure 2.1 illustrates these concepts.

Given a network topology $\mathcal{G}(\mathcal{N}, \mathcal{E})$, a *path* p in the network is a sequence of links $p = (e_1, \dots, e_k)$, such that end node of link e_i is the initial node of link e_{i+1}. The initial node of a path p is the initial node of its first link, and we denote it as $a(p)$. The end node of path p, denoted as $b(p)$, is the end node of its last link. For instance, in Fig. 2.1 $p = (e_2, e_3)$ is a path with $a(p) = n_1$ and $b(p) = n_3$. If a path traverses a link e, we denote it as $e \in p$. If a node n is the initial, final or intermediate node of a path, we denote it $n \in p$. Note that a path may traverse a node or a link more than once. When a path p starts and ends at the same node ($a(p) = b(p)$) we call it a cycle. A *simple path* is a path which does not contain any cycle, and thus does not traverse any node more than once. In Fig. 2.2, where each connection between two nodes represents two links, one in each direction, the path $p = (e_{23}, e_{34}, e_{42})$ is a cycle. The path $p' = (e_{12}, e_{23}, e_{34}, e_{42})$ is not a simple path, since it contains the cycle p.

Given a topology $\mathcal{G}(\mathcal{N}, \mathcal{E})$, we say that two nodes $n, n' \in \mathcal{N}$ are connected, when $\mathcal{G}$ contains at least one path between them. In communication networks, we commonly work with *connected topologies*. Formally, we say that a topology $\mathcal{G}(\mathcal{N}, \mathcal{E})$ is connected, when all the

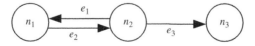

Figure 2.1   Example. Topology with $|\mathcal{N}| = 3$ nodes, $\mathcal{N} = \{n_1, n_2, n_3\}$, $|\mathcal{E}| = 3$ links $\mathcal{E} = \{e_1, e_2, e_3\}$. End nodes of e_1 are $a(e_1) = n_2, b(e_1) = n_1$. Outgoing links of node n_2 are $\delta^+(n_2) = \{e_1, e_3\}$, and incoming links of n_2 are $\delta^-(n_2) = e_2$

---

[1] Note that this representation puts no restriction in the number of links between two nodes.

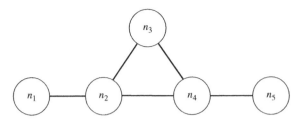

Figure 2.2   Example topology

possible node pairs in the topology are connected between them. For instance, the topology in Fig. 2.1 is not connected, since there is no path from node n_3 to node n_2 (nor to node n_1), while the topology in Fig. 2.2 is connected.

We define a *multipath* or *multicast tree* p, originated at node $a(p)$ and ended at the set of nodes $b(p) \in \mathcal{N} - a(p)$, as a set of $|b(p)|$ paths $p_n, n \in b(p)$. Each path p_n starts at the common origin node of the tree $a(p)$, and ends at $n \in b(p)$. We use $e \in p$ and $n \in p$ to denote that a link e and a node n belong to the multipath. The paths in a multipath should form a directed tree, so that path p_n is the only sequence of links of the multipath that permits connecting $a(p)$ and $n \in b(p)$. This means that each node $n \in p$ has one and only one incoming link $e \in p$, but the tree origin node $a(p)$ that has none. Also, it holds that the number of links in a multicast tree equals its number of nodes minus one. Figure 2.3 clarifies the concept of what is and what is not a multipath or multicast tree.

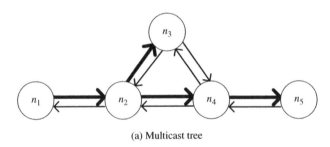

(a) Multicast tree

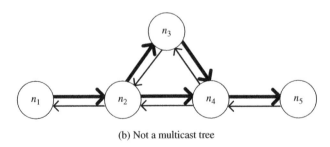

(b) Not a multicast tree

Figure 2.3   (a) Multicast tree p with origin $a(p) = n_1$ and destination nodes $b(p) = \{n_3, n_5\}$. (b) Not a multicast tree. Note that n_4 has two input links

2.3   Installed Capacities

Each link $e \in \mathcal{E}$ has associated a real number $u_e \geq 0$ representing its capacity, that is, the amount of average traffic it can transfer. To simplify the notation, it is sometimes convenient to represent the capacities of all network links as a vector $u = \{u_e, e \in \mathcal{E}\}$, where each coordinate contains the capacity of its correspondent link.

Link capacities can be measured in any traffic unit that is appropriate to the problem. In packet-switching networks, bits-per-second is the standard capacity unit. Commonly, packet-switching technologies do not permit the designer choosing each link capacity arbitrarily, but restrict them to a discrete set of possible capacity candidates. For instance, in SDH networks, link capacities can be limited to be integer multiples of 155 Mbps, the capacity of a single STM-1 circuit. When the link capacities are constrained to belong to a discrete set, we refer to them as *modular capacities*. In connection-based networks dimensioned using loss systems, the link capacity is commonly given as the (integer) number of simultaneous connections that the link can carry. Thus, link capacities in this case are inherently modular.

2.4   Traffic Demands

Traffic offered to the network is composed of a set $\mathcal{D}$ of *traffic demands*. Each demand $d \in \mathcal{D}$ represents a traffic flow. We write $a(d)$ to denote the set of one or more nodes generating the flow d, and $b(d)$ is the set of one or more nodes where the flow is targeted. No node can simultaneously be source and destination of a demand $(a(d) \cap b(d) = \emptyset)$. We use h_d to indicate the offered demand intensity of d. In general, h_d is measured in the same units as the link capacities u. Vector $h = \{h_d, d \in \mathcal{D}\}$ denotes in a compact form the traffic offered by all the network demands.

The meaning of flow intensity h_d values is slightly different in packet-switched and circuit-switched networks:

- In packet-switching networks, demands correspond to packet flows with a given *average* intensity (e.g., measured in bits-per-second or in packets-per-second). Thus, instantaneous traffic injected by the source can fast and randomly fluctuate around its average according to different patterns, oscillating from time intervals when the traffic is above its average (traffic bursts) or below it. The maximum instantaneous rate of a source is usually called its *peak rate*. The *effective bandwidth* of a source is a value between its average rate and its peak rate, defined as the amount of link capacity that a source requires in each traversing link, to statistically satisfy its Quality of Service (QoS) threshold. Different procedures exist to estimate the effective bandwidth of a source under different QoS conditions. Describing them is out of the scope of this book, some seminal papers in the topic are [1, 2]. In general, the more stringent the flow QoS is, and the longer its bursts are, the closer that the effective bandwidth is to the peak rate. When the intensity h_d of a flow d is a measure of its effective bandwidth, the network design should enforce that the sum of the effective bandwidths traversing a link is below the link capacity. When h_d values are the flow average intensities, network designs typically limit the maximum average link utilization to a value between 50 and 90%, to grossly guarantee the flow QoS.
- In circuit-switching networks, a demand d is a source of connection requests arriving randomly to the network, with average arrival frequency of λ_d connections per time unit (e.g.,

per second). Each connection can have a random duration with an average time (e.g., in seconds) given by μ_d^{-1}. A carried connection occupies in all its traversed links a deterministic amount of capacity s_d. When all the demands generate connection requests with the same size (typically $s_d = 1, \forall d$), the traffic is qualified as *single-class*. In contrast, in *multi-class traffic*, requests from different demands can have different s_d sizes. In either case, we use h_d to denote the average traffic intensity, which is given by:

$$h_d = s_d \frac{\lambda_d}{\mu_d}$$

The h_d values are given in the same units as s_d values. Usually, s_d and link capacity u_e are measured using integer numbers, such that link capacities are considered the number of simultaneous connections a link can carry. In these cases, the link capacities (u_e) and traffic intensities (h_d) are said to be measured in Erlangs, in honor to Agner Krarup Erlang, who greatly contributed to teletraffic theory in early twentieth century.

2.4.1 Unicast, Anycast, and Multicast Demands

The notation $a(d)$ and $b(d)$ as the *set* of initial and end nodes of a demand d, permits us modeling in a consistent form the main types of traffic demands in the networks:

1. *Unicast demands (1 – 1)*: A demand d is of the unicast type when it has a single initial and end node (i.e., $|a(d)| = 1$, $|b(d)| = 1$). Unicast demands are the most usual form of traffic: each demand is originated in a particular node and is targeted to other particular node. When the offered traffic is composed of one unicast demand per each node pair, the offered traffic is frequently represented using a *traffic matrix*: a square matrix with as many rows and columns as nodes, such that coordinate (i, j) of the matrix contains the traffic offered (h_d) for the demand d initiated at node i and destined to node j. Figure 2.4 illustrates with an example the generation of a traffic matrix M from a list of demands.
2. *Anycast demands (j – k)*: A demand d is of type anycast $j - k$ when (i) the demand can be originated in one or several nodes, among a set of j source nodes ($|a(d)| = j$) and (ii) the traffic should be delivered to one or more nodes, chosen among a set of k different destination

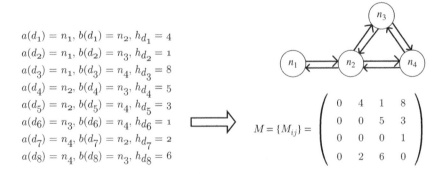

$$a(d_1) = n_1, \, b(d_1) = n_2, \, h_{d_1} = 4$$
$$a(d_2) = n_1, \, b(d_2) = n_3, \, h_{d_2} = 1$$
$$a(d_3) = n_1, \, b(d_3) = n_4, \, h_{d_3} = 8$$
$$a(d_4) = n_2, \, b(d_4) = n_3, \, h_{d_4} = 5$$
$$a(d_5) = n_2, \, b(d_5) = n_4, \, h_{d_5} = 3$$
$$a(d_6) = n_3, \, b(d_6) = n_4, \, h_{d_6} = 1$$
$$a(d_7) = n_4, \, b(d_7) = n_2, \, h_{d_7} = 2$$
$$a(d_8) = n_4, \, b(d_8) = n_3, \, h_{d_8} = 6$$

$$M = \{M_{ij}\} = \begin{pmatrix} 0 & 4 & 1 & 8 \\ 0 & 0 & 5 & 3 \\ 0 & 0 & 0 & 1 \\ 0 & 2 & 6 & 0 \end{pmatrix}$$

Figure 2.4   Example. Traffic matrix M is created from the list of demands at the left

nodes. Most frequent cases correspond to anycast $1 - k$ and anycast $k - 1$ demands. For instance, let us suppose we are using a cloud computing service that has the information spread in a set of k different server nodes $(n_1 \dots n_k)$. In this situation, a user in node n that wants to submit a file to the cloud, can be modeled as an anycast $1 - k$ source, if there is the flexibility to freely choose one server to upload the file to. In its turn, a user in node n willing to download a file that is mirrored in k servers can be modeled as a $k - 1$ anycast demand, as long it is possible to freely choose the server where to download the file.

3. *Multicast demands ($1 - k$)*: A demand d is of the multicast $1 - k$ type, when it consists of a traffic flow originated in a single node ($|a(d)| = 1$), and such that k exact *copies* of the traffic flow should be delivered, one for each k destinations in $b(d)$ ($|b(d)| = k$). Note that the need of delivering a copy of the traffic to each destination is a difference with respect to anycast $1 - k$ demands: in anycast demands, the traffic should be delivered to *any* of the target nodes, but not to all of them. Common examples of multicast sources are multimedia distribution services, on-line games, or mirroring among servers, where a single source wants to simultaneously update information in a set of destination nodes.

4. *Broadcast demands*: A broadcast demand d is a subtype of multicast demand, where an origin node $a(d)$ wants to deliver a copy of the traffic to all the rest of the nodes in the network, but itself ($b(d) = \mathcal{N} - a(d)$).

2.4.2   Elastic versus Inelastic Demands

In previous definitions, traffic demands represent random packet or connection arrivals, which can have fast fluctuations with respect to the average on the short time scale. The concepts of elastic and inelastic demand distinguish how the *average* flow intensity varies at *longer time-scales*.

In *elastic* traffic demands, the average flow intensity can vary along time, adapting its rate to the perceived state of the network. The typical example of an elastic source is an *elephant* TCP connection: a TCP connection that is always willing to transmit at higher rate if allowed, and whose actual rate is periodically adjusted by the TCP congestion control mechanism to adapt to perceived network conditions. In Chapter 6 we model the rate control problem for elastic demands. Different techniques for devising decentralized algorithms that enable a fast adaptation to network conditions, will be presented in Part II of the book.

In *inelastic* demands, the average offered flow intensity is supposed to be constant in long time scales. Typical cases are multimedia sources that generate a flow rate with very particular and constant patterns, and do not react to congestion signals. Inelastic demands are also considered in offline network design problems where the demand volumes are supposed to be known in advance (e.g., coming from a traffic forecast or a monitoring process).

2.5   Traffic Routing

Traffic routing is the process by which the traffic demands are carried from their initial to their end nodes. This is represented in different forms for different demand types:

- *Unicast demands.* The routing of an unicast demand d is realized through one or more *paths*, each of them with the same initial and end nodes as d. When the demand is routed through

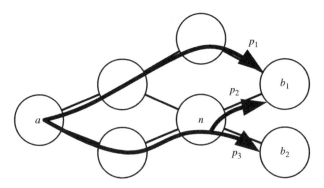

Figure 2.5 Bifurcated multicast example. A fraction f of the traffic is delivered through multicast tree $\{p_1, p_3\}$ and node a forwards two copies of the traffic, while a fraction $1 - f$ is delivered through the tree $\{p_2, p_3\}$ and node n is the one that forwards two copies of this traffic

two or more paths (e.g., 30% of the traffic in one path, 70% in other), we say that it is *balanced* or *bifurcated*.

- *Anycast demands*. The routing of an anycast demand d is realized through one or more *paths*. Each path starts in *any* node belonging to $a(d)$, and end in *any* node of $b(d)$. Similarly to the unicast case, if the demand is carried by two or more paths, we say that it is balanced or bifurcated.
- *Multicast demands*. The traffic of a multicast demand d, with origin node $a(d)$ and the set of destination nodes $b(d)$, can be routed through one or more *multicast trees* starting in $a(d)$ and ending in all nodes in $b(d)$. Again, if more than one multicast tree is used to carry the demand traffic, we say that the demand is balanced or bifurcated. This happens in the example of Fig. 2.5 when a fraction f of the traffic is delivered through a multicast tree, and a fraction $1 - f$ through other. Note that multicast routing requires from the network nodes the capability of making copies of the traffic in an input link, to potentially more than one output link. For instance, in Fig. 2.5, nodes a and n require such a copy capability.

Chapter 4 is dedicated to model multiple routing problem variants, while Part II provides several examples of routing algorithms.

References

[1] F. P. Kelly, "Effective bandwidths at multi-class queues," *Queueing Systems*, vol. 9, no. 1–2, pp. 5–15, 1991.
[2] A. I. Elwalid and D. Mitra, "Effective bandwidth of general Markovian traffic sources and admission control of high speed networks," *IEEE/ACM Transactions on Networking (TON)*, vol. 1, no. 3, pp. 329–343, 1993.

3

Performance Metrics in Networks

3.1 Introduction

In this chapter we introduce the main performance metrics in network design that will be part of the optimization models along the book. Link and end-to-end delay estimations are provided for packet switching networks, and several models are presented for evaluating blocking probability in circuit switching networks. Average number of hops and network congestion metrics are defined. Then, cost models are introduced that capture the economies of scale discounts appearing in network equipment acquisitions. Network availability is used for evaluating network resilience, under failure risks modeled using the Shared Risk Group (SRG) concept, and we present a method to estimate network availability in general protection or restoration schemes. Finally, we provide a rigorous definition of fairness in the allocation of resources (e.g., bandwidth in the links) among competing entities, and its connection with utility functions.

We pursue, when possible, simplified estimations of the performance metrics to ease its utilization in network models. In such case, their convexity properties are analyzed, since this is a property that can be exploited when designing optimization algorithms.

3.2 Delay

In packet-switching networks, randomness in the sources reflect in randomness in the packet flows observed in the links. This is the origin of buffering delays in the nodes and even packet drops when these buffers become full. Keeping such degrading effects under a satisfactory threshold requires an appropriate network design. For this, we will provide estimations for the average delay (i) traversing a link, (ii) in an end-to-end route, and (iii) a figure which captures the average delay in the network.

3.2.1 Link Delay

Given a link e, T_e denotes the average delay of the packets traversing it. To estimate T_e we assume a simple link model, shown in Fig. 3.1. Packets arriving to a node (and also packets

Optimization of Computer Networks – Modeling and Algorithms: A Hands-On Approach,
First Edition. Pablo Pavón Mariño.
© 2016 John Wiley & Sons, Ltd. Published 2016 by John Wiley & Sons, Ltd.
Companion Website: www.wiley.com/go/PavonMarinoSol16

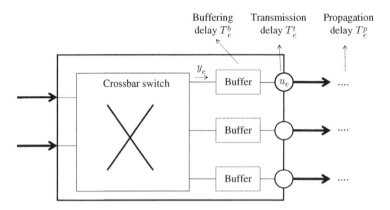

Figure 3.1   Node and link model

generated at the node), are supposed to be instantaneously switched by a crossbar, to its output port. There, they are stored in a buffer of infinite size waiting for their turn to be transmitted. According to this model, the link delay is the sum of the buffering delay, transmission delay and link propagation delay:

$$T_e = T_e^b + T_e^t + T_e^p$$

- Average buffering delay (T_e^b), represents the time spent by the packet in the buffer. This time depends on the link capacity u_e, the average intensity y_e, and other statistical properties of the traffic.
- Average transmission delay T_e^t is the average time needed to transmit the packet when it leaves the buffer. This is given by the ratio between the average packet length in bits (L), and the link transmission rate u_e: $T_e^t = L/u_e$.

  Example: A 500 bytes packet (a good approximation for the average packet size in the Internet), has a transmission time of 0.4 ms in a 10 Mbps link, and 0.4 μs in a 10 Gbps link.
- Propagation delay T_e^p is the time needed by the electromagnetic signal to reach the link end node. This is given by the ratio between the link distance d_e and the propagation velocity: $\approx$ 300,000 km/s in wireless links, and $\approx$ 200,000 km/s in wired technologies.

  Example: In a 100 m link in a Local Area Network (LAN) the propagation delay is in the order of 0.5 μs, while in 500 km Wide Area Network (WAN) link, it is $\approx$ 2.5 ms.

The estimation of transmission and propagation delays do not pose any particular difficulties. In contrast, a precise estimation of the average buffering delay would require the full statistical characterization of the traffic from all the sources traversing the link. This is actually far from being possible in real networks, even for the simplified node architecture assumed in Fig. 3.1.

Instead of pursuing a precise queueing (buffering) delay estimation, network design is commonly based on simplified expressions that capture the main trends governing this delay, but simple enough to be introduced in network-wide mathematical models. We focus on two buffering delay estimations: the Poisson traffic model and the self-similar traffic model.

3.2.1.1 Poisson Traffic Model

The Poisson traffic model, assumes that the offered traffic to each link is a packet flow, where the time between two consecutive packets are independent samples of a negative exponential distribution. This assumption is supported by the Palm–Khintchine theorem ([1], p. 160), which states that under mild assumptions, the multiplexing in a link of a large number of independent renewal packet sources (inter-arrival times in each source are i.i.d), asymptotically approximates a Poisson source. If the packet sizes are also independent samples of a general distribution, the average buffering delay is given by the Pollaczek–Khinchine M/G/1 formula (3.1):

$$T_e^b = \frac{L}{u_e} \frac{\rho_e}{2(1 - \rho_e)} (1 + CV_L^2) \tag{3.1}$$

where CV_L is the coefficient of variation (standard deviation divided by its mean) of the packet length, u_e the link capacity in bps, and ρ_e the average link utilization $\rho_e = y_e/u_e$. If packet sizes have an exponential distribution ($CV_L^2 = 1$), previous expression results in the well-known M/M/1 formula (e.g., see [2] for details):

$$T_e^b = \frac{L}{u_e} \frac{\rho_e}{1 - \rho_e} \tag{3.2}$$

3.2.1.2 Self-Similar Traffic Model

Delay estimations given by Poisson traffic models are optimistic approximations with respect to what is monitored in real networks. In particular, the probability of finding long bursts decays exponentially in Poisson models, which contradicts empirical traffic observations. The reason is that in real traffic interarrival and packet transmission times are not independent processes, and the user traffic behavior can be better approximated with patterns of infinite variance.

Self-similar models are widely accepted as a more realistic form of characterizing the network traffic. In self-similar traffic models, traffic is bursty at different scales: there are short bursts and short low traffic intervals, but also the probability of having long bursts and long low-traffic intervals is significant. Long bursts tend to saturate the links, and to worsen the buffering delays and the probability of packet drops. There are many types and flavors of self-similar traffic models, which intend to accurately capture these effects. The interested reader can find a comprehensive view of the topic in [3]. Frequently, buffering delay estimations under self-similar traffic are quite complex, and do not result in a closed formula expression. In (3.3) we show an exception to this: an approximation to the buffering delay presented in [4] for the case of an infinite size queue, fed with self-similar traffic generated using a FBM (Fractional Brownian Motion) model:

$$T_e^b = (L/u_e) \frac{\rho_e^{1/2(1-H)}}{(1 - \rho_e)^{H/(1-H)}} \tag{3.3}$$

parameter $H \in [0.5, 1)$ in (3.3) is the so-called Hurst parameter, which characterizes the degree of self-similarity of the traffic. High H parameters $H \approx 1$ appear in traffic models with high levels of self-similarity, while $H = 0.5$ characterizes traffic that is not self-similar. Different traffic measurements in the Internet have reported traces with Hurst parameters between 0.6 and 0.9. Note that approximation (3.3) converts in (3.2) formula when $H = 0.5$.

3.2.1.3 Discussion

Figure 3.2 plots the link delays associated to (3.3) for different link utilizations ρ and Hurst parameters H. As can be seen, link delays are small and approximately constant at low link utilizations and then increase abruptly when the utilization exceeds a threshold. The threshold is around $\rho = 0.5$ for very bursty traffics ($H = 0.9$), and $\rho = 0.95$ for non-self-similar traffic ($H = 0.5$). These results support a common practice in network planning, consisting in dimensioning the network links such that utilization for the average expected traffic is below 50–60%. Note that in this case, buffering delays can be negligible in high-bit-rate links, compared to user reaction times, or to propagation delays. For instance, in a 10 Gbps link of 500 km, buffering delays can be in the order of microseconds, and can be neglected with respect to propagation delays, which are in the order of milliseconds.

Observing expression (3.3), we notice also that, for the same link utilization, average delays are inversely proportional to link capacities. This means that, for example, the average delay of a 10 Mbps link loaded at a 50% is 1000 times larger than the delay of a 10 Gbps link at a 50% load. Thus, the optimization of the buffering delay figures in a network is a force towards concentrating the network traffic in a small number of large-capacity links, instead of spreading it in a high number of small ones. As will be shown in Chapter 5 and Chapter 7, and is made evident along the book, this trend competes with other opposite ones, so that optimum network design is often the one that finds the right balance among multiple trade-offs.

Finally, according to estimation (3.3), average delay is proportional to the average packet size. Thus, ideally, dividing by two the packet sizes would halve the average buffering delay. However, the packet lengths are not commonly a figure to optimize in network design, but an input provided by the architecture of protocols in the network. Moreover, reducing the packet length would have two negative effects: (i) each packet still needs to carry a header with control information so, for example, duplicating the number of packets for the same data, means duplicating the overhead, and (ii) the cost of the packet switching nodes grows with the

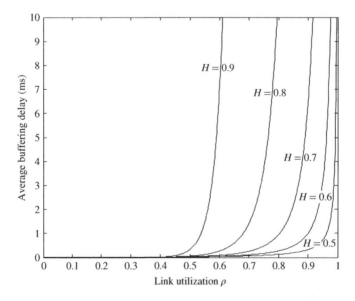

Figure 3.2   Average buffering delay estimation (3.3). $L = 500 \times 8$ bits, $u_e = 100$ Mbps

number of packet headers per second to process: for example reducing the packet length by two, roughly means multiplying the rate of headers to process by two.

3.2.2 End-to-End Delay

The average end-to-end of a traffic traversing a path p is given by T_p:

$$T_p = \sum_{e \in p} T_e$$

If the traffic is multicast, and p is the multicast tree routing it, the end-to-end delay can be different for different destination nodes $n \in b(p)$. In general, we are interested in the worst-case or maximum average delay, experienced by the traffic going to the different destinations:

$$T_p = \max_{n \in b(p)} \left\{ \sum_{e \in p_n} T_e \right\} \tag{3.4}$$

3.2.3 Average Network Delay

The average network delay is defined as the average end-to-end delay suffered by a traffic unit (e.g., a packet) chosen randomly in the network. This figure can be useful as an objective function to minimize in network design, since it condenses in a scalar number the delay performance of the network. In this section, we estimate network delay for non-multicast traffic, following the method in [5]. Let P be the set of paths carrying traffic, x_p the amount of traffic carried by path p, associated with demand $d(p)$, and h_d the total amount of traffic carried of demand $d \in D$. The average network delay T is the weighted delay for the different paths in the network:

$$T = \sum_p \frac{x_p}{\sum_d h_d} T_p$$

Assuming that all paths of all demands are treated equally (i.e., there are no priority paths that observe smaller link delays), we have that:

$$T = \frac{1}{\sum_d h_d} \sum_p x_p \sum_{e \in p} T_e = \frac{1}{\sum_d h_d} \sum_e T_e \sum_{p \in P_e} x_p$$

Where in the second equality, the order of the summatories are exchanged, so P_e is the set of paths that traverse link e. Denoting y_e as the total amount of traffic in link e ($y_e = \sum_{p \in P_e} x_p$) we have the final expression for average network delay.

$$T = \frac{1}{\sum_d h_d} \sum_e y_e T_e \tag{3.5}$$

An expression for average network delay with multicast traffic is obtained in Exercise 3.1.

3.2.4 Convexity Properties

It can be shown by direct inspection of the second derivatives, that link delay formulas for M/M/1 (3.2), M/G/1 (3.1), and self-similar traffic (3.3) are strictly increasing and strictly convex expressions with respect to the traffic in the link (y_e variables). Logically, same properties apply with respect to utilization variables ρ_e. In addition, aforementioned formulas are also strictly decreasing and strictly convex with respect to capacity variables (u_e). However, joint convexity with respect to (y_e, u_e) does not hold. See Exercise 3.2 for the derivations. Figure 3.2 helps to illustrate the strictly increasing and convex evolution of the link delay with respect to ρ_e values.

From previous results, convexity of the end-to-end delay T_p in a path p, $T_p = \sum_{e \in p} T_e$ is guaranteed, as a function of y_e or u_e, since the sum of convex functions is also convex. Note that strict convexity with respect to u or y variables does not hold unless the path traverses all the links in the network. Similarly, expression (3.4) for worst-case multicast tree delay is a convex function of u_e or y_e or ρ_e variables, since the maximum of a convex function is also convex.

Finally average network delay expression T (3.5) is a strict convex function with respect to u_e, y_e or ρ_e variables, for any of the studied T_e estimations. See Exercise 3.3 for the derivations.

3.3   Blocking Probability

According to our notation model, in circuit-switching networks a demand d is the source of connection requests that arrive randomly to the network and, if carried, occupy a fixed and deterministic amount s_d of bandwidth in each traversed link. The average intensity in Erlangs of such demand h_d is given by:

$$h_d = s_d \frac{\lambda_d}{\mu_d}$$

Where λ_d is the mean number of connection requests per time unit, and μ_d^{-1} the average connection holding time. Link capacity values (u_e) are integer units (e.g., number of channels), and also the amount of bandwidth consumed by a connection s_d is integer. When a request does not find a path with enough available capacity, it is blocked.

In this section, we will treat both the single-class and multiclass traffic cases. Recall that in *single-class* traffic, all the demands generate connection requests with the same size $s_d = 1$, while in *multiclass* networks, requests from different demands can have different s_d sizes. We will first describe a simplified model to estimate the blocking probabilities at the link, path, and network level. The benefits of this model are its simplicity, and the opportunity to provide reasonably simple estimations, for which a convexity analysis can be presented. This is obtained at a cost of estimation accuracy. In the last part of this section, other more complex and more accurate models are presented.

3.3.1   Link Blocking Probability

Figure 3.3 illustrates the link model considered for estimating the blocking probabilities of connections traversing a link e. The traffic intensity of each demand d in the link is noted y_{de}, $y_e = \sum_d y_{de}$ is the total offered link traffic, and $\rho_e = y_e/u_e$ the normalized link load. We assume that connection arrival times are Poisson, and connection holding times are independent samples of any distribution. We also assume the so-called *complete sharing* link model [6], which means that a request is blocked only if there are not enough idle channels in the link. In these conditions, if the traffic is single-class, the blocking probability B_e is the same

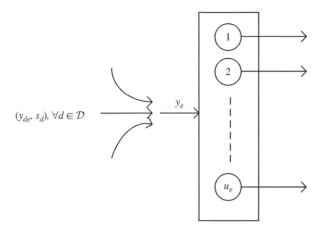

Figure 3.3 Node and link blocking model

for all the demands, and given by the celebrated Erlang-B formula:

$$B_e = E_B[y_e, u_e] = \frac{y_e^{u_e}/u_e!}{\sum_{k=0}^{u_e} y_e^k/k!} \tag{3.6}$$

In the multiclass context, demands with different s_d values can observe different blocking probabilities $B_e(d)$. These can be computed for the complete sharing model using the Kaufman–Roberts recursion (proposed independently in [7] and [8]), described in Algorithm 1. Note that, logically, all the demands with the same bandwidth requirements s_d perceive the same blocking.

Algorithm 1 Kaufman–Roberts recursion

1: $g(c) = 0, c < 1; g(0) = 1$
2: **for all** $c = 1, \ldots, u_e$ **do**
3:    $g(c) = \frac{1}{c} \sum_d y_{de} g(c - s_d)$
4: **end for**
5: $G = \sum_{c=0}^{u_e} g(c)$
6: $B_e(d) = \frac{1}{G} \sum_{i=0}^{s_d-1} g(C - i), \quad \forall d \in \mathcal{D}$
7: **return** $\{B_e(d), d \in \mathcal{D}\}$

3.3.1.1 Discussion

Figure 3.4 plots the blocking probabilities in the single-class traffic case, for a link e with different link capacities (u_e) and load values (ρ_e). The first observation is that blocking probabilities always increase with link offered load. For a given link utilization, blocking probabilities decrease with link capacity. For instance, let us assume a 1 Gbps link, which is offered an average of 600 Mbps, consisting of arrivals of 100 Mbps connections. In this case, the link capacity equals 10 units (10 simultaneous connections), utilization is 60%, and the resulting blocking probability is $E_B(0.6, 10) \approx 4\%$. In its turn, if half of the traffic (300 Mbps)

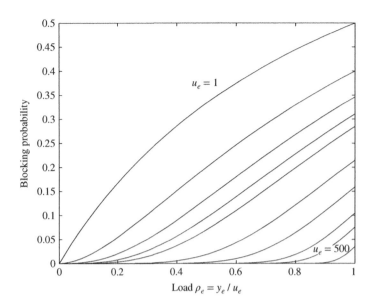

Figure 3.4   Blocking probability (single-class, Erlang-B). $u_e \in \{1, 2, 3, 4, 5, 10, 20, 50, 100, 500\}$

was offered to a link with half capacity (500 Mbps), the resulting blocking would be higher: $E_B(0.6, 5) \approx 11\%$. A similar effect could be obtained if we reduce the *granularity* of the connections. For instance, if 600 Mbps of traffic are offered to a 1 Gbps link, but with connection sizes of 10 Mbps instead of 100 Mbps, the link can now accommodate 100 simultaneous connections ($u_e = 100$). This reduction in connection granularity, results in a much reduced blocking probability $E_B(0.6, 100) \approx 6 \times 10^{-7}$.

As a general rule, similar trends favoring the concentration of the traffic in links of larger capacities, and reducing the granularity, also hold in multiclass networks. However, some anomalous behaviors, like the blocking probability of a traffic class being reduced when the traffic of the class is increased, can occur when the link capacity is of a comparable size with respect to the connection bandwidths s_d. We will briefly present some examples on this in Section 3.3.4.

3.3.2   Demand and Network Blocking Probability

The demand blocking probability B_d provides the probability of rejecting a connection request of a given demand, while the average network blocking B is the weighted average of the fraction of traffic rejected by the network:

$$B = \frac{1}{\sum_d h_d} \sum_d h_d B_d \tag{3.7}$$

The exact calculation in a network of the demand blocking probabilities, managing the interplay of different demands traversing different links, accepting one or more possible (alternate) routes for the connection, becomes intractable for all but simple cases. The derivation of usable estimations has received a lot of attention in the literature. Section 3.3.3 will provide a short review of some of them, the interested reader is recommended the compendiums [6] and [9].

In this section we present a simplified approach, valid for single and multiclass networks, where traffic can be *unicast, anycast, multicast,* or any combination of them. We make the following assumptions:

1. *Poisson traffic model*: Connection requests associated to any demand follow a Poisson arrival process.
2. *Load sharing* model: If a demand d is assigned more than one possible path $p \in P_d$ (or multicast tree for a multicast demand), each connection arrival randomly and independently chooses one path out of P_d. If the path has not enough capacity, the connection is rejected (i.e., with no re-trial in other path). As a result, a path p associated to demand $d(p)$ can be treated as an independent demand of load $h_p = h_{d(p)} z_p$, where z_p is the path selection probability.
3. *Link independent Poisson arrivals*: Connection requests associated to a link are supposed to be a Poisson process, independent from other links.
4. *Low-blocking regime*: The load and capacity conditions in the network are supposed to be such that the blocking probabilities are moderate (e.g., below 5%).

Given a path p, link independence assumption permits estimating the path blocking probability B_p as:

$$B_p = 1 - \prod_{e \in p}(1 - B_{ep}) \tag{3.8}$$

where B_{ep} is the probability of not finding enough idle resources in link e for connections of the size s_d (being d the demand associated to p). Note that in single-class networks, we have $B_{ep} = B_e, \forall p$. In the low-blocking regime, path blocking (3.8) can be approximated as:

$$B_p = 1 - \prod_{e \in p}(1 - B_{ep}) \approx \sum_{e \in p} B_{ep}$$

Previous approximation is pessimistic, and its accuracy proportional to the product of blockings (the smaller the blockings the better). Finally, we assume that all the links in a path p sum a quantity equal to h_p to the offered traffic. Therefore, in the multiclass case, Algorithm 1 can be used to compute the B_{ep} values, taking as link offered traffic the sum of the traffics in the traversing paths. According to this approximation, the average network blocking probability B can be obtained as:

$$B = \frac{1}{\sum_p h_p} \sum_p h_p B_p = \frac{1}{\sum_p h_p} \sum_p h_p \sum_{e \in p} B_{ep}$$

In the single-class case, Erlang-B formula (3.6) is used to compute B_e blockings and the link offered traffic is given by $y_e = \sum_{p \in P_e} h_p$. The average network blocking expression can be further simplified, resulting in:

$$B = \frac{1}{\sum_p h_p} \sum_p h_p \sum_{e \in p} B_{ep} = \frac{1}{\sum_p h_p} \sum_e \sum_{p \in P_e} h_p B_e = \frac{1}{\sum_p h_p} \sum_e y_e B_e \tag{3.9}$$

3.3.3 Other Blocking Estimations

In this section we provide an overview of more complex blocking estimations appearing in the literature. Again, the reader is referred to classical books like [6] or [9] for an in-depth survey.

3.3.3.1 Load Sharing: Reduced Load Approximation

Reduced load approximations are probably the most popular approximations in loss networks. Let us assume a load sharing model for a single-class or multiclass network where each demand d is a source of connection requests associated to a single path p_d. The main idea behind reduced load approach is that connection blocking probabilities B_d for the path p_d can be approximated by:

$$B_d = 1 - \prod_{e \in p_d}(1 - B_{ed}) \tag{3.10}$$

where B_{ed} is the probability that a connection of demand d is not accepted on link e. In other words, (3.10) says that blocking on different links are approximately independent. The most common reduced load approximation is the one from Kelly (see [10] for a complete theoretical background), which uses path blocking probabilities of the form:

$$B_d = 1 - \prod_{e \in p_d}(1 - B_e)^{s_d} \tag{3.11}$$

This means that a request for *one unit of capacity* in link e is blocked with probability B_e, and we make the approximation that all these blocking events are independent. According to this, the traffic offered to link e will be Poisson and the level of carried traffic $h_d(1 - B_d)$ will be $\sum_d h_d \prod_{e' \in p_d}(1 - B_{e'})^{s_d}$. The B_e values are determined by solving the following system of $|\mathcal{E}|$ equations involving the Erlang-loss (E_B) formula:

$$B_e = E_B\left(\frac{1}{1 - B_e}\sum_d h_d \prod_{e' \in p_d}(1 - B_{e'})^{s_d}, u_e\right), \quad e \in \mathcal{E} \tag{3.12}$$

Equations (3.12) simply state that B_e values in all the links should be consistent with the appropriate level of carried traffic of all demands.

Kelly proved [11] that there is a unique solution to (3.12), and that this solution is asymptotically exact for large networks (when capacities and traffics are multiplied by a factor $k \to \infty$). The blocking estimation has shown to be accurate for many topologies of interest in the single-class case, while its accuracy can be degraded when multiclass traffic appears.

The *Erlang fixed point method* depicted in Algorithm 2 is the most popular approach to find the B_e values solving (3.12), by using a repeated substitution iteration.

Algorithm 2 Erlang fixed point method

1: $k = 0$, Choose $B_e^0 \in [0, 1], \forall e \in \mathcal{E}$
2: **for all** $e \in \mathcal{E}$ **do**
3: $B_e^{k+1} = E_B\left(\frac{1}{1 - B_e^k}\sum_d h_d \prod_{e' \in p_d}\left(1 - B_{e'}^k\right)^{s_d}, u_e\right)$
4: **end for**
5: $G = \sum_{c=0}^{u_e} g(c)$
6: $B_e(d) = \frac{1}{G}\sum_{i=0}^{s_d-1} g(C - i), \quad \forall d \in D$
7: **return** $\{B_e(d), d \in D\}$

In single-class networks, the Erlang fixed point iteration is empirically known to converge in a few iterations to the correct B_e values. Actually, formal convergence guarantees are that it does not diverge, but may oscillate without improvements in some hand-picked scenarios (e.g., see [12]). In multi-class networks, the method has no convergence guarantees, and its use is discouraged.

3.3.3.2 Alternate and Adaptive Routing

The obvious drawback of load sharing model is that if the path selected by a demand is unavailable, the request is lost, although other paths may be free. For this reason, pure load sharing is not used in real networks, and its interest is mostly theoretical. The improvement is to permit choosing the path from a given set, taking into account the state of the network.

One example is the so-called *alternate routing*, where each demand has a precomputed and ordered list of paths. When a connection request arrives, the first path available in the list is chosen, and the request is blocked only if all are unavailable. Other examples fall into the so-called *adaptive routing* techniques. In these schemes, when more than a path is available for a connection, the one to choose does not depend on a pre-defined order, but on the current network state. This is the case of the so-called *Least Congested Routing* (LCR). In LCR, a measure of congestion is computed for each connection admissible path. The path congestion is defined as the idle capacity in the traversed link with less idle capacity. Then, LCR routes the connection through the path with lower congestion.

Estimating the blocking probabilities in alternate and adaptive routings is difficult, and closed-form expressions are not available but in trivial cases. In its turn, a plethora of estimations based on numerical methods have been presented, for different routing flavors. The reader is referred to [6] for an exhaustive compilation.

Alternate and adaptive routing techniques bring into discussion new effects in the network: random oscillations between different network states, and the need of admission control techniques to address them. To illustrate this, let us imagine a network where each demand has two alternate paths, a short and a long one. In some circumstances, for the same average offered traffic, two stable states can appear in the network: a high blocking state and a low blocking state.

- In the high blocking state, most of the demands use the long alternate path, which occupy more links. This situation becomes stable since subsequent connection requests find their short paths occupied by connections following alternate paths of other demands.
- In the low blocking state, most of the connections use the short path, occupying a similar amount of link resources as in the previous state, but carrying more connections.

Oscillations between these two states can occur because of random fluctuations of the arrivals process: a network in a low blocking state after a burst of arrivals can enter in the high blocking state, and stay there even when the burst disappears. This type of network instabilities have been studied in several works (e.g., see [6] for further details), and are known to occur in just some regimes of offered traffics. In order to avoid such instabilities, the so-called *trunk reservation* technique is used. This consists of the following: (i) a link with available capacity always accepts a connection if it follows its primary (short) path, but (ii) if the connection is routed through an alternate (long) path, the link should have at least $T > 1$ idle capacity units to accept it. The result is that, when the links are heavy loaded, connections are only routed through their short paths, or rejected, so no wasteful alternate routes are accepted. By an appropriate dimensioning of the trunk reservation factor T, it is possible to eliminate network instabilities and reduce the network blocking at high loads.

3.3.4 Convexity Properties

It is well known that Erlang-B blocking (3.6) is an increasing function with respect to the link utilization ρ_e or link traffic y_e, and a decreasing function of the link capacity u_e. In [13], it was shown that Erlang-B is a convex function of the link capacities, when they take integer values. Extensions of the Erlang-B for continuous values of u_e (of interest in some blocking estimation methods), have been shown to be also convex [14].

The convexity of the Erlang-B as a function of the traversed traffic y_e or the link utilization ρ_e, has been studied in several works. In [15], it is shown that if a link has one unit of capacity ($u_e = 1$), the blocking probability is strictly concave function of y_e (or ρ_e). In contrast, if $u_e > 1$, the blocking is a strictly convex if the link utilization is below a given threshold $\rho_e < \rho^*$, and strictly concave if $\rho_e > \rho^*$. Figure 3.4 illustrates the inflection point. Note that in those utilization regimes where the blocking is below 10%, the Erlang-B function is always convex. Finally, it holds that, for any utilization regime, the link throughput $y_e(1 - B_e)$ is strictly concave in y_e or ρ_e, and thus the amount of blocked traffic $y_e B_e$ is a strictly convex function of ρ_e or y_e. Note that according to this, the average single-class network blocking estimation (3.9) is a strictly convex function with respect to $\{y_e, e \in \mathcal{E}\}$ variables.

Unfortunately, the multiclass case does not enjoy many of the monotonicity and convexity properties of its single-class counterpart (see [16, 17] for details). Figure 3.5 helps us to illustrate this, for two Poisson traffic classes traversing a link:

- In Fig. 3.5a, we see how *increasing* the link capacity for a fixed traffic can *increase* the blocking probability of some connections. This happens since the extra capacity can reduce the blocking of some *wide* traffic classes, leaving less room to the rest.
- In Fig. 3.5b the traffic of the demand with $s_d = 2$ is increased, keeping constant the load of a traffic class with $s_d = 3$. As can be seen, *increasing* the traffic of a class can actually *decrease* the blocking of such class. This is because the extra traffic can increase the blocking of other classes, which then gives extra capacity to reduce the blocking of the first.

3.4 Average Number of Hops

The average number of hops $\bar{n}$ is a popular measure to characterize the routing. It is defined as the average number of links traversed by a packet or a traffic connection in the network.

$$\bar{n} = \frac{\sum_p l_p x_p}{\sum_p x_p}$$

Where l_p is the number of links in path p, or, in multicast trees, the average number of links traversed among each destination. If the traffic is just unicast and/or anycast, but not multicast, and all the offered traffic is carried, we can further simplify the previous expression:

$$\bar{n} = \frac{\sum_e y_e}{\sum_d h_d} \tag{3.13}$$

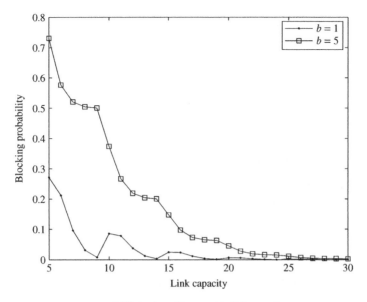

(a) Behavior with respect to link capacity

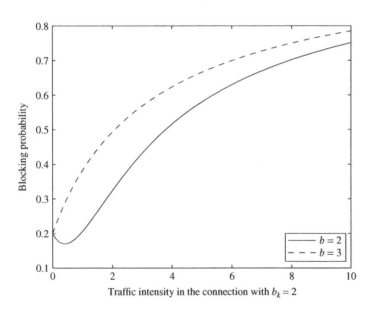

(b) Behavior with respect to link traffic

Figure 3.5   Example. Evolution of blocking probability for two traffic classes $k = 1, 2$ of connection sizes s_k, traversing a link (Poisson assumption). (a) $u_e = \{5 \dots 30\}$, $s_1 = 1, s_2 = 5, y_1 = 1\ E, y_2 = 5\ E$. (b) $u_e = 7, s_1 = 2, s_2 = 3, y_1 = [0, 20], y_2 = 3$

Which means that the metric $\bar{n}$ can be obtained by using just aggregated information of the traffic in the links and the demands. Note that the numerator in (3.13) is obtained by rewriting the sum per path as a sum per link:

$$\sum_p l_p x_p = \sum_e \sum_{p \in P_e} x_p = \sum_e y_e$$

It is easy to see that metric $\bar{n}$ (3.13) is a linear (and thus convex and concave) function with respect to the traffic in the links.

3.5 Network Congestion

In packet switched networks, the network congestion cg is defined as the utilization in the link of the network, which has a higher utilization (so-called, the *bottleneck link*):

$$cg = \max_{e \in \mathcal{E}} \frac{y_e}{u_e} \tag{3.14}$$

In other words, if a network has a congestion of 0.7, means that at least one link has a utilization of 70%, and the rest have a 70% utilization or less. The other common way to represent the network congestion is the *worst-case unused bandwidth* (u_u) metric. This is defined as the unused bandwidth ($u_e - y_e$) in the link of the network with less unused bandwidth:

$$u_u = \min_{e \in \mathcal{E}} u_e - y_e \tag{3.15}$$

Note that according to the M/M/1 formula, the average transmission plus buffering delay in a link is given by $\frac{L}{u_e - y_e}$, and thus is inversely proportional to the link unused bandwidth. Thus, designs maximizing the worst-case unused bandwidth u_u, simultaneously minimize the worst-case average link delay in the network.

Expression (3.14) is a convex (but not strictly convex) function of y_e and also of u_e variables, since $\frac{y_e}{u_e}$ is a convex function with respect to both, and the pointwise maximum of a set of convex functions is also convex. Similarly, expression (3.15) is concave, as a function of y_e or u_e, since $u_e - y_e$ is linear (and thus concave), and the pointwise minimum of a set of concave functions is itself concave.

3.6 Network Cost

The cost of the network is naturally one of the major aspects subject to optimization. A precise determination of the cost of a design is a complex task, and in realistic scenarios, it is commonly not possible to find a close expression for it. In this section, we will describe a cost model that intends to be simple, but flexible enough to capture the main trends in cost evolution and being applicable in different contexts.

We denote as C the cost of a network, which is assumed to be given by the sum of a cost per link ($c(e)$), and a cost per node ($c(n)$):

$$C = \sum_e c(e) + \sum_n c(n) = \sum_e c_f(e) + c_v(e, u_e) + \sum_n c_f(n) + c_v(n, u_n)$$

Both link and node costs are separated as the sum of a fixed cost (c_f) and a variable cost (c_v). Fixed costs in links and nodes are applied only if the link/node exists, whatever the link capacity (u_e) or node capacity (which we denote as u_n). In turn, variable costs c_v are increasing functions of the link/node capacity. Both fixed and variable costs can have multiple forms, depending on the network technology and, for example, the particular cost structure of the network operator. Several illustrative examples follow:

- In optical backbone networks, links are optical fibers deployed between major cities. Usu-ally, the companies building railways or highways infrastructures in the country are the owners of these fibers. They rent them to the telco carriers and the telco is responsible of adding all the necessary transmission equipment to the fiber[1]. The renting cost is in general a linear function of link length d_e ($c_f(e) = \alpha + \beta d_e$). A rented fiber can host a quite variable number of channels (e.g., up to 160 40G channels), and the cost of using a channel is low compared to the fiber renting cost. In this context, $c_f(e)$ costs are dominant with respect to $c_v(e)$ costs.
- Corporate networks of private companies are usually composed of small to medium size IP routers, connected by virtual circuits hired to the network carrier. In this case, circuits costs given by the carrier tariffs strongly depend on the link rate, and are dominant with respect to the router costs.

The dependence of the variable costs with respect to the link/node capacity often follows the *economies of scale law* or the *law of diminishing returns*: as the capacity grows, (i) the absolute cost grows, but (ii) the cost per capacity unit decreases. As an example, this explains why buying a 10 Gbps link is cheaper than buying *ten* 1 Gbps links. On some occasions, economies of scale appear as a lower list price per capacity unit in higher capacity equipment, or take the form of a discount in the link/node cost if more capacity units are acquired.

The economies of scale law is reflected in c_v functions that are (i) increasing, and (ii) con-cave functions with respect to the link/node capacities. Figure 3.6 illustrates this, plotting the (normalized) prices of several wholesale Internet access links in Spain as published in p. 92 of [18]. Clearly, we see that costs can be approximated by an increasing and concave curve. In particular, a least-square fit of the costs is given by:

$$c(e) = 0.575 - 0.42u_e + 0.843\sqrt{u_e}$$

3.7 Network Resilience Metrics

Network resilience is a term describing the ability of a network to provide an acceptable ser-vice level even in the presence of failures or attacks. This is a critical aspect in network design, given the economic importance of the activities that rely on communication networks today. Service Level Agreements (SLA) that carriers sign with their clients often include different measures relative to network resilience. The most important among them is the *service avail-ability*, defined as the percentage of time during which the service should be operative during the observation period (e.g., 1 year). Typical availabilities appearing in SLAs are between

---

[1] Since the carrier *lights* the fiber adding the optical transmission equipment, the fiber rented is usually called *dark fiber*.

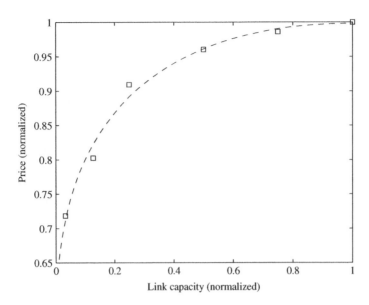

Figure 3.6　Concave costs example. Normalized wholesale Internet access prices (GigADSL) for different rates in Spain (p. 92, [18])

99.9% and 99.999%. The latter (popularly called *five nines*) is considered a premium class performance in standard (non-critical) services. Note that a five nines availability means that the service can be discontinued at most for about 5 minutes per year.

　　SLAs include penalties if the performance metrics are not met and it is important for network operators and service providers to precisely estimate the availability guarantees they can offer. This requires an assessment of the failure risks that threaten the network: risk of software and hardware malfunctions, cable cuts, power outages, and so on. As an example, Table 3.1 shows an estimation study in [19] for different network resources. For each equipment type, availability is extracted as a function of two separated estimations:

- Mean Time Between Failures (MTBF): Average time elapsed between two consecutive equipment failures.
- Mean Time To Repair (MTTR): Average time needed to repair a failure. During this time, the resource is unavailable.

The availability A of an equipment is related to MTBF and MTTR by the formula:

$$A = \frac{MTBF - MTTR}{MTBF}$$

The successful provision of any actual network service involves the cooperation of multiple resources, which can suffer individual or coordinated failures. For instance, Fig. 3.7 shows an

Table 3.1 Typical MTBF MTTR and availability values [19]

Equipment	MTBF	MTTR	Availability
Web server	$10^4 - 10^6$ h	1 h	99.99–99.9999%
IP interface card	$10^4 - 10^5$ h	2 h	99.98–99.998%
IP router itself	$10^5 - 10^6$ h	2 h	99.998–99.9998%
ATM switch	$10^5 - 10^6$ h	2 h	99.998–99.9998%
SONET/SDH DXC or ADM	$10^5 - 10^6$ h	4 h	99.996–99.9996%
WDM OXC or OADM	$10^5 - 10^6$ h	6 h	99.994–99.9994%
Long distance cable (per 1000 km)	50–200 days	hours-days	99–99.75%

Figure 3.7 Simple service availability example

example of a web connection whose successful completion requires the correct operation of three links, two routers, and the web server. The resulting availability of the service A_s, if the up/down state of each resource was statistically independent, would be given by:

$$A_s = 0.999 \times 0.99995 \times 0.992 \times 0.9995 \times 0.998 \times 0.99999 \approx 0.9889$$

This example helps us to illustrate that, in service provisions subject to realistic link and node MTBF/MTTR values, the availability figures obtained can be very poor. Thus, the only way to improve these numbers is the adoption of fast and automatic recovery actions that reroute the traffic affected by a failure, such that the service is not disrupted during the (usually long) reparation time of the failing resources. We distinguish two main strategies:

- Protection recovery: In this case, the recovery actions are pre-planned and typically pre-signaled in the network such that failure reactions can be very fast (e.g., from tens of ms to 1 s). The most popular protection systems are illustrated in Fig. 3.8. In the *1+1 protection* (Fig. 3.8a), a traffic connection through a (primary) path p is backed by a secondary or back up path p', which is disjoint to p. The source node sends two copies of the traffic, one for each path. In case of failure in the primary path, the receiver is reconfigured to read the traffic coming from the backup path. As a variation on this, in *1:1 protection* systems, the secondary path can be idle or transmit low priority traffic when the primary path is up. When p fails, any low priority traffic is pre-empted, and the secondary path is used to carry the original traffic. In both 1+1 and 1:1 protection schemes, the capacity reserved in the primary and secondary paths are the same, and the backup capacity is dedicated to protect a particular connection. In its turn, in the *M:N* protection

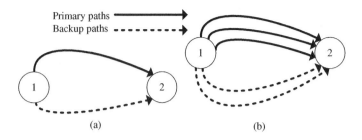

Figure 3.8   (a) 1+1 protection (dedicated), (b) 2:3 protection (shared)

schemes, a set of M backup paths are allocated to protect N primary connections. Any primary connection can be allocated any backup path. Thus, the capacity in the backup links is shared among the primary connections. Figure 3.8b shows a 2:3 shared protection scheme.

- Restoration recovery: In restoration schemes, the recovery actions are not pre-planned, but decided after the failure/reparation is detected. This permits the maximum flexibility to decide on the most efficient recovery adapted to current network state, at a cost of a slower reaction.

Typically, network design is constrained to guarantee a minimum availability to the traffic demands. In the following subsections, we provide two different models to estimate such metrics. Both models are based on the definition of the Shared Risk Groups (SRG) in the network.

3.7.1   Shared Risk Groups

We refer to *risks* as the possible causes that can create failures in a network. For instance, a risk could be the malfunction of a particular router, the risk of fire in a particular site affecting all the equipment located there, or the risk of accidentally cutting a duct between two cities carrying a multitude of links. In this context, we define a *Shared Risk Group* (SRG) f, associated to a particular risk, as the set of network nodes and network links that simultaneously fail if the associated risk is materialized. Given the one-to-one relation between risks and SRGs, we will use both terms interchangeably throughout the text. We denote as $\mathcal{F}$ the set of all risks/SRGs identified in a network.

Note that an SRG can be composed of one or more nodes and/or links. And that a single node or link can belong to zero, one, or more SRGs. The example in Fig. 3.9 helps us to illustrate this. In this network, we identify three risks of failure, associated to three different ducts that can be accidentally cut. Note that links 1–2 and 1–3 share the f_1 duct so if it is cut, both links fail simultaneously. Also note that link 1–2 belongs to two SRGs and link 2–3 to none.

Given a particular risk f, we denote $MTBF_f$, $MTTR_f$ and A_f as its MTBF, MTTR and availability. The estimation of such values is typically an input to network design, coming from a risk assessment study. For instance, Net2Plan gives full flexibility to configure the SRGs in the network, their MTBF/MTTR values and the associated failing nodes and links in an arbitrary form.

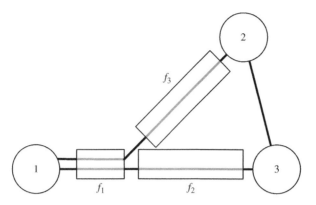

Figure 3.9   Example. Three SRGs f_1, f_2, f_3 associated to three ducts that is estimated that can be accidentally cut

3.7.2   Simplified Availability Calculations

In this simplified model, two assumptions are made:

- Independent risks: The events of failure in two SRGs are statistically independent. Thus, the probability that two SRGs f_1, f_2 are currently working correctly is given by $A_{f_1} A_{f_2}$.
- Disjoint SRGs: There is no link nor node in the network that belongs to more than one SRG. This permits simplifying the notation and write A_e and A_n to refer to the availability of the SRG to which e and n belong, respectively. If a link/node does not belong to any SRG, it will never fail, and thus has an availability of one.

The independent and disjoint SRG assumptions ease the calculation of the availability of services that require the simultaneous correct operation of multiple nodes and links. For instance, let us assume a service represented by a traffic demand d, whose traffic is carried through a particular path p. We assume that the chances of a node failure are neglected and thus only link failures are considered. The availability of a demand d is equal to A_p, the availability of the path, equal to the probability of having all the traversing links working correctly. This is given by:

$$A_d = A_p = \prod_{e \in p} A_e$$

If the primary path p is protected with a 1+1 scheme by (disjoint) backup path p', the demand availability is given by:

$$A_d = 1 - (1 - A_p)(1 - A_{p'}) = 1 - \left(1 - \prod_{e \in p} A_e\right)\left(1 - \prod_{e \in p'} A_e\right)$$

3.7.3   General Model

In this section, we describe a general model which drops the SRG disjointness assumption of the simplified model. In this case, for example since two links/nodes can share a risk, we cannot consider the probability of having both simultaneously active as the product of their availabilities.

To cope with this difficulty we consider a set S of possible network states in which the network can be. A state $s \in S$ is defined by the set $down(s)$ of risks that are in failure, and the set $up(s) = F - down(s)$ of risks that are not. We denote as s_0 to the no-failure state, when no SRGs are down. By applying the SRG statistical independence assumption, the probability π_s of finding the network in state s is given by:

$$\pi_s = \prod_{f \in up(s)} A_f \prod_{f \in down(s)} (1 - A_f)$$

In particular, the fraction of time in which the system is in the no-failure state is given by:

$$\pi_{s_0} = \prod_{f \in F} A_f$$

Note that π_s values can be computed independently of the recovery system applied (if any), and only depends on the set of SRGs identified in the network and their availabilities. Also note, however, that the number of possible states in a system grows with $2^{|F|}$.

Given a particular network state s, we denote $h_d(s)$, as the amount of traffic of demand d that is carried in the network when, starting from a no-failure situation, all the SRGs in $down(s)$ fail in any particular order. That means $h_d(s)$ is a measure of the surviving traffic of demand d in state s. We assume that in the no-failure situation all the traffic is carried, such that $h_d(s_0) = h_d$.

Interestingly, $h_d(s)$ values can be easily computed for any network state, by tracking the reactions of the recovery mechanism, whatever its complexity is. In particular, this methodology can be applied to both protection and restoration schemes.

A limitation of the method is that the $h_d(s)$ values are computed considering that the network starts in a no-failure situation, and then failures in $down(s)$ occur one by one. However, in some recovery schemes, the $h_d(s)$ values can differ depending on the particular order in which failures occur, or be different if the network did not start in a no-failure situation, but, for example, in a state with more failures than s, where some SRGs are repaired. The only situation when $h_d(s)$ computations are exact are when the surviving traffic of a demand only depends on the state of the network and not on how this state is reached. This holds, for instance, in networks with no recovery strategy and in networks with dedicated protection schemes.

The computation of π_s and $h_d(s)$ values is enough to compute many availability performance metrics in the network:

- Availability of demand d (A_d): The fraction of *time* in which the 100% of the traffic of the demand is carried:

$$A_d = \sum_{s:h_d(s)=h_d} \pi_s$$

- Traffic survivability of demand d (A_d^H): The weighted average fraction of traffic of demand d carried, in all the network states:

$$A_d^H = \frac{\sum_s \pi_s h_d(s)}{h_d}$$

- Network availability (A): The fraction of *time* in which the 100% of the traffic of the network is carried:

$$A = \sum_{s:h_d(s)=h_d,\forall d\in D} \pi_s$$

- Network traffic survivability (A^H): The weighted average fraction of traffic carried, in all the network states:

$$A^H = \frac{\sum_s \pi_s \sum_d h_d(s)}{\sum_d h_d}$$

See that all previous metrics require the enumeration of all the network states in S, and computing π_s and $h_d(s)$ values for each of them. However, it is possible to obtain approximate values if we enumerate only a subset $S' \subset S$ of the possible states, such that the probability π_{ne} of finding the network in a non-enumerated state $s \notin S'$ is small enough.

We illustrate this with an example. Let us suppose a network with 50 SRGs identified, all of them with availability $A = 0.999$. The number of possible network states is given by 2^{50}, and the computation of the metrics enumerating all these states is considered prohibitive. If we consider only single-SRG failures, the number of states enumerated is 51, and the π_{ne} is given by:

$$\pi_{ne} = 1 - 0.999^{50} - 50 \times (1 - 0.999)0.999^{49} \approx 0.0012$$

The first substraction is the no-failure probability, and the second, the single failure probability. The availability estimations when not all the states are enumerated, become a pessimistic estimation, and it holds that:

$$A_d(S') \leq A_d \leq A_d(S') + \pi_{ne}$$
$$A_d^H(S') \leq A_d^H \leq A_d^H(S') + \pi_{ne}$$
$$A(S') \leq A \leq A(S') + \pi_{ne}$$
$$A^H(S') \leq A^H \leq A^H(S') + \pi_{ne}$$

That is, considering all the network states would increase the availability in, at most, π_{ne}, which would occur in the (rare) case in which all the traffic was carried in the non-enumerated states. In our example, if we are designing a system with a 0.999 availability target in mind (0.001 unavailability), it is better to enumerate a number of states such that the possible error π_{ne} becomes at least one order of magnitude lower than the maximum allowed unavailability. In our case, we satisfy this by just enumerating both single and double failures (1276 states), such that:

$$\pi_{ne} = 1 - 0.999^{50} - 50 \times (1 - 0.999)0.999^{49} - \binom{50}{2}(1 - 0.999)^2 0.999^{48} \approx 0.000019$$

As a final remark, the *availability report* functionality of Net2Plan implements the model described in this section to estimate the availability and survivability performances of a network. Availability estimations can be also produced using the built-in simulator that tests the recovery mechanism under failure and repairing events randomly created according to the SRG information.

3.8   Network Utility and Fairness in Resource Allocation

Many network design problems are versions of allocation problems, in which resources have to be assigned to different entities, under several constraints. For instance, in congestion control problems, traffic flows are assigned the rate they are allowed to inject in the network. The allocation of different flows is coupled, since the sum of the traffic of the demands traversing each link cannot exceed its capacity. Another example is the assignment of transmission power to the different users of a cellular network. In this case, assigning more power to a user means increasing the bandwidth of its connection with the base station, but generating more interference that may degrade the bandwidth of other users.

The NUM (*Network Utility Maximization*) model is a framework for addressing this type of problems, as an application to communication networks of the social welfare maximization principle in economics, in which resources are allocated to maximize the sum of the well-being perceived by the individuals. Let $\mathcal{A}$ be a set of users (e.g., demands in a bandwidth assignment problem, transmitters in a cellular network, ...), to which we have to assign resources. Given a user $a \in \mathcal{A}$, we should decide the amount x_a of resources assigned to it. For this, we define the *utility function of user a*, $U_a(x_a)$, a function that returns the utility (which can be interpreted as a "profit") that user a perceives depending on the amount of resources x_a granted. Utility functions are always non-decreasing, meaning that assigning more resources to a user (higher x_a) is always perceived as better (higher $U_a(x_a)$).

In the general form of the NUM problem (3.16), resource allocation $x = (x_a, a \in \mathcal{A})$ is targeted to maximize the sum of the utilities perceived by all the users, subject to a general set of constraints $x \in \mathcal{X}$:

$$\max_x \sum_a U_a(x_a), \quad \text{subject to:} \quad x \in \mathcal{X} \tag{3.16}$$

Different shapes of the utility function U_a result in different allocations when NUM framework is applied. In this section, we are interested in showing the connection between the particular form of the utility function, and the *fairness* of the resulting resource allocation. In this context *fair* means avoiding those allocations where some users are granted a high amount of resources (high x_a), while others, comparatively, suffer starvation (low x_a).

3.8.1   Fairness in Resource Allocation

Intuitively, fairness in resource allocation means avoiding those assignments where some users are granted a high amount of resources (high x_a), while others, comparatively, suffer starvation (low x_a). However, formalizing the essence of what a fair resource allocation is, is not an easy task, and fairness has been defined in a number of different ways. The definitions have been mostly presented in the context of the problem of congestion control in data networks, and we generalize them here for any resource allocation.

One of the most common fairness notions is *max-min fairness*. An allocation x is said to be max-min fair if the resources granted to any user a_1 cannot be increased without decreasing the resources of some other user a_2 which in x received less bandwidth ($x_{a_2} \leq x_{a_1}$). As a side effect, this policy tries to maximize the resources granted to the user with minimum resources allocated (and this motivates the name *max-min fairness*).

In [20], Kelly proposed the concept of proportional fairness. A vector x^* is proportionally fair if, for any other feasible allocation x, the aggregate of the proportional change of x with respect to x^* is negative:

$$\sum_a \frac{x_a - x_a^*}{x_a^*} \leq 0, \quad \forall x \text{ feasible}$$

That is, the percentages of increases/decreases with respect to any other allocation should result in a negative sum.

In [21], Mo and Walrand extended the notion of proportional fairness. Let $w = (w_a, a \in \mathcal{A})$ be a vector of positive weight coefficients and α a non-negative number. A resource allocation a^* is said to be (α, w)-proportionally fair if for any other feasible allocation x it holds that:

$$\sum_a w_a \frac{x_a - x_a^*}{x_a^{*\alpha}} \leq 0, \quad \forall x \text{ feasible} \tag{3.17}$$

The w_a values can be used to give more importance to the resources allocated to some users. If all users are equal for the system ($w_a = 1, \forall a \in \mathcal{A}$), classical fairness notions are produced for some α values. In particular, 0-proportionally fair solutions ($\alpha = 0$) are those that maximize the total amount of resources granted $\sum_a x_a$. When x_a means the amount of bandwidth assigned to a source a in a network, this solution is the one which maximizes the network throughput. Actually, as we will see later in this section and in Chapter 6, maximum throughput solutions can be arbitrarily unfair, granting all the resources to some users and zero to others. If $\alpha = 1$ we have the Kelly's notion of proportional fairness. In addition, it can be shown that max-min fairness solutions are obtained when $\alpha \to \infty$ [21].

There is no consensus on which particular value of α is best suited for being "fair enough" in a resource allocation context. Actually, this decision is clearly problem dependent. Lower values of α tend to produce solutions where the total amount of resources $\sum_a x_a$ is higher, but with larger differences between the resources x_a allocated to different users (more "unfair"). In turn, higher α values reduce the difference between users, commonly at a cost of a lower aggregated throughput.

3.8.2 Fairness and Utility Functions

The importance of the definition (3.17) of (w, α) fairness is that, if appropriate utility functions U_a are used, the optimum solutions of NUM resource allocation problems, are also (w, α)-fair. This connection was presented in [21], for a basic version of the NUM problem for bandwidth assignment. The following proposition extends its application to a more general set of resource allocation problems.

Proposition 3.1 Let us define the generalized NUM problem for resource allocation, with decision variables (x, y). Vector $x = \{x_a, a \in \mathcal{A}\}$ represents the resources to assign to the users, while y represents any arbitrary set of auxiliary variables. We define a generalized resource allocation problem as:

$$\max \sum_a U_a(x_a) \quad \text{subject to } (x, y) \in \mathcal{X} \tag{3.18}$$

where $\mathcal{X}$ is an arbitrary non-empty, closed, convex set. Let the utility functions in (3.18) be:

$$U_a(x_a) = \begin{cases} w_a x_a & \text{if } \alpha = 0 \\ w_a \log x_a & \text{if } \alpha = 1 \\ w_a \frac{x_a^{1-\alpha}}{1-\alpha} & \text{if } \alpha > 0, \alpha \neq 1 \end{cases} \tag{3.19}$$

Then, it holds that a resource allocation (x, y) is (α, w)-proportionally fair if, and only if, it is the optimum solution of (3.18).

Proof. The proof is based on a classical optimality condition in convex optimization [[22], Prop. 3.1] that states that given $\mathcal{X}$ a non-empty, closed and convex set, and F a convex function in $\mathcal{X}$, a vector $x^* \in \mathcal{X}$ is an optimal solution of the problem $\min_{x \in \mathcal{X}} F(x)$ if and only if, $(x - x^*) \nabla F(x^*) \geq 0$, for every $x \in \mathcal{X}$. Applying this property to problem (3.18) we have that a solution (x^*, y^*) is optimal to (3.18) if and only if it is (w, α)-fair according to (3.17).

In the following example, we illustrate the different allocations that different notions of fairness produce. We focus on a network of two links and three demands like the one in Fig. 3.10. Rate allocation is performed solving the NUM problem:

$$\max \sum_d U_a(h_d), \quad \text{subject to:}$$

$$h_0 + h_1 \leq 1$$

$$h_0 + h_2 \leq 1$$

where $U_a(h_d)$ is an utility function of the form (3.19) with all w coefficients equal to 1. In our example, assigning more bandwidth to demand d_0 consumes more link capacity (since it traverses two links), than granting bandwidth to d_1 or d_2 (that only traverse 1). Figure 3.11 shows the allocations for different α values. As we can see, small α values tend to allocate a reduced amount of bandwidth to d_0 and more to d_1 and d_2. Higher α results in a more equal bandwidth distribution, at a cost of reducing the throughput. As $\alpha \to \infty$ the solution approximates to the max-min fairness, which in this example assigns the same rate to all demands (although this does not hold in general networks).

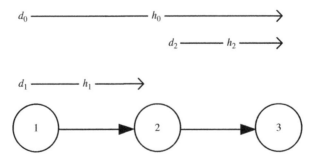

Figure 3.10 Example. Bandwidth allocation example. Demands d_0, d_1, d_2 are assigned a rate h_1, h_2, h_3, respectively. Link capacities are equal to one

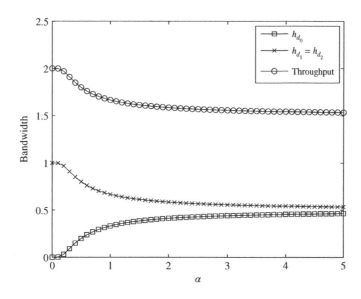

Figure 3.11 α-Fair allocations in Fig. 3.10, for different α values

3.8.3   Convexity Properties

Figure 3.12 plots utility functions of the form (3.19) for different $\alpha \geq 0$ values. It is easy to see that utilities are concave functions and strictly concave for $\alpha > 0$. Being concave means that a sort of diminishing returns effect occurs in resource allocation, that is, increasing the amount of resource granted to a user from z to $z + 1$ means a higher increase in utility than increasing a unit of resource from $z + 1$ to $z + 2$. Interestingly, the sum of the utilities $\sum_a U_\alpha(x_a)$ is a concave function of vector x and NUM problems for which the set of feasible solutions $\mathcal{X}$ is convex are convex optimization problems that can be solved efficiently.

NUM modeling will be applied in several contexts throughout the book, such as the fair assignment of transmission power or access probabilities to the links in wireless networks, or the fair rate assignment to the flows subject to congestion control.

3.9   Notes and Sources

Multiple models for packet traffic exist in the literature, too many to cite them all. We refer to classical books on the topic like [2, 5, 23, 24]. Markovian modulated processes, where a source behavior is dependent on an internal state for which the transition probabilities are known and regression models, where the traffic intensity depends on a (usually linear) relation with previous inputs and outputs, are probably the most popular. It has been shown that self-similarity and long-range dependence (related to the chances of long bursts) can appear in the aggregation of such sources.

No closed-form formulas ara available for sophisticated traffic models. In the scope of network-wide optimization models, the interest on precise delay predictions becomes less and less important as bit rates increase, making buffering delays negligible with respect to,

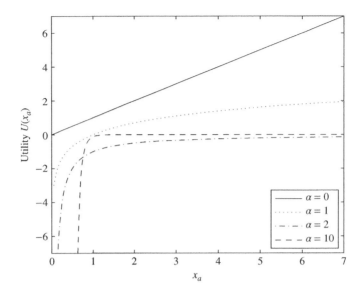

Figure 3.12 α-Utility functions $U_\alpha(x_a)$ ((3.19), $\alpha = 0, 1, 2, 10$, $w_a = 1$)

for example, propagation delays. Expressions (3.1) and (3.3) capture the main trends in packet delay and their simplicity makes them appealing for network models. A recent result producing simple approximations in self-similar models is [25].

Apart from the references in the text, a plethora of results exist for blocking probability models in loss networks, many of them presented in teletraffic symposiums for which proceedings are not easy to access. Fortunately, excellent compilations like [9] or [6] make them available to a wider audience.

The basic introduction to network recovery has been extracted from [19]. The general model presented for estimating the availability of arbitrary network recovery schemes is original to the best of the author's knowledge, although it is established on standard and simple principles.

Utility maximization in the context of computer networks has been first introduced in [20]. Max-min fairness has its routes in economics and philosophy. A seminal application in computer networks is the classic Bertsekas and Gallager book [26] (Section 6.5.2), [27] contains a more recent survey. Proportional fairness was introduced by Kelly [20] in the context of network communications, and corresponds to the log utilities that enforce the Nash bargaining solution in economics. α-fair utilities also have their roots in the economics [28]. The first application to resource allocation in communication networks appears in [21]. Proposition 3.1 generalizing the connection of fairness and utility functions in [21] is original, although established in very simple properties of convex programs. Utility maximization has become a useful tool to address multiple network problems, as will be shown throughout the book. Some books covering the topic are [29, 30], and [31]. A path breaking contribution for understanding the interactions among network layers in the framework of network utility maximization is [32].

3.10   Exercises

3.1   Let us define the average end-to-end delay in a multicast tree p as the average among the end-to-end delay to each destination:

$$T_p = \frac{1}{|b(p)|} \sum_{n \in b(p)} \sum_{e \in p_n} T_e$$

Using this expression, find a simple formula for the average network delay in a network fed with multicast traffic. Show that the resulting formula is not convex with respect to x_p variables.

3.2   Show that expressions (3.1), and (3.3) for the average link buffering delay are strictly increasing and strictly convex functions of ρ_e. Show that they are also strictly decreasing and strictly convex functions with respect to capacity variables.

3.3   Let T_e by a link delay estimation for link e, being a strictly increasing and strictly convex function with respect to y_e variables and strictly convex with respect to u_e variables. Show that the average network delay expression T given by (3.5) is also strictly convex with respect to u_e, y_e, or ρ_e variables.

3.4   We want to compare two options: (i) a 10 Mbps link fed by a 6 Mbps flow and (ii) two 5 Mbps links fed by 3 Mbps flows each. Average packet size is 500 bytes. Compute the M/M/1 average delay in both cases and comment on the sentence: "network designs targeted to minimize average network delays tend to concentrate traffic into fewer links of higher capacity". Repeat the computations when the average packet length is 64 bytes. What are the trade-offs network architects face when optimizing the packet sizes?

3.5   We want to compare two options: (i) a 10 Mbps link fed by an average load of 6 Mbps of random connection arrivals (occupying 1 Mbps of capacity each), and (ii) two 5 Mbps links fed by an average load of 3 Mbps of random connection arrivals (occupying 1 Mbps of capacity each). Compute the blocking probability in each case. Comment on the sentence: "network designs targeted to minimize blocking probabilities tend to concentrate the traffic into fewer links of higher capacity".

3.6   Let e be a 10 Mbps link, fed by Poissonian connection request arrivals where six connections per minute arrive in average, connection average duration is 1 min, and each connection occupies 1 Mbps. Compare the blocking probability obtained with the cases: (i) the number of connection arrivals per minute is doubled and the average connection duration is halved, (ii) the number of connection arrivals per minute is doubled, but each connection occupies 500 Kbs. Is there any benefit in fractioning the traffic in lower size connections? Why?

3.7   Net2Plan online algorithm `Online_evProc_generalProcessor` in the repository implements three connection admission control rules for incoming connections: (i) alternate routing among the k-shortest paths, (ii) load sharing using a route selection probability proportional to the routes carried traffic, and (iii) least-congested routing

(LCR) selection rule among the k-shortest paths. Use Net2Plan simulation function-alities in the network and traffic in `example7Nodes.n2p` file, setting link capac-ities to 20 units, to compare the performance of the three options. Use the built-in generator for exponential interarrival times with connection size equal to one. Use $k = 10$ in alternate and LCR routings. In load sharing, the routes carried traffic are set before calling the simulation to those that minimize the average network blocking using `Offline_fa_xpFormulations` algorithm. Simulate 10^6 events with a transitory of 10^5 events.

3.8 Repeat the previous experiments restricted to alternate routing and LCR. Now use dif-ferent connection sizes for different demands chosen arbitrarily. Find the link capacity (common to all the links) that make the simulated blocking fall below 10^{-3} in the two cases.

3.9 Let $\mathcal{G}(\mathcal{N}, \mathcal{E})$ be a network, and let $\{y_e, e \in \mathcal{E}\}$ be the known traffic carried by each link e. The network cost C is given by:

$$C = 35.4 + \sum_e 2u_e + 12u_e^\alpha$$

where $\alpha \in (0, 1)$ is a given constant. Show that C is a strictly concave function with respect to $u = \{u_e, e \in \mathcal{E}\}$ variables. Show that C is a strictly convex function with respect to $\rho = \{\rho_e = y_e/u_e, e \in \mathcal{E}\}$ variables.

3.10 Figure 3.13 shows a ring network, the ducts where the links are placed, and each duct length. Each duct is assigned a risk of accidental cutting, which would break all the traversing links. Ducts suffer an average of two cuts per year, per 1000 km in length, and have an average repair time of 12 h. (i) Identify the SRGs, their associated links and nodes, and their availability, (ii) compute the availability of a demand from node 1 to node 4, with a primary route in link 1–4, and 1+1 protected through path 1–2–3–4, and (iii) use Net2Plan availability report to estimate the availability and survivability of all the demands, and the network availability and survivabilities.

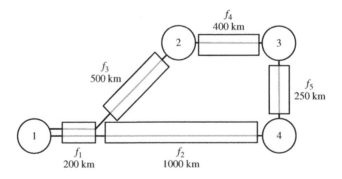

Figure 3.13 Figure Exercise 3.10

3.11 We want to design a shared protection system $M : N$, to protect $N = 2$ connections. All the $M + N$ routes are SRG disjoint and have an availability $A_p = 0.999$. Indicate the number of backup routes M needed to provide a 0.99999 availability to the connections.

3.12 Net2Plan online algorithm `Online_evProc_ipOspf` implements the reactions to network failures of an IP network using the OSPF protocol. Use Net2Plan availability report, with this algorithm as a provisioning algorithm, and `Online_evGen_generalGenerator` as a failure generator, to compute the minimum integral link capacity (common to all links) that ensures an availability of 0.999 to the `example7Nodes.n2p` network. The identified network risks are one SRG per bidirectional link with a MTTF of 1 year and MTTR of 1 day, and assuming random exponential distributions..

3.13 Repeat the previous exercise, using the Net2Plan per SRG failure analysis report, to obtain the minimum integral link capacity, common to all the links that ensures a 100% availability under single SRG failure.

3.14 A set of users $\mathcal{A}$ is allocated resources in the network solving the NUM model $\max_{x \in \mathcal{X}} \sum_a U(x_a)$, where $U(x)$ are (w, α) utility functions in (3.19), and x_a reflects the bandwidth assigned to user a measured in bps. Does the optimum sum of utilities change if we solve a modified NUM problem where x_a is measured in Gbps? Does the allocation change?

References

[1] D. P. Heyman and M. J. Sobel, *Stochastic Models in Operations Research: Stochastic Optimization.* Courier Dover Publications, 2003, vol. 2.

[2] L. Kleinrock, *Queueing Systems.* 1975, vol. I: Theory, (Published in Russian, 1979. Published in Japanese, 1979. Published in Hungarian, 1979. Published in Italian 1992.).

[3] K. Park and W. Willinger, *Self-Similar Network Traffic and Performance Evaluation,* 1st edn. New York, NY, USA: John Wiley & Sons, Inc., 2000.

[4] I. Norros, "A storage model with self-similar input," *Queueing Systems,* vol. 16, no. 3-4, pp. 387–396, 1994.

[5] L. Kleinrock, *Queueing Systems.* Wiley Interscience, 1976, vol. II: Computer Applications.

[6] A. Girard, *Routing and dimensioning in circuit-switched networks,* ser. Addison-Wesley series in electrical and computer engineering: Telecommunications. Addison-Wesley, 1990.

[7] J. Kaufman, "Blocking in a shared resource environment," *Communications, IEEE Transactions on,* vol. 29, no. 10, pp. 1474–1481, Oct. 1981.

[8] J. Roberts, "A service system with heterogeneous user requirements," in *Performance of Data Communications Systems and Their Applications,* vol. 29, no. 10. Amsterdam, The Netherlands: North-Holland, 1981, pp. 423–431.

[9] K. Ross, *Multiservice loss models for broadband telecommunication networks,* ser. Telecommunication networks and computer systems. Heidelberg, Germany: Springer, 1995.

[10] F. P. Kelly, "Loss networks," *The Annals of Applied Probability,* vol. 1, no. 3, pp. 319–378, 08 1991.

[11] F. P. Kelly, "Blocking probabilities in large circuit-switched networks," *Advances in Applied Probability,* pp. 473–505, 1986.

[12] W. Whitt, "Blocking when service is required from several facilities simultaneously," *AT&T Technical Journal,* vol. 64, no. 8, pp. 1807–1856, 1985.

[13] E. Messerli, "Proof of a convexity property of the Erlang B formula," *The Bell System Technical Journal,* vol. 51, no. 951, p. 553, 1972.

[14] A. Jagers and E. Van Doorn, "On the continued Erlang loss function," *Operations Research Letters,* vol. 5, no. 1, pp. 43–46, 1986.

[15] A. Harel, "Convexity properties of the Erlang loss formula," *Operations Research,* pp. 499–505, 1990.

[16] P. Nain, "Qualitative properties of the Erlang blocking model with heterogeneous user requirements," *Queueing Systems,* vol. 6, no. 1, pp. 189–206, 1990.

[17] K. W. Ross and D. D. Yao, "Monotonicity properties for the stochastic knapsack," *Information Theory, IEEE Transactions on,* vol. 36, no. 5, pp. 1173–1179, 1990.

[18] (2008) Informe comisión del mercado de las telecomunicaciones (Spain), mtz 2008/626.

[19] J.-P. Vasseur, M. Pickavet, and P. Demeester, *Network recovery: Protection and Restoration of Optical, SONET-SDH, IP, and MPLS*. Elsevier, 2004.

[20] F. Kelly, "Charging and rate control for elastic traffic," *European transactions on Telecommunications*, vol. 8, no. 1, pp. 33–37, 1997.

[21] J. Mo and J. Walrand, "Fair end-to-end window-based congestion control," *IEEE/ACM Transactions on Networking (ToN)*, vol. 8, no. 5, pp. 556–567, 2000.

[22] D. P. Bertsekas and J. N. Tsitsiklis, *Parallel and Distributed Computation: Numerical Methods*. Athena Scientific, 1997.

[23] R. Cooper, Introduction to Queuing Theory, 2nd edn. 1981.

[24] D. Gross, J. F. Shortle, J. M. Thompson, and C. M. Harris, *Fundamentals of Queueing Theory*, 4th ed. New York, NY, USA: John Wiley & Sons, Inc., 2008.

[25] J. Chen, H. Bhatia, R. Addie, and M. Zukerman, "Statistical characteristics of queue with fractional brownian motion input," *Electronics Letters*, vol. 51, no. 9, pp. 699–701, 2015.

[26] D. Bertsekas and R. Gallager, *"Data networks. 1992," PrenticeHall, Englewood Cliffs, NJ*, 1992.

[27] D. Nace and M. Pióro, "Max-min fairness and its applications to routing and load-balancing in communication networks: a tutorial," *Communications Surveys & Tutorials, IEEE*, vol. 10, no. 4, pp. 5–17, 2008.

[28] J. W. Pratt, "Risk aversion in the small and in the large," *Econometrica: Journal of the Econometric Society*, pp. 122–136, 1964.

[29] S. Shakkottai, S. G. Shakkottai, and R. Srikant, *Network Optimization and Control*. Now Publishers Inc, 2008.

[30] R. Srikant, *The Mathematics of Internet Congestion Control*. Springer Science & Business Media, 2012.

[31] R. Srikant and L. Ying, *Communication Networks: An Optimization, Control, and Stochastic Networks Perspective*. Cambridge University Press, 2013.

[32] M. Chiang, S. H. Low, R. Calderbank, and J. C. Doyle, "Layering as optimization decomposition," *Proceedings of IEEE*, 2006.

4

Routing Problems

4.1 Introduction

Routing in communication networks is the particular set of rules that decide the path followed by the traffic units from their origin to destination nodes. The two main technological strategies to forward the traffic are *flow-based* routing and *destination-based* routing:

- In flow-based routing, network nodes are able to identify the demand d of the arriving traffic and make a different per-demand routing decision. Flow-based routing is character- istic of connection-oriented technologies like MPLS, ATM, frame-relay or SONET/SDH, where a flow completes a connection establishment stage before any data is carried. During this stage, the network decides and signals the flow route and configures the internal rout- ing tables of the traversed nodes accordingly. After that, the flow data frames are attached enough control information in their headers so that the intermediate nodes can enforce the routing previously defined.
- In destination-based routing, the network nodes apply the same routing decision to all the traffic units targeted to the same destination. This happens irrespective of the particular flow or demand the traffic belongs to or, for instance, the particular initial node of the traf- fic. IP and Ethernet networks are the main representatives of destination-based routing, as IP routers and Ethernet switches make forwarding decisions observing the packet IP desti- nation address and MAC destination address, respectively.

  Destination-based routing is typical of connectionless network layers, where a source can inject traffic (IP datagrams, Ethernet frames) without any prior connection establishment. The nodes' routing tables should be configured beforehand, associating to each destination the output link or links to forward the traffic. When a destination is assigned more than one output link, a splitting rule sets the fraction of traffic forwarded to each.

Naturally, destination-based routing offers less flexibility than flow-based routing. For instance, in the network of Fig. 4.1, the routing of demands d_1 and d_2 would be feasible in an MPLS network, where each demand is carried through a virtual circuit. However, this routing is not implementable in an IP network, since node 3 would have to distinguish between the

Optimization of Computer Networks – Modeling and Algorithms: A Hands-On Approach,
First Edition. Pablo Pavón Mariño.
© 2016 John Wiley & Sons, Ltd. Published 2016 by John Wiley & Sons, Ltd.
Companion Website: www.wiley.com/go/PavonMarinoSol16

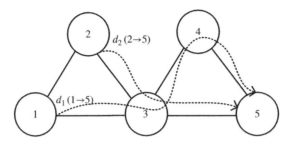

Figure 4.1   Example. Routing not valid in networks implementing destination-based routing

traffic of demand d_1 and d_2, which have the same destination node. Still, as we will see later, destination-based routing can be as bandwidth efficient as flow-based routing: any traffic that can be carried in a form can be carried in the other not needing extra capacity in the links.

This chapter models the problem of deciding the connection routes in flow-based networks and setting the routing tables in destination-based networks. First, we introduce the three main modeling strategies that lead to the flow-path, flow-link, and destination-link formulations. In the latter case, we elaborate on the case of shortest-path based routing in IP networks governed by the OSPF or IS-IS protocols. For this case, the forwarding follows the Equal-Cost Multi-Path (ECMP) rule, which determines that if a node has more than one shortest path to a destination, the traffic is split in *equal* fractions among all the output links in the shortest paths. Finally, we expose a comprehensive set of problem variants such as the routing of any-cast and multicast traffic, routing in the presence of protection and restoration schemes, or routing under variable traffic (multi-hour routing).

4.2   Flow-Path Formulation

The two characteristics of flow-path routing formulations are:

- Each demand $d \in D$ is associated (as an input to the problem) a set of *candidate paths* P_d, which are the only ones admissible for carrying traffic of d. Given a path p, $d(p)$ denotes the (unique) demand that p is associated to. We denote as $P = \bigcup_d P_d$ to the set of all candidate paths in the network, and $P_e \subset P$ is the subset of those paths traversing link e.
- There is a decision variable x_p for each candidate path $p \in P$ determining its carried traffic:

$$x_p = \{\text{Amount of traffic of demand } d(p) \text{ carried by } p\}, \quad \forall p \in P$$

In this section, we use as a case study the flow-path formulation (4.1) that finds the minimum congestion routing by maximizing u_u, the worst-case unused bandwidth in the links.

$$\max_{x, u_u} \ u_u \quad \text{subject to:} \tag{4.1a}$$

$$\lambda_d : \sum_{p \in P_d} x_p = h_d, \quad \forall d \in D \tag{4.1b}$$

$$\pi_e : \sum_{p \in P_e} x_p \le u_e - u_u, \quad \forall e \in \mathcal{E} \tag{4.1c}$$

$$v_p : x_p \ge 0, \quad \forall p \in P \tag{4.1d}$$

We use $\mathcal{N}$ to denote the set of nodes, $\mathcal{E}$ the set of links, and $u_e \ge 0$ the capacity of link e. The offered traffic is composed of a set of *unicast* demands D. Constraints (4.1b) state that all the traffic is carried: the sum of the traffic in the candidate paths of a demand d, equals its offered traffic h_d. Constraints (4.1c) mean that link traffic (the sum of the traffic in the traversing paths) does not exceed its capacity, minus the amount u_u. That is, u_u is the maximum bandwidth we can remove from *all* the network links, while still keeping the problem feasible[1]. Finally, (4.1d) prohibit a negative amount of carried traffic in a path.

An analogous flow-path formulation is obtained if we use $\hat{x}_p$ variables representing *fractions* of traffic carried in each path:

$$\hat{x}_p = \{\text{Fraction} \in [0, 1] \text{ of traffic of demand } d(p) \text{ (with respect to } h_{d(p)}) \text{ carried by } p\},$$

$$\forall p \in P \tag{4.2}$$

Then, the traffic actually carried by a path p is given by: $h_{d(p)}\hat{x}_p$. As we will see throughout the book, using $\hat{x}_p$ variables can sometimes ease the modeling of problem constraints, like, for example, non-bifurcated routing. The rewriting of formulation (4.1) using new decision variables is straightforward and left as an exercise.

4.2.1 Optimality Analysis

Problem (4.1) is linear and thus enjoys strong duality property. In this section, we show how the application of KKT optimality conditions and other general optimization results in the problem gives a significant insight on what the optimum routing looks like.

The Lagrange function of (4.1), put as a minimization problem, is given by:

$$L(x, u_u, \lambda, \pi, v) =$$

$$= -u_u + \sum_d \lambda_d \left(h_d - \sum_{p \in P_d} x_p \right) + \sum_e \pi_e \left(\sum_{p \in P_e} x_p - u_e + u_u \right) - \sum_p v_p x_p =$$

$$= u_u \left(\sum_e \pi_e - 1 \right) + \sum_p x_p \left(\sum_{e \in p} \pi_e - \lambda_{d(p)} - v_p \right) + \sum_d h_d \lambda_d - \sum_e u_e \pi_e \tag{4.3}$$

The term reorganization in the last equality eases the computation of the Lagrange minimization optimality conditions:

$$\frac{\partial L}{\partial u_u} = 0 \Rightarrow \sum_e \pi_e = 1 \tag{4.4a}$$

$$\frac{\partial L}{\partial x_p} = 0 \Rightarrow \sum_{e \in p} \pi_e = \lambda_{d(p)} + v_p, \forall p \in P \tag{4.4b}$$

---

[1] In the summation $\sum_{e \in P_e} x_p$ throughout the book we assume that if a path p traverses a link more than once, its x_p variable is summed the appropriated number of times.

The complementary slackness conditions are:

$$\pi_e \left(\sum_{p \in P_e} x_p - u_e + u_u \right) = 0, \forall e \in \mathcal{E} \tag{4.5a}$$

$$v_p x_p = 0, \quad \forall p \in P \tag{4.5b}$$

4.2.1.1 Multipliers Interpretation

We interpret $\pi_e \geq 0$ as a weight associated to each link. Then:

- Observing (4.4b), and since $v_p \geq 0$, the weight of any path p using π_e as link weights ($\sum_{e \in p} \pi_e$) is always *greater or equal* than $\lambda_{d(p)}$, the optimum multiplier for the path demand. If the path p carries traffic ($x_p > 0$), then $v_p = 0$ (4.5b), and the path weight is exactly $\lambda_{d(p)}$. That is:
  - If a path p carries traffic, it is a shortest path among $P_{d(p)}$, using π_e values as link weights. $\lambda_{d(p)}$ is the shortest path weight.
  - If a path p is *not* a shortest path, it will *not* carry traffic.
  - It is still possible to be a shortest path and not carry traffic ($v_p = 0, x_p = 0$).
- If $\pi_e > 0$, applying (4.5a) we get the result that the traffic in the link equals $u_e - u_u$ and thus the link is a bottleneck (has the minimum unused bandwidth among the network links). In contrast, if a link e is not a bottleneck, $\pi_e = 0$.

Summing up, if we use optimal π_e multipliers as link weights, the optimum routing will be one that *only* carries traffic using shortest paths among P. However, note that in our problem more than one shortest path can exist for a demand and in that case we cannot deduce from the optimum multipliers how the traffic should be split among them to achieve optimality. Actually, although we know that at least one splitting yields to the optimal routing, other splittings among the shortest paths could even violate link capacity constraints. Example 10.1 in Chapter 10 illustrates this situation in the context of an adaptive routing (dual) algorithm that iterates to find the optimum multipliers as an indirect form to optimize the routing. In Chapter 10 we show the practical importance of having strictly convex objective functions to avoid these difficulties.

4.2.1.2 Traffic Bifurcation in the Optimum

Since problem (4.1) is linear, we can use Prop. B.2 (the Fundamental Theorem of Linear Programming) to find a bound to the number of paths carrying traffic in an optimum solution.

Proposition 4.1 If problem (4.1) is feasible, there is an optimum solution for which, at most, $|D| + k$ paths carry traffic, where k is the number of bottleneck links with exactly u_u units of unused capacity.

Proof. According to Theorem B.2 in Appendix B, if a linear problem of the form $\min_x c^T x, s.t. Ax = b, x \geq 0$, is feasible, there is an optimum solution with a number of non-zero coordinates lower or equal than the number of equality constraints. We can reformulate

problem (4.1) converting link capacity inequality constraints into an equality as in (4.6) by
adding $|\mathcal{E}|$ slack variables $s_e \geq 0$:

$$\sum_{p \in P_e} x_p + s_e - u_u = u_e, \forall e \in \mathcal{E} \tag{4.6}$$

If a link e is not a bottleneck, $s_e > 0$. Then, the number of non-zero x_p variables is limited to
$|D| + |\mathcal{E}| - (|\mathcal{E}| - k) = |D| + k$.

Proposition 4.1 means that we can find an optimum solution to (4.1), where at most $|\mathcal{E}|$
demands are bifurcated, the worse (and rare) case when all the links are a bottleneck. Interest-
ingly, in many networks $|D| >> |\mathcal{E}|$ and an optimum routing exists where most of the demands
are not bifurcated. For instance, in a network with $|\mathcal{N}| = 100$ nodes, a demand between each
node pair ($|D| = 100 \times 99 = 9900$) and an average of six links per node (a reasonable number
in backbone networks) ($|\mathcal{E}| = 600$), at most $\approx 6\%$ of the demands would be bifurcated.

4.2.1.3 Sensitivity Analysis

Optimal multipliers to problem (4.1) provide us sensitivity information on how the optimum
routing would change if we modified or *perturbed* the problem constraints (see Appendix B
for details). In this section, we perturb the original problem as follows:

- The offered traffic of a demand d is augmented in z_d units (h_d is replaced by $h_d + z_d$ in
  (4.1)).
- The capacity of a link e is augmented in z_e units (u_e is replaced by $u_e + z_e$ in (4.1)).

Let $p^*(z)$ be the optimum unused bandwidth u_u, when the original problem is perturbed
by $z = \{z_d, d \in D, z_e, e \in \mathcal{E}\}$ values. $p^*(0)$ is the optimum u_u unused capacity in the original
problem. Applying the sensitivity results in Appendix B, we have that π^* and λ^* optimum
multipliers of the original problem are subgradients of the perturbation function p^* in the origin
$z = 0$ and thus it holds that:

$$p^*(z) \leq p^*(0) + \sum_e \pi_e z_e - \sum_d \lambda_d z_d$$

Then:

- Increasing the capacity of one link ($z_e > 0$) leaving the rest unchanged can increase the
  unused bandwidth u_u in, at most, $\pi_e z_e$ units. Reducing the link capacity ($z_e < 0$) worsens u_u
  in $\pi_e z_e$ units or more. Note that if $\pi_e = 0$, increasing the link capacity provides no congestion
  improvement ($p^*(z) \leq p^*(0)$).
- Decreasing a demand offered traffic ($z_d > 0$) leaving the rest unchanged, will improve the
  u_u value in at most $\lambda_d z_d$ units. Then, if $\lambda_d = 0$, reducing the demand yields to no congestion improvement. In its turn, increasing the offered traffic ($z_d < 0$), will worsen the idle
  bandwidth in at least $\lambda_d z_d$ units.

Note that since the objective function of (4.1) is linear, the dual function may be
non-differentiable at some points. For this reason, the optimum multipliers may be non
unique, and there is no possibility to interpret the multipliers as the partial derivatives of the
perturbation function in $z = 0$.

4.2.2 Candidate Path List Pre-Computation

Flow-path formulation forces the designer to pre-compute a list of admissible paths for each demand. This approach can have a number of advantages. Practical routings often have particular constraints, as avoiding too long paths, paths with loops, paths traversing specific nodes, and so on. The candidate path list is a good tool for applying these policies by just not including the invalid paths in the candidate list. However, if we want to compute an optimal routing that accepts any path as admissible, we are forced to populate the list with *all* the paths in the network, which can be an infinite number. Even, if we restrict to just the set of loopless paths, this number can still grow exponentially with the number of network links.

Example 4.1 In a network of $|\mathcal{N}| = 30$ nodes and one link between each node pair, the number of paths $|\mathcal{P}|$ traversing a maximum of two links is given by:

$$|\mathcal{P}| = 30 \times 29 + 30 \times 29 \times 28 = 25{,}320$$

which is a reasonable amount of paths for solving a linear flow-path formulation, with standard computing facilities. In turn, the number of candidate paths would be impractical if we just restrict them to the loopless paths (not traversing a node twice):

$$|\mathcal{P}| = 30 \times 29 + 30 \times 29 \times 28 + 30 \times 29 \times 28 \times 27 + \ldots \approx 7.2 \times 10^{32}$$

The next sections elaborate on two main workarounds to the issue of oversized candidate path lists: (i) limiting the list to the top paths (e.g., top k paths) in an elaborated ranking of paths and (ii) the application of *Candidate Path List Augmentation* (CPLA) techniques.

4.2.3 Ranking of Paths Elaboration

This technique consists of restricting the candidate path list to the k minimum cost paths for each demand, k being a design parameter. By doing so, the size of the candidate path list is limited to $k|D|$. The rationale behind this approach is that in many design case studies, a restricted candidate list (e.g., $k = 10$) has shown to be enough to produce optimal, or close to optimal, results.

Let $w_e > 0$ be a constant cost applied to every link in the network, and n and n' two nodes. We consider the cost of a path as the sum of the costs of its traversing links. According to this, the k-minimum cost path problem finds a ranking of k paths $(p_1, \ldots, p_k)$, from n to n' such that:

$$\sum_{e \in p_1} w_e \leq \sum_{e \in p_2} w_e \leq \ldots \leq \sum_{e \in p_k} w_e$$

and there is no other path from n to n' with a cost lower than p_k. Note that when $w_e = 1, \forall e$, the paths are ranked according to the number of traversed links.

The case $k = 1$ is the well-known shortest path problem, which can be solved in $\mathcal{O}(|\mathcal{E}| + |\mathcal{N}| \log |\mathcal{N}|)$ using the celebrated Dijkstra algorithm [1]. Several other algorithms have been proposed for the general case $k > 1$ (see [2] and references therein). It is of special interest the algorithm proposed by Yen [3], which computes the k-shortest *and loopless* paths between two nodes. That is, those paths with loops are excluded for the ranking[2]. Net2Plan tool provides methods to elaborate and handle candidate path lists, including different variants of the k-minimum cost path rankings.

Finally note that although there is no *a priori* guarantee of optimality of the resulting routing when a restricted candidate list is used, it may be possible to make an *a posteriori* optimality check. Moreover, if the check does not certify that optimality has been reached, it suggests a set of paths that could be added to the candidate path list, with good chances of improving the solution. This is the essence of the CPLA technique described in Section 4.2.4.

4.2.4   Candidate Path List Augmentation (CPLA)

Candidate Path List Augmentation (CPLA), is an application to the flow-path formulation of the so-called *column generation* method for linear programs (e.g., see [4] or [5] for details). CPLA technique consists of solving the original flow-path problem with the set of admissible paths P, by sequentially solving flow-path formulations for a restricted candidate path list $P^R \subset P$, with a much lower number of paths. The rationale behind CPLA technique is that in many practical problems, in the optimum, the number of paths carrying traffic of a demand is small (e.g., usually one, up to three).

The CPLA technique has been frequently used for linear problems, the reader is referred to [6] for examples. In this book, we present a generalization of this technique, valid for a somewhat wider set of problems. Let (4.7) be a generalized flow-path routing problem, for a candidate path list P. Decision variables x_p are the flow-path variables, while $z = \{z_1, \ldots, z_K\}$ denotes an arbitrary set of decision variables. Constraints (4.7b–d) are the standard flow-path constraints, while functions g_{je} are problem-specific convex constraints.

$$\min_{x,z} \sum_e f_e(x, z), \quad \text{subject to:} \tag{4.7a}$$

$$\lambda_d : \sum_{p \in P_d} x_p = h_d, \quad \forall d \in D \tag{4.7b}$$

$$\pi_e : \sum_{p \in P_e} x_p \le u_e, \quad \forall e \in \mathcal{E} \tag{4.7c}$$

$$\upsilon_p : x_p \ge 0, \quad \forall p \in P \tag{4.7d}$$

$$w_j : \sum_e g_{je}(x, z) \le 0, \quad j = 1, \ldots, J \tag{4.7e}$$

Note that this definition includes non linear convex problems. We are interested in comparing the optimal solution of problem (4.1), when a restricted candidate path list is used ($P = P^R$) (*restricted problem*), and when all the admissible paths are included in the list ($P = P^A \supseteq P^R$) (*original problem*). Given a solution x to the restricted problem, we denote as $a(x)$ its extension to the original problem given by:

$$a(x)_p = \begin{cases} x_p, & \text{if } p \in P^R \\ 0, & \text{if } p \in P^A - P^R \end{cases}$$

---

[2] A trivial but useful ranking variant is admitting for each demand the minimum cost path and all the paths with a cost up to, for example, 50% greater than it.

That is, both x and $a(x)$ reflect the same routing, and the variables $a(x)_p$ for those paths which are not in the restricted problem are set to zero. We make the following assumptions:

- Assumption 1: f_e and g_{je} are convex differentiable functions and both restricted and augmented problem versions enjoy the property of strong duality.
- Assumption 2: f_e and g_{je} are solely functions of z and of $y_e = \sum_{p \in P_e} x_p$, the total amount of traffic traversing a link. This means that:

$$f_e(x, z) = f_e(a(x), z), \quad \forall (x, z) \text{ feasible} \tag{4.8a}$$

$$g_{je}(x, z) = g_{je}(a(x), z), \quad \forall (x, z) \text{ feasible} \tag{4.8b}$$

$$\frac{\partial f_e}{\partial y_e}(y, z) = \frac{\partial f_e}{\partial x_p}(x, z) = \frac{\partial f_e}{\partial x_p}(a(x)x, z), \quad \forall (x, z) \text{ feasible} \tag{4.8c}$$

$$\frac{\partial g_{je}}{\partial y_e}(y, z) = \frac{\partial g_{je}}{\partial x_p}(x, z) = \frac{\partial g_{je}}{\partial x_p}(a(x)x, z), \quad \forall (x, z) \text{ feasible} \tag{4.8d}$$

We remark that functions like average path or network delay, Erlang-B blocking probability, average number of hops and so on. satisfy previous assumptions and can fit into these definitions.

Proposition 4.2 Let $s = (x^*, z^*, \pi^*, \lambda^*, w^*)$ be an optimal primal-dual solution to the restricted version of problem (7.1) (set of paths $\mathcal{P}^R$). Then, under the conditions and assumptions described in this section, it holds that $(a(x^*), z)$ is a primal optimal solution to the augmented problem (set of paths $\mathcal{P}^A \supseteq \mathcal{P}^R$), if and only if:

$$\lambda_d^* = \min_{p \in \mathcal{P}_d^A} \sum_{e \in p} c_e^*, \quad \forall d \in D \tag{4.9}$$

where c_e^* is a cost associated to each link given by:

$$c_e^* = \pi_e^* + \frac{\partial f_e}{\partial y_e}(x^*, z^*) + \sum_j w_j^* \frac{\partial g_{je}}{\partial y_e}(x^*, z^*) \tag{4.10}$$

Previous condition means that, if we use c_e^* as the cost per link, λ_d^* multipliers for each demand, must be the cost of the shortest paths among all the paths in $\mathcal{P}_d^A$.

Proof. The Lagrangian minimization optimality conditions in the restricted problem are, after reorganizing terms, and applying $v_p \geq 0$:

$$\sum_{e \in p} \left(\frac{\partial f_e}{\partial x_p}(x^*, z^*) + \pi_e^* + \sum_j w_j^* \frac{\partial g_{je}}{\partial x_p}(x^*, z^*) \right) \geq \lambda_{d(p)}^*, \quad \forall p \in \mathcal{P}^R \tag{4.11a}$$

$$\sum_{e \in p} \frac{\partial f_e}{\partial z_k}(x^*, z^*) + \sum_j w_j^* \frac{\partial g_{je}}{\partial z_k}(x^*, z^*) = 0, \quad \forall k = 1, \dots, K \tag{4.11b}$$

When a path p carries traffic, expression (4.11a) is an equality. By applying (4.9) and (4.8) we have that expression (4.11a) holds also in the augmented problem, for all $p \in \mathcal{P}^A$. In addition, (4.11b) is satisfied in the augmented problem since both f_e and g_{je} functions just depend on the total traffic in the links, which does not change when we augment the path lists. Then, it is easy to see that solution $s^A = (a(x^*), z^*, \pi^*, \lambda^*, w^*)$ is an optimal solution for the augmented problem, since the rest of the optimality conditions hold: (i) the $(a(x^*), z)$ solution is feasible since (x^*, z^*) is, (ii) π^* multipliers are non-negative, and (iii) complementary slackness holds since it holds in the restricted problem.

For the necessary condition, let us now suppose that an augmented solution is optimal and all the paths in $\mathcal{P}^A - \mathcal{P}^R$ carry zero traffic. Then, it must be also optimal for the restricted problem. Applying (4.11a) to the augmented problem, it is easy to see that (4.9) must hold.

Proposition 4.2 permits certifying if an optimal solution to the restricted problem is optimal to the original one: if the shortest path cost for each demand d using c as link costs equals λ_d, the optimum has been reached. If the solution is not optimal, a good heuristic attempt to get closer to the optimum is augmenting the set $\mathcal{P}^R$, adding the shortest paths for the demands not passing the test. Although it is not guaranteed that these new paths will enter in the optimal solution in any future iteration, empirical tests support this approach [6]. Algorithm 3 summarizes the pseudocode of the CPLA technique described.

Algorithm 3 CPLA Algorithm

1: Initialize $\mathcal{P}_d^R$ sets in any form (e.g., k-minimum cost paths).
2: **repeat**
3:     Solve problem for path lists $\mathcal{P}_d^R$: π^*, λ^* are the optimal multipliers.
4:     **for all** $d \in D$ **do**
5:         p_d is the shortest path in $\mathcal{P}^A$ with link weights c_e^* as in (4.10)
6:         **if** $\sum_{e \in p_d} c_e^* < \lambda_d^*$ **then**
7:             $\mathcal{P}_d^R = \mathcal{P}_d^R \bigcup \{p_d\}$
8:         **end if**
9:     **end for**
10: **until** no new path is added in the loop

Alternatively, it is possible to modify the stopping condition in the CPLA, terminating the algorithm when the cost is close enough to a lower bound to the optimal cost of the original problem. This can be done in some cases by computing in each iteration a value of the dual function in the augmented problem. Exercise 4.6 shows an example of this.

4.3   Flow-Link Formulation

Let $\mathcal{G}(\mathcal{N}, \mathcal{E})$ be a network, and D be the set of offered demands, composed solely of *unicast* traffic. In flow-link formulation, the routing is represented using the so-called flow-link decision variables:

$$x_{de} = \{\text{Amount of traffic of demand } d \text{ that traverses link } e\}, \quad \forall d \in D, e \in \mathcal{E} \quad (4.12)$$

In (4.13) we show the equivalent flow-link formulation to (4.1), for finding the routing that maximizes the worse case unused link bandwidth u_u.

$$\max_{x, u_u} u_u \quad \text{subject to:} \quad (4.13a)$$

$$\lambda_{nd} : \sum_{e \in \delta^+(n)} x_{de} - \sum_{e \in \delta^-(n)} x_{de} = \begin{cases} h_d, & \text{if } n = a(d) \\ -h_d, & \text{if } n = b(d), \quad \forall d \in D, n \in \mathcal{N} \\ 0, & \text{otherwise} \end{cases} \tag{4.13b}$$

$$\pi_e : \sum_d x_{de} \le u_e - u_u, \forall e \in \mathcal{E} \tag{4.13c}$$

$$v_{de} : x_{de} \ge 0, \forall d \in D, e \in \mathcal{E} \tag{4.13d}$$

Constraints (4.13b) are the so-called *flow conservation constraints* that guarantee that all the traffic is carried in a form that *could be* a unicast routing. They will be explained in further detail in next subsection. Constraints (4.13c) mean that link carried traffic does not exceed its capacity minus the unused bandwidth. Finally, non-negativity constraints (4.13d) prohibit having a negative amount of traffic of a demand in a link.

An alternate flow-link formulation for the same problem can be obtained using fractional $\hat{x}_{de}$ decision variables:

$$\hat{x}_{de} = \{ \text{Fraction} \in [0, 1] \text{ of traffic of demand } d \text{ (with respect to } h_d) \text{ that traverses link } e \},$$

$$\forall d \in D, e \in \mathcal{E} \tag{4.14}$$

Again, formulation (4.13) can be rewritten taking into account that $x_{de} = h_d \hat{x}_{de}$. In particular, note that flow conservation constraints in this case become independent of the h_d values:

$$\sum_{e \in \delta^+(n)} \hat{x}_{de} - \sum_{e \in \delta^-(n)} \hat{x}_{de} = \begin{cases} 1, & \text{if } n = a(d) \\ -1, & \text{if } n = b(d), \quad \forall d \in D, n \in \mathcal{N} \\ 0, & \text{otherwise} \end{cases} \tag{4.15}$$

4.3.1   Flow Conservation Constraints

One flow conservation constraint (4.13b) exists for each node n and each demand d, stating that (see Fig. 4.2):

- The sum of the demand traffic that leaves n through its output links: $\sum_{e \in \delta^+(n)} x_{de}$ $\cdots$
- minus the sum of the demand traffic that enters n through its input links: $\sum_{e \in \delta^-(n)} x_{de}$ $\cdots$
- is: (i) h_d if node n is the origin of the demand (and thus produces an excess of h_d units of traffic of this demand), (ii) $-h_d$ if node d is the end node of demand d (and thus is a sink consuming h_d units of traffic), or (iii) zero in the rest of the nodes (what gets in, gets out), since intermediate nodes do not produce nor consume demand traffic and just forward it.

The $|\mathcal{N}| \times |D|$ flow conservation constraints can be rewritten in a compact matrix form using the concept of *incidence matrix* of a graph. Let us define the incidence matrix for the set of links $\mathcal{E}$, as a matrix C_{ne} with as many rows as nodes $|\mathcal{N}|$ and as many columns as links $|\mathcal{E}|$:

$$C_{ne} = \{1 \text{ if } n = a(e), -1 \text{ if } n = b(e), 0 \text{ otherwise}, \quad \forall n \in \mathcal{N}, e \in \mathcal{E}\}$$

Similarly, we define the incidence matrix for the set of demands D as a matrix C_{nd} with as many rows as nodes $|\mathcal{N}|$ and as many columns as demands $|D|$:

$$C_{nd} = \{1 \text{ if } n = a(d), -1 \text{ if } n = b(d), 0 \text{ otherwise}, \quad \forall n \in \mathcal{N}, d \in D\}$$

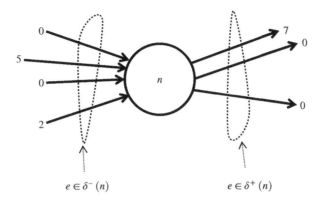

$$e \in \delta^- (n) \qquad\qquad\qquad e \in \delta^+ (n)$$

Figure 4.2   Flow conservation constraint example for a demand d and node n. x_{de} values are plotted next to incoming and outgoing links of the node. The node is not a source nor destination of the demand: the traffic of the demand entering and leaving the node are the same $(5 + 2 = 7)$

If we consider x_{de} and $\hat{x}_{de}$ variables as matrices of size $|\mathcal{D}| \times |\mathcal{E}|$, the flow conservation constraints (4.13b) and (4.15) can be rewritten as matrix equality constraints[3]:

$$C_{ne} x_{de}^T = C_{nd} \mathbf{diag}(h_d), \quad C_{ne} \hat{x}_{de}^T = C_{nd}$$

Where **diag** is the diagonal matrix with elements h_d in its diagonal and zero elsewhere.

4.3.2   Obtaining the Routing from x_{de} Variables

Flow conservation constraints are *necessary conditions* for a solution to be a routing fully satisfying the traffic demands. However, we may encounter some difficulties when trying to extract the path routing (x_p information) from the x_{de} variables. Figure 4.3 helps us to illustrate the two main problems:

- *Ambiguity in the routing*. More than one x_p routing can be compatible with x_{de} variables. For instance, in Fig. 4.3a the following two routings are consistent with x_{de} variables depicted: (i) routing five traffic units in each path 1-2-4-6-8 and 1-3-4-7-8, and (ii) routing five traffic units in each path 1-2-4-7-8 and 1-3-4-6-8.
- *Existence of isolated cycles*. Fig. 4.3b is an example of a x_{de} solution that contains an isolated cycle of traffic (nodes 2-6-5). This solution satisfies flow conservation constraints, but is not a real routing. Note that solutions with cycles are never optimal in problems where the objective function sums a strictly positive cost to each x_{de} variable. This does not hold in (4.13), so in this problem the solver could return a solution with isolated cycles, which should be removed offline.

Net2Plan tool provides a conversion function which produces a x_p routing from any feasible x_{de} variables. Ambiguities are resolved arbitrarily, and cycles (isolated or not) are eliminated so that the resulting routing consumes the same or less link capacity than x_{de} solution.

---

[3] Net2Plan incorporates a built-in library for computing incidence matrices, used in the flow-link formulations with JOM.

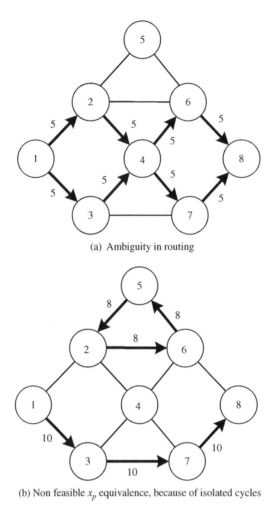

(a) Ambiguity in routing

(b) Non feasible x_p equivalence, because of isolated cycles

Figure 4.3   Example. Potential problems when converting a feasible x_{de} solution into a x_p routing, for a demand d with $h_d = 10$, from node 1 to node 8. The number in each link illustrates its x_{de} value

4.3.3   Optimality Analysis

The Lagrangian function of problem (4.13) put as a minimization problem is given by:

$$L(x, u_u, \lambda, \pi, v) =$$

$$= -u_u + \sum_{nd} \lambda_{nd} \left(C_{nd} h_d - \sum_e C_{ne} x_{de} \right) + \sum_e \pi_e \left(\sum_d x_{de} - u_e + u_u \right) - \sum_{de} v_{de} x_{de} =$$

$$= u_u \left(\sum_e \pi_e - 1 \right) + \sum_{de} x_{de} (\lambda_{b(e)d} - \lambda_{a(e)d} + \pi_e - v_{de}) + \sum_d h_d (\lambda_{a(d)d} - \lambda_{b(d)d}) - \sum_e u_e \pi_e$$

$$(4.16)$$

The Lagrange minimization and complementary slackness optimality conditions are:

$$\lambda_{a(e)d} - \lambda_{b(e)d} + v_{de} = \pi_e, \quad \forall d \in D, e \in \mathcal{E}, \quad \sum_e \pi_e = 1 \tag{4.17a}$$

$$\pi_e \left(\sum_d x_{de} - u_e + u_u \right) = 0, \quad \forall e \in \mathcal{E}, \quad v_{de} x_{de} = 0, \quad \forall d \in D, e \in \mathcal{E} \tag{4.17b}$$

We can interpret $\pi_e \geq 0$ multipliers as a weight associated to each link, similar to what happened in the flow-path formulation. In turn, λ_{nd} is interpreted as a potential of node n with respect to demand d, such that the traffic is transferred only from nodes of higher to lower (or equal) potential. Then:

- Let p_d be any set of links composing a path between demand d end nodes. According to (4.17b) we have:

$$\sum_{e \in p_d} \pi_e = \sum_{e \in p_d} \lambda_{a(e)d} - \lambda_{b(e)d} + v_{de} = \lambda_{a(d)d} - \lambda_{b(d)d} + \sum_{e \in p_d} v_{de}$$

Since $v_{de} \geq 0$, the sum of the weights of the traversed links is greater or equal than the difference of potential between the demand end nodes. If p_d carries traffic of d, then $x_{de} > 0$ and $v_{de} = 0$ for the path links, according to (4.17b). In summary, the traffic of each demand is only routed through shortest paths between demand end points, taking π_e as the link weights and the shortest path weight equals the end nodes difference of potential for that demand.
- If $\pi_e > 0$, applying (4.17b) we have that the link carries $u_e - u_u$ units of traffic, and thus is a bottleneck link. Similarly, if a link e is not a bottleneck, then $\pi_e = 0$.

Note that these conclusions are analogous to that of the flow-path formulation (4.1).

4.4 Destination-Link Formulation

We consider a network $\mathcal{G}(\mathcal{N}, \mathcal{E})$, with offered traffic given by a traffic matrix h of size $|\mathcal{N}| \times |\mathcal{N}|$, where $h_{st}, s \neq t$ is the traffic that source node s generates, targeted to node t. In destination-based routing, nodes take forwarding decisions depending solely on the traffic destination node, for example independently on the traffic ingress node or any notion of end-to-end demand. Destination-link formulation is a suitable form of modeling such situations, characterized by x_{te} decision variables:

$$x_{te} = \{\text{Amount of traffic targeted to node } t, \text{ that traverses link } e\}, \quad \forall t \in \mathcal{N}, e \in \mathcal{E}$$

A destination-link formulation for the maximum worst-case unused bandwidth u_u is shown in (4.18).

$$\max_{x, u_u} u_u \quad \text{subject to:} \tag{4.18a}$$

$$\sum_{e \in \delta^+(n)} x_{te} - \sum_{e \in \delta^-(n)} x_{te} = \begin{cases} h_{nt}, & \text{if } n \neq t \\ -\sum_s h_{st}, & \text{if } n = t \end{cases}, \quad \forall t, n \in \mathcal{N} \tag{4.18b}$$

$$\sum_t x_{te} \le u_e - u_u, \quad \forall e \in \mathcal{E} \tag{4.18c}$$

$$x_{te} \ge 0, \quad \forall t \in \mathcal{N}, e \in \mathcal{E} \tag{4.18d}$$

Constraints (4.18b) are a variation of the flow conservation constraints adapted to destination based routing, while (4.18cd) are the standard link capacity and non-negativity constraints.

Figure 4.4 helps us to illustrate the flow conservation constraints (4.18b). When a node n is not the destination of the traffic (Fig. 4.4a), the difference between what gets out and what gets in targeted to t, is exactly the amount h_{nt} of traffic that n generates to t. The constraint for node $n = t$ (Fig. 4.4b), means that the difference between what leaves the node (which should be zero in general) and what is received is just $-h_t = -\sum_s h_{st}$: the total amount of traffic that the rest of the network nodes generate, targeted to t. Note that this constraint is a linear combination of the previous ones and is thus redundant.

The optimality analysis for the destination-link formulation (4.18) is similar to the one for flow-link formulation, and left as an exercise. Again, multipliers π_e of the link capacity constraints (4.18c) are link weights according to which the optimum routing is a shortest path routing. Also, $\pi_e > 0$ indicates that the link is a bottleneck.

Finally note that destination-link formulation has $|\mathcal{N}| \times |\mathcal{E}|$ variables and just $|\mathcal{N}|^2$ flow conservation constraints, which is usually much less than the $|D| \times |\mathcal{E}|$ variables and $|\mathcal{N}| \times |D|$ conservation constraints in flow-link formulation. This usually results in a significant reduction of the solver time for finding an optimal solution numerically.

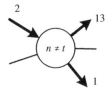

(a) Case $n \ne t$. Node n produces $h_{nt} = 13 + 1 - 2 = 12$ traffic units to t

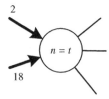

(b) Case $n = t$. The total network traffic targeted to t is $h_t = \sum_{st} h_{st} = 18 + 2 = 20$

Figure 4.4   Example. Flow conservation constraints for destination-link formulation

4.4.1   Obtaining the Routing Tables from x_{te} Variables

In destination-based routing networks, the routing tables in the nodes define the fraction of traffic to forward to each output link. This is represented by the f_{te} variables as follows:

f_{te} ={Fraction $\in [0, 1]$ of the traffic in node $n = a(e)$ targeted to t, that is forwarded to e},

$\quad \forall t \in \mathcal{N}, e \in \mathcal{E}$

In particular, the routing table in node n is reflected by the f_{te} values for the outgoing links of n. Obtaining the routing tables from the destination-link variables x_{te} is easy, by computing for each node the ratio of the traffic in their outgoing links, with respect to the own generated traffic plus the one received through the input links:

$$f_{te} = \frac{x_{te}}{h_{a(e)t} + \sum_{e \in \delta^-(a(e))} x_{te}}, \quad \forall t \in \mathcal{N}, e \in \mathcal{E} \tag{4.19}$$

If a node does not generate nor receive traffic to a destination t, both numerator and denominator in previous expression are zero. This indetermination can be arbitrarily solved by assigning a value $\hat{x}_{te} = 1$ to one outgoing link of the node (e.g., the one in the shortest path in number of hops to t) and zero to the rest.

4.4.2   Some Properties of the Routing Table Representation

Let $\mathcal{G}(\mathcal{N}, \mathcal{E})$ be a connected topology, and f_{te} a set of routing tables in it. We say that routing tables are *well-defined*, when for any offered traffic matrix (assuming links of infinite capacity), all the traffic units reach their destination node in a *finite or infinite* number of hops. If the number of hops is always finite, which means that the routing has no cycles, we say that the routing is *well-defined in the strict sense*.

Figure 4.5 helps us to illustrate previous definitions. When the routing tables are *not* well defined (Fig. 4.5a), it may happen that for some destination node t, the routing has cycles such that all the traffic to t entering such cycles never leaves them. When the cycles are open in the sense that a fraction of the traffic leaves them and eventually reaches the destination, the routing is well-defined, but not in the strict sense. For instance, in Fig. 4.5b, a traffic unit to node 4, arriving to node 2 could enter in a cycle between nodes 2-5, or among nodes 2-3-5. If the routing decision in each node is made randomly, the probability of not reaching the destination after k hops is positive for any $k \to \infty$. Finally, when the routing has no cycles (e.g., Fig. 4.5c), the destination node is always reached in a finite number of hops.

Naturally, we are interested in routings that are strictly well-defined (loopless). We try now to characterize them. First, we base our approach in a modified network version, where for each destination t two new nodes are added: a success node n_s and a loss node n_l. Both are sink nodes, meaning that the traffic arriving to them stays in them:

- The *success node* n_s represents the traffic that arrives to target node t that is not forwarded again to the network. Usually, *all* the traffic targeted to t that reaches t stays in it, but the

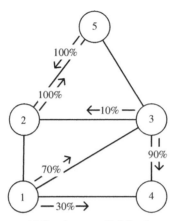

(a) Routing not well-defined

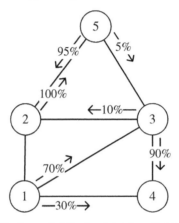

(b) Routing well-defined, not strictly well-defined

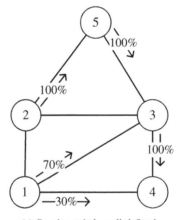

(c) Routing strictly well-defined

Figure 4.5   Routing table examples for destination node 4

node n_s permits us modeling situations where this does not happen when (maybe by a configuration error) the node t is part of a routing cycle.

- The *loss node* n_l receives the traffic targeted to node t, that any node drops and thus will never arrive to the destination. Dropping of traffic occurs in reality, for example, when a node receives traffic to a destination t, but has not configured any entry in the routing tables for it.

To represent the routing, we define the $|\mathcal{N}| + 2 \times |\mathcal{N}| + 2$ routing matrix for destination t, and we denote it as $P^{(t)}$. The coordinate (i, j) of $P^{(t)}$ contains the fraction of the traffic arriving to node i, that is forwarded to node j, $i, j \in \mathcal{N} \bigcup \{n_s, n_l\}$. This information is structured as shown below:

$$
P^{(t)} = \left(\begin{array}{c|c|c}
Q_{|\mathcal{N}|\times|\mathcal{N}|} & l_{|\mathcal{N}|\times 1} & s_{|\mathcal{N}|\times 1} \\ \hline
0 & 1 & 0 \\ \hline
0 & 0 & 1
\end{array} \right)
\tag{4.20}
$$

- Matrix Q contains the routing information among the nodes in $\mathcal{N}$, coming from f_{te} routing coefficients:

$$
Q_{ij}^{(t)} = \sum_{e : a(e)=i, b(e)=j} f_{te}, \quad \forall i, j \in \mathcal{N}
\tag{4.21}
$$

- Coordinate s_i indicates the fraction of traffic in node i that is forwarded directly to the success node. In general:

$$
s_i = \begin{cases} 0 & \text{if } i \neq t \\ 1 & \text{if } i = t \end{cases}
$$

but coordinate $s_t < 1$ when the network has loops that involve the target node t.
- Coordinate l_i contains the fraction of traffic arriving to node i, targeted to t, that node i drops. Traffic dropping can be caused by the absence of an appropriate entry in the routing table of a node. If these misconfigurations do not occur, $l_i = 0, \forall i \in \mathcal{N}$.

Note that routing matrices $P^{(t)}$ are stochastic matrices: (i) all their elements are non-negative and (ii) their rows add up to 1. Thus, they can be seen as a transition matrix of a discrete time Markov chain, with one state per network node, and $P_{ij}^{(t)}$ representing the transition probability between states. In our analogy, we interpret $P_{ij}^{(t)}$ as the probability that a traffic unit (e.g., a packet) in node i at time k, is forwarded to node j (and thus is there at time $k + 1$). States n_s and n_l are absorbing states of the matrix, since when they are reached, the chain stays in them eternally. This corresponds to a traffic unit that arrives the destination t (state n_s) or is dropped (state n_l). Also in this analogy, matrix $Q^{(t)}$ contains the forwarding decisions in the network.

The Markov chain representation helps us to extract some interesting properties of the routing tables.

Proposition 4.3 The routing represented by $\{P^{(t)}, t \in \mathcal{N}\}$ matrices is well-defined, if and only if the matrix $(I - Q^{(t)})$ is invertible.

Proof. Matrix $(I - Q^{(t)})$ is not invertible for a destination t if and only if it has an eigenvalue equal to $\lambda = 1$ (recall that the condition for being an eigenvalue of a matrix A is that det $(A - \lambda I) = 0$). Then, there is an eigenvector $v \neq 0$ such that $v = vQ^{(t)}$. The eigenvector represents an amount of traffic in the nodes that persists unchanged along time, eternally being routed in a closed cycle.

Proposition 4.4 The routing represented by $\{P^{(t)}, t \in \mathcal{N}\}$ matrices is well-defined in the strict-sense (and thus has no loops), if and only if the diagonal values in the so-called fundamental matrices $M^{(t)}$:

$$M^{(t)} = \sum_{k=0}^{\infty} (Q^{(t)})^k = (I - Q^{(t)})^{-1} \qquad (4.22)$$

are all equal to one, for all destination nodes $t \in \mathcal{N}$.

Proof. The matrix $M^{(t)}$ is the so-called *fundamental matrix of the Markov chain* associated to node t. A well-known property of this transition matrix is that its coordinate M_{ij} is the expectation of the number of times that the chain is in state j, assuming it starts in state i. If $i = j$, $M_{ii}^{(t)} \geq 1$ since the starting situation counts. Then, the chain never gets back to node i after it leaves it if and only if $M_{ii}^{(t)} = 1$. In our analogy, this means that the routing has no cycles if and only if $M_{ii}^{(t)} = 1$. Finally, a well-defined routing without cycles is strictly well-defined. ∎

The following proposition provides a relation to extract x_{te} destination-link information (and thus the amount of traffic traversing network links) from the routing tables f_{te}, valid also when the routing is well-defined, but has open loops.

Proposition 4.5 Let f_{te} be a well-defined routing, and h_{st} the offered traffic matrix. Then, the x_{te} values:

$$x_{te} = \sum_{i} h_{it} M_{ia(e)}^{(t)} f_{te}$$

are non-negative, and satisfy the flow conservation constraints (4.18b).

Proof. The product $h_{it} M_{ia(e)}^{(t)}$ is the amount of traffic originated in node i and targeted to t, that appears at node $a(e)$. By summing for all the possible origins i we have the total amount of traffic targeted to t, that reaches node $a(e)$. Thus:

$$\sum_{i} h_{it} M_{ia(e)}^{(t)} = h_{a(e)t} + \sum_{e \in \delta^-(a(e))} x_{te}, \quad \forall t \in \mathcal{N}, e \in \mathcal{E}$$

Then, a fraction of traffic given by f_{te} is routed to e. Resulting x_{te} values satisfy (4.19), and thus are non-negative and compatible with the flow-conservation constraints. ∎

Other metrics that can be easily extracted from the fundamental matrices are:

- The fraction of the traffic originated in i and targeted to t that successfully arrives to the destination is given by $M_{it}^{(t)} s_t$, which is also the i-th coordinate of vector $M^{(t)} s$.
- The fraction of the traffic originated in i and targeted to t that is dropped is $1 - M_{it}^{(t)} s_t$, which is also the i-th coordinate of vector $M^{(t)} l$.
- The expected number of hops in the network of a packet targeted to t, initiated in node i, is given by the i-th coordinate of the vector $M^{(t)} 1$, where 1 is the column vector of length $|\mathcal{N}|$ full of ones.

In Net2Plan, the routing in each network layer can be configured according to two forms. In so-called *source-routing*, end-to-end routes are assigned to demands (alike to x_p information). In *hop-by-hop routing*, the routing is defined by setting the routing tables f_{te}. In this latter case, Net2Plan detects open and closed loops and estimates routing statistics using the expressions in this section.

4.4.3 Comparing Flow-Based and Destination-Based Routing

Let $\mathcal{G}(\mathcal{N}, \mathcal{E})$ be a network, with offered traffic given by a traffic matrix h_{st}. We consider two possible forms of carrying the traffic: using flow-based routing (e.g., using MPLS tunnels), and destination-based routing (e.g., using an IP network with static routing tables). Trivially, any destination-based routing x_{te} could be reproduced with flow-based routing (x_{de}, x_p), by just appropriately defining the flows. We are interested in this section in assessing the opposite question: which are the limitations that destination-based routing imposes?

The previous question was partially answered in Section 4.1 (e.g., the example in Fig. 4.1), where it was shown that some routings which are possible in flow-based networks, could never be exactly reproduced with destination-based forwarding. However, the following property shows that for any feasible routing in a flow-based network, we could find a feasible destination-based routing, carrying the same offered traffic with same link capacities.

Proposition 4.6 Let $\mathcal{G}(\mathcal{N}, \mathcal{E})$ be a network and h_{st} a traffic matrix with the offered traffic between nodes $s, t \in \mathcal{N}, s \neq t$. Let $\mathcal{D}$ denote the set of demands created from this matrix: one demand between each node pair. Then, it holds that for every x_{de} solution carrying the offered traffic $\mathcal{D}$ that is feasible according to the flow-link constraints (4.13b–d), we can create a x_{te} solution feasible according to the destination-link constraints (4.18b–d), carrying the same traffic matrix h_{st} in the same network and link capacities.

Proof. By making $x_{te} = \sum_{d:b(d)=t} x_{de}$, it is easy to see that x_{te} values satisfy all the constraints (4.18b–d).

4.5 Convexity Properties of Performance Metrics

Assessing the convexity of a routing optimization problem may require determining if several performance metrics are convex functions of the routing variables. This task is eased by the fact that, for any network link e, the amount of traffic traversing the link is a linear expression with respect to the routing variables in any formulation:

$$y_e = \sum_{p \in P_e} x_p, \quad \text{in flow-path formulations}$$

$$y_e = \sum_d x_{de}, \quad \text{in flow-link formulations}$$

$$y_e = \sum_t x_{te}, \quad \text{in destination-link formulations}$$

As shown in Section A.2.4 of Appendix A, if a function $f(y)$ is convex (concave) with respect to y, the function $f(Ax + b)$ is convex (concave) with respect to x, although strict

Table 4.1 Convexity of some functions and constraints with respect to routing variables

Function/constraint	Convex?
$B = \sum_e y_e B_e$	yes (uni/any/multicast), single-class[4]
$T = \sum_e y_e T_e$	yes (uni/any cast), no multicast (see Exercise 3.1)
$\sum_{e \in p} B_e \leq B^{max}$	yes if single-class, $u_e > 1, \forall e, B^{max} < 15\%$
$\sum_{e \in p} y_e B_e \leq h_p B^{max}$	yes if single-class (h_p: traffic offered to path p)
$\sum_{e \in p} T_e \leq T^{max}$	yes
$\bar{n} = \sum_e y_e / \sum_d h_d$	yes
$\max_e y_e / u_e$	yes

convexity/concavity is not kept unless A is full-rank. From this, we have that all performance metrics that are convex (concave) functions with respect to the set of variables $\{y_e, e \in \mathcal{E}\}$, are also convex (concave) as functions of the routing variables x_p, x_{de} or x_{te} in their respective formulations. Strict convexity or concavity with respect to the routing variables does not hold in general, even if the function is strictly convex or concave with respect to $\{y_e, e \in \mathcal{E}\}$.

Example 4.2 The average network delay expression for unicast traffic $T = \frac{1}{\sum_d h_d} \sum_e \frac{y_e}{u_e - y_e}$ is a strictly convex function with respect to y_e variables, taking h_d and u_e as constants. Thus, it is a convex function with respect to x_p, x_{de}, or x_{te} variables in routing problem formulations. Strict convexity with respect to routing variables would only hold in the case (of no practical interest), when there is a one-to-one relation between possible routings and y_e values. This happens, for instance, in a flow-path problem with one path per link.

In Table 4.1 we summarize the convexity properties of some performance metrics and typical constraints that can be introduced in routing problems. Proofs can be easily derived, and are omitted or referred to exercises.

4.6   Problem Variants

In this section, several routing problem variants are presented, including in some cases their formulations, which serve as modeling examples.

4.6.1   Anycast Routing

Let us assume a routing problem with offered traffic composed of a set D of anycast demands. Each demand $d \in D$, has a set of possible origin nodes $a(d)$ and possible destination nodes $b(d)$. Carrying the anycast demand means that all the demand traffic routed from any node in $a(d)$ to any node in $b(d)$, should sum up the offered traffic h_d.

---

[4] Formulations including the Erlang-B can be solved in JOM with the built-in `erlangB` function.

4.6.1.1 Flow-Path Formulation

Modeling anycast routing problems in a flow-path formulation (e.g., (4.1)) just needs populating the (pre-computed) set of candidate paths P_d for every anycast demand d, with all the accepted paths from any node in $a(d)$, to any node in $b(d)$. Once P_d path lists are computed, the same formulation (e.g., (4.1)) returns the optimal anycast routing.

4.6.1.2 Flow-Link Formulation

A convenient form of modeling anycast routing problems in flow-link formulations such as (4.13) is performing a graph transformation of the problem. This consists of adding two extra nodes to the network for each anycast demand d, so-called *source* node a_d and *sink* node b_d. A link of infinite capacity connects a_d with each node in $a(d)$, and each node in $b(d)$ with the sink b_d. Then, the anycast demand d, is replaced by a standard unicast demand d', starting in a_d and ending at b_d. Figure 4.6 illustrates this procedure. The original network is composed of nodes $\{1, \ldots, 9\}$. The network is offered an anycast demand d, with origins $a(d) = \{1, 2, 3\}$ and destinations $b(d) = \{6, 8\}$. The unicast routing solution obtained in the transformed problem, plotted in the figure, means that four units of traffic are routed from node 1 to node 6, and six units of traffic from node 2 to node 8.

4.6.1.3 Destination-Link Formulation

A similar approach as in the previous case permits adding anycast traffic demands to a destination-link formulation like (4.18). Note that in this case, the two nodes added for each

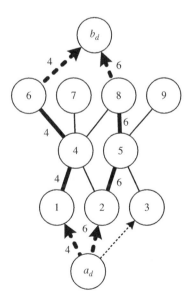

Figure 4.6 Anycast → unicast graph transformation example

anycast demand means enlarging with two more rows and columns per demand the matrix of offered traffic h_{st}, which would have just one non-zero value in the coordinate (a_d, b_d).

4.6.2   Multicast Routing

A multicast traffic demand d has a single initial node $a(d)$, multiple end nodes $b(d)$, and a traffic volume h_d, such that exactly h_d units of traffic have to be delivered to *each* node in $b(d)$. If all the traffic of a multicast demand is carried through a single tree, the demand is not bifurcated. In contrast, if more than one tree is used (e.g., 30% of h_d in one tree and the remaining 70% in other), the demand is said to be bifurcated.

4.6.2.1   Flow-Path Formulation

Multicast routing problems can be modeled in a flow-path formulation exactly like unicast routing problems (e.g., (4.1)), by just populating the P_d sets of candidate paths with *multicast trees* from $a(d)$ node to destinations $b(d)$, instead of paths. Then:

- Pre-computation of the candidate path lists involves solving the k-minimum cost multicast tree problem. In its simpler version ($k = 1$), the problem is called the *Steiner tree problem for directed graphs*, which is known to be $\mathcal{NP}$-complete. The general versions $k > 1$ are also naturally $\mathcal{NP}$-complete. Thus, there is no algorithm that guarantees computing minimum cost multicast trees in worse case polynomial time, which can be an issue in large networks. In Exercise 4.11 we present an algorithm for producing k-shortest multicast trees solving a sequence of k ILP formulations. This scheme is integrated in Net2Plan to ease the development of multicast routing algorithms.
- The CPLA technique described in Section 4.2.4 and Algorithm 3 can be applied to multicast routing problems. The only change occurs in line 1 and 5 of the algorithm: shortest path computations should be replaced by the equivalent minimum cost multicast tree computations, which now involves solving an $\mathcal{NP}$-complete problem.

It is important to remark that, even though multicast routing flow-path formulations can be linear (e.g., (4.1)), and thus solvable in polynomial time, the computation of the candidate path list involves $\mathcal{NP}$-hard computations, and the overall multicast routing problem versions are $\mathcal{NP}$-complete. An exception to this occurs with *broadcast* demands: where the destination nodes are all but the origin: $b(d) = \mathcal{N} - \{a(d)\}$. In this case, the computation of the minimum cost broadcast tree gets the name of the Minimum Spanning Tree (MST) problem, which can be solved in polynomial time with, for example the Edmond's algorithm in [7]. Also, we can compute the k-minimum cost spanning trees in polynomial time, for example with Gabow *et al.* algorithm in [8].

4.6.2.2   Flow-Link Formulation

Flow-link formulations of the multicast problem involve integer (in general 0–1) decision variables[5]. As an example, (4.23) presents a linear flow-link formulation for finding the

---

[5] Since the optimal multicast routing problem even for a single demand is $\mathcal{NP}$-complete, no LP nor convex formulations for the problem can exist (unless $\mathcal{P} = \mathcal{NP}$).

non-bifurcated multicast routing which maximizes the unused bandwidth in the bottleneck link (u_u). Decision variables are:

u_u = {Amount of unused link bandwidth in the bottleneck link}

$\hat{x}_{de}$ = {1 if multicast tree of demand d traverses link e, 0 otherwise}, $\quad \forall d \in D, e \in \mathcal{E}$

$\hat{x}_{det}$ = {1 if multicast tree of demand d traverses link e, in the path to t, 0 otherwise},

$\quad \forall d \in D, e \in \mathcal{E}, t \in b(d)$

$\max u_u \quad$ subject to: $\hfill$ (4.23a)

$$\sum_{e \in \delta^+(n)} \hat{x}_{det} - \sum_{e \in \delta^-(n)} \hat{x}_{det} \begin{cases} = 1, & \text{if } n = a(d) \\ = -1, & \text{if } n = t \\ 0 & \text{otherwise} \end{cases} , \quad \forall d \in D, t \in b(d), n \in \mathcal{N} \quad (4.23b)$$

$$\hat{x}_{det} \le \hat{x}_{de}, \quad \forall d \in D, t \in b(d), e \in \mathcal{E} \hfill (4.23c)$$

$$\sum_d h_d \hat{x}_{de} \le u_e - u_e, \quad \forall e \in \mathcal{E} \hfill (4.23d)$$

Constraints (4.23b) are the flow conservation constraints for the individual paths of each multicast tree. (4.23c) means that a link e belongs to a tree d if it appears in any path to any destination in $b(d)$. Constraints (4.23d) are the standard link capacity constraints, so that u_u contains the maximum amount of capacity that is unused in all the links. Finally, 0–1 integrality of $\hat{x}_{de}$ and $\hat{x}_{det}$ variables mean that the routing is non-bifurcated, and thus each demand is carried by one and only one multicast tree.

It is interesting to remark that constraints (4.23b,c) are necessary conditions for a valid multicast routing, but not sufficient. In particular, these constraints do not avoid the existence of loops similarly to the unicast case, or unnecessary links in the trees, which should be removed offline. Extra constraints can help to avoid some (but not all) of these situations. For instance, making the number of incoming links of a tree in a node $n(\sum_{e \in \delta^-(n)} x_{de})$ equal 0 for the tree initial node and ≤ 1 for the rest. Finally, these difficulties do not appear if the objective function sums a positive cost coefficient for each $\hat{x}_{de}$ variable, since then solutions with unnecessary links are never optimal. Note that this is not the case in (4.23a), although it is possible to replace it in practice by $\bar{u} - \epsilon \sum_{de} \hat{x}_{de}$, being $\epsilon > 0$ a small enough value.

4.6.3 Non-Bifurcated Routing

In non-bifurcated routing, the traffic of each demand d is carried in one and only one path (or multicast tree if the demand is multicast). In the general case, finding the optimal non-bifurcated routing is an $\mathcal{NP}$-complete problem. As shown in the next subsections, it can be modeled using 0–1 integer constraints in the three formulations, flow-path, flow-link, and destination-link.

4.6.3.1 Flow-Path Formulation

It is relatively easy to add the non-bifurcated constraint to a flow-path formulation like (4.1) by just rewriting it using the $\hat{x}_p$ variables (4.2). These variables represent the fraction $\in [0, 1]$ of traffic of the demand, that is carried by the path. Then, the constraint $\hat{x}_p \in \{0, 1\}$, that forbids x_p to take fractional values, makes the routing non-bifurcated.

4.6.3.2 Flow-Link Formulation

Similar to the previous case, it is possible to rewrite the problem using the variables $\hat{x}_{de}$ (4.14) that contain the fraction $\in [0, 1]$ of the traffic of the demand, that traverses link e. Adding the constraints $\hat{x}_{de} \in \{0, 1\}$ makes the routing non-bifurcated.

4.6.3.3 Destination-Link Formulation

In a non-bifurcated destination-based routing, a node n handles all the traffic to a destination node t, forwarding it to a single outgoing link. Adding this type of constraints requires some rewriting. Below (4.24) we present a formulation for the non-bifurcated variant of the maximum worst-case unused bandwidth routing problem (4.18). Decision variables are:

$u_u = \{$Amount of unused link bandwidth in the bottleneck link$\}$

$x_{te} = \{$Amount of traffic targeted to node t, that traverses link $e\}$, $\quad \forall t \in \mathcal{N}, e \in \mathcal{E}$

$f_{te} = \{1$ if traffic targeted to node t traverses link $e, 0$ otherwise$\}$, $\quad \forall t \in \mathcal{N}, e \in \mathcal{E}$

$$\max \ u_u \quad \text{subject to:} \tag{4.24a}$$

$$\sum_{e \in \delta^+(n)} x_{te} - \sum_{e \in \delta^-(n)} x_{te} = \begin{cases} h_{nt}, & \text{if } n \neq t \\ -\sum_s h_{st}, & \text{if } n = t \end{cases} \quad , \forall t, n \in \mathcal{N} \tag{4.24b}$$

$$\sum_t x_{te} \leq u_e - u_u, \quad \forall e \in \mathcal{E} \tag{4.24c}$$

$$x_{te} \geq 0, \quad \forall t \in \mathcal{N}, e \in \mathcal{E} \tag{4.24d}$$

$$x_{te} \leq M f_{te}, \quad \forall t \in \mathcal{N}, e \in \mathcal{E} \tag{4.24e}$$

$$\sum_{e \in \delta^+(n)} f_{te} \leq 1, \quad \forall t, n \in \mathcal{N}, t \neq n \tag{4.24f}$$

The objective function and constraints (4.24b–d) are equal to the original problem (4.18). Constraints (4.24e) forces that $f_{te} = 1$ when a link e carries traffic to t ($x_{te} > 0$). M is a big constant, larger than any value x_{te} can take (e.g. $M = \sum_n h_{nt}$). Then, 0–1 integrality of f_{te} variables and (4.24f) a node to use at most one output link to forward traffic to t, making the routing non-bifurcated.

4.6.4 Integral Routing

In some network design problems, traffic demands are composed of an integer number of traffic units, for example number of channels or connections to be established. In such cases, although bifurcating the traffic of a demand is possible, the routing must carry an integer number of traffic units in each path. This constraint is called *integral routing*. Figure 4.7 illustrates it with an example.

Finding the optimum integral routing is an $\mathcal{NP}$-complete problem in the general case. Modeling it is quite easy using the flow-path, flow-link and destination-link formulations: (i) the offered traffic h_d (or h_{st}) should be given as the integer number of traffic units to carry, (ii) then, routing variables x_p, x_{de}, or x_{te} represent the number of units carried and are restricted to integers.

4.6.5 Destination-Based Shortest Path Routing

In general destination-based routing networks, the optimal routing tables f_{te} computed (e.g., first solving (4.18, then applying (4.19)), should be signaled by separately communicating to each node the set of output links and their forwarding percentages for each destination. This process has two main shortcomings: (i) the complexity of signaling all this information and (ii) the hardware impossibility for common equipment to apply arbitrary forwarding percentages to each output link and destination.

Shortest path routing is an attempt to address both difficulties. In shortest path networks, each network link e has associated a cost $c_e > 0$. Link costs can be offline decided by a centralized authority, or more or less dynamically adjusted by link end nodes. In any case, a signaling protocol should propagate the c_e costs of all the links in the network, to all the network nodes. In IP networks, where shortest path routing is popularly applied, this is commonly the task of the so-called link-state protocols like OSPF or IS-IS. Once a node knows the link costs c_e, it autonomously devises its routing table by applying two rules:

- *Shortest path rule*: Each node n forwards the traffic to each destination node t, only using the shortest path (or paths) from n to t, computed using c_e values as the cost per link.
- *Splitting rule*: In case that more than one shortest path exists (thus with equal cost), an unambiguous splitting rule is applied to decide which fraction of traffic is forwarded through each output link belonging to a shortest path.

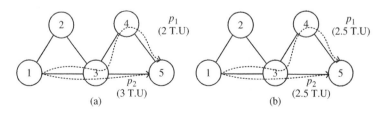

Figure 4.7 Example. Routing of demand d with $h_d = 5$ traffic units (T.U). (a) integral routing and, (b) not an integral routing, since one of the paths does not carry an integer amount of traffic

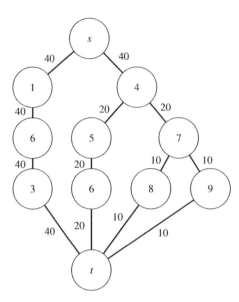

Figure 4.8   ECMP splitting rule example. All the links have cost $c_e = 1$. 80 units of traffic are delivered from s to t, nodes 1, 4, and 7 split the traffic equally between two links

The most popular splitting rule in IP networks is the Equal-Cost Multi-Path (ECMP) scheme. Let $sp(n, t)$ denote the set of outgoing links of n that belong to at least one shortest path to t. Then, the ECMP rule in n splits the traffic to t *equally* among all the links in $sp(n, t)$. Figure 4.8 helps us to illustrate the ECMP technique, for a case where 80 traffic units are generated in s, targeted to t.

IP routers can implement ECMP in different forms. The simpler one distributes one packet at a time to each output link in $sp(n, t)$ using a round robin pointer. This would produce a very finely balanced distribution of traffic. However, it would also cause IP packets of the same TCP transport connection to follow different routes in the network, creating packet out-of-sequence events at the TCP receiver end. This is something to avoid, since TCP could interpret packet disorders as packet losses and then react unnecessarily reducing its rate. For this reason, many routers implement ECMP using hash-based splitting: each packet origin and destination address (and maybe transport ports) are hashed, producing a number, for example between 0 and 15. Then, a map between hash values and the shortest path's output links determines where the packet is forwarded. Since all the packets of the same transport connection produce the same hash value, they will follow the same path in the network. However, depending on the traffic, resulting splitting ratios can be very unbalanced: for example if a single TCP connection produces the 100% of the traffic to a destination, only one output link would be used.

The constraints imposed by the shortest path and the splitting rules make it difficult or even impossible to reproduce optimal routings achievable in general flow-based forwarding. One of the first works to explore this for the ECMP rule was [9]. It showed that properly selecting c_e weights, which are constrained to be integer numbers, could produce close to optimal solutions in many practical cases, although for some topologies the performance could differ substantially from the optimum. The problem of finding the best OSPF weights for a network setting is $\mathcal{N}P$-hard [9]. Since this is a traffic engineering problem of practical interest in IP networks, we have selected it in a case study in Chapter 12 and will present several heuristic algorithms for it.

Interestingly, as we have seen in Section 4.2.1 and Section 4.3.1, we can find a set of link costs c_e, such that the optimal routes of a flow-based or destination-based routing problem are shortest paths according to them[6]. This establishes that the shortest path rule is not in essence a major obstacle for routing performance, and that the impossibility of matching with IP/OSPF the optimum routings achievable in general destination-based routing, comes from the ECMP splitting rule.

4.6.6   SRG-Disjoint 1+1 Dedicated Protection Routing

Let $\mathcal{G}(\mathcal{N}, \mathcal{E})$ be a network, $\mathcal{D}$ the set of offered traffic demands, and $\mathcal{F}$ the set of risks or Shared Risk Groups (SRG) identified for them (see Section 3.6.1 for the definition and notation of SRGs). In this section, we model the problem of finding a 1+1 dedicated protection routing where primary and backup paths should be SRG disjoint. By doing so, we guarantee that all the traffic is carried in the presence of a single SRG failure.

We model this routing problem using a flow-path formulation that maximizes the worst-case unused bandwidth u_u in the links (3.15). For each demand d, we denote as P_d its set of admissible paths. Logically, a necessary condition for finding a valid routing is that for each demand, at least two SRG-disjoint paths exist in P_d. This requires that demand end nodes do not belong to any SRG.

The decision variables are:

$u_u = \{\text{Amount of unused link bandwidth in the bottleneck link}\}$

$\hat{x}_p = \{1 \text{ if path } p \text{ is the primary path of demand } d \ (p), 0 \text{ otherwise}\}, \quad \forall p \in P$

$\hat{x}'_p = \{1 \text{ if path } p \text{ is the backup path of demand } d(p), 0 \text{ otherwise}\}, \quad \forall p \in P$

The routing formulation is given by (4.25):

$$\max_{\hat{x}, \hat{x}'} u_u \quad \text{subject to:} \tag{4.25a}$$

$$\sum_{p \in P_d} \hat{x}_p = 1, \quad \forall d \in D \tag{4.25b}$$

$$\sum_{p \in P_d} \hat{x}'_p = 1, \quad \forall d \in D \tag{4.25c}$$

$$\sum_{p \in P_e} h_{d(p)}(\hat{x}_p + \hat{x}'_p) \leq u_e - u_u, \quad \forall e \in \mathcal{E} \tag{4.25d}$$

$$\sum_{p \in P_f \cap P_d} \hat{x}_p + \hat{x}'_p \leq 1, \quad \forall d \in D, f \in \mathcal{F} \tag{4.25e}$$

---

[6] In the routing problems addressed in those sections, these link costs were the optimal multipliers π_e of the link capacity constraints.

With constraints (4.25b, c), we force each demand to have one primary and one backup path associated. In (4.25d), the traffic in each link is constrained to be lower or equal to the capacity minus the worst-case unused bandwidth u_u. Constraints (4.25e) make that for every demand d and SRG f, no two primary-backup paths associated to the demand ($\in P_d$) and affected by the same failure (which we denote as $\in P_f$), can simultaneously carry traffic. This is true since $\hat{x}_p$ and $\hat{x}_p$ are constrained to be 0–1 integers.

An alternate flow-path formulation for the problem can be obtained by populating the candidate path lists, not with paths, but with *SRG-disjoint 1+1 path-pairs*. For instance, from P_d we compute the set P_d^{1+1} of path pairs (p_1, p_2), such that both p_1 and p_2 belong to P_d and are SRG-disjoint. Then, we apply a standard flow-path formulation using P_d^{1+1} as "path" lists. Note that in this case, we do not need the routing to be non-bifurcated to enforce SRG-disjointness. However, the general problem of finding the minimum cost SRG-disjoint path pair is $\mathcal{N}P$-hard, and so is the problem of enumerating the k-shortest SRG-disjoint path pairs. Still, there are polynomial algorithms for some problem variants, like the Suurballe's algorithm for minimum cost link-disjoint paths [10, 11].

4.6.7   Shared Restoration Routing

In contrast to dedicated 1+1 protection, in shared restoration routing the link bandwidth can be shared by different backup paths. In other words, depending on the particular failure state (failed nodes and links in the network), different primary paths may be using it. Designing the routing in such a restoration scheme usually involves first defining a set S of possible failure states of interest, such that the possibility π_s of finding the network in failure state s cannot be neglected. Usually, set S includes at least the no-failure state (denoted s_0), and the single-failure states, when just one SRG is down. The reader is referred to Section 3.6.3, for more information on how to enumerate failure states in the network, and estimate their π_s probabilities from the MTBF and MTTR values of the SRGs.

In this section we present a flow-link formulation for finding the different $|S|$ non-bifurcated routings (one for each possible network state considered), that optimizes the network congestion by maximizing the unused bandwidth in the bottleneck, in any network state. Decision variables are:

$u_u = \{$Amount of unused link bandwidth in the bottleneck link$\}$

$\hat{x}_{des} = \{1$ if traffic of demand d is carried in link e during network state s, 0 otherwise$\}$,

$$\forall d \in D, e \in \mathcal{E}, s \in S$$

The routing formulation is given by (4.26):

$$\max_{\hat{x}} u_u \quad \text{subject to:} \tag{4.26a}$$

$$\sum_{e \in \delta^+(n)} \hat{x}_{des} - \sum_{e \in \delta^-(n)} \hat{x}_{des} = \begin{cases} 1, & \text{if } n = a(d) \\ -1, & \text{if } n = b(d), \\ 0, & \text{otherwise} \end{cases} \quad \forall d \in D, n \in \mathcal{N}, s \in S \tag{4.26b}$$

$$\sum_d h_d x_{des} \leq u_e - u_u, \forall e \in \mathcal{E}, s \in S \tag{4.26c}$$

$$x_{des} = 0, \forall d \in D, s \in S, e \in \mathcal{E}(s) \tag{4.26d}$$

Constraints (4.26b) and (4.26c) just force all the routings in all the network states to follow the standard flow conservation and link capacity constraints. (4.26d) means that during a network state s, the failed links represented by $\mathcal{E}(s)$ (including all the input and output links of a node, if such node is failed), cannot carry traffic.

In (4.26), the routings can be totally different from one state to the other. This means for instance, that the routing of a demand d could change if a link fails, even if the failed link is not traversed by the demand. Frequently, restoration in real networks do not permit such flexibility, and only the demands directly affected by a failure can be rerouted. This limitation can be incorporated by adding the constraint (4.27):

$$- \sum_{e' \in \mathcal{E}(s)} x_{de's_0} \leq x_{des} - x_{des_0} \leq \sum_{e' \in \mathcal{E}(s)} x_{de's_0}, \quad \forall d \in D, e \in \mathcal{E}, s \in S - s_0 \tag{4.27}$$

Expression $\sum_{e' \in \mathcal{E}(s)} x_{de's_0}$ is the number of links carrying traffic of the demand, that are affected by the failure s. If this number is zero, the constraint (4.27) forces that $x_{des} = x_{des_0}$, such that the routing must be the same as the one in the non-failure state. If not, any 0–1 combination for the variables x_{des} and x_{des_0} is allowed, and thus the routing can be different.

The reader is referred to Exercise 4.15 to see an example of flow-path formulation for a similar problem.

4.6.8   Multi-Hour Routing

In many real networks, the offered traffic follows periodic (daily/weekly) patterns easy to forecast. For instance, the traffic in the morning is usually different than that at the non-working hours in the evening, and different than what we find at midnight. These variations can be modeled as a *multi-hour traffic demand*: a sequence of demands (or of traffic matrices), each estimating the offered traffic at contiguous time intervals. We use index t, $t = 1, \ldots, T$, for identifying the time interval. Then, for a given set of demands D, we denote h_{dt} as the amount of traffic offered by demand d during time interval t. The number of time intervals used in multi-hour depends on the time granularity of our traffic estimations. Typically, we can find design ranges from $T = 2$ (e.g., *day traffic* and *night traffic*), to more complex studies with, for example one traffic interval for each hour of the week ($T = 7 \times 24 = 168$).

The target of multi-hour design is take benefit of the knowledge of traffic fluctuations to devise more efficient routings that make a better use of network resources. Two main multi-hour routings variants exist: (i) *oblivious* or *static* routing and (ii) *dynamic* routing.

4.6.8.1   Oblivious Routing

In the oblivious routing case, the routing is static, and is never changed along the time intervals. This means that the fraction of traffic $\hat{x}_p$ of demand $d(p)$ that is carried in a path p, is not changed, or in a destination-based routing networks like IP, that the routing tables f_{te} are never modified. Naturally, since the *fractions* of traffic routed in each path are unchanged, the amount of traffic in the links fluctuates as the offered traffic changes. For instance, if a demand d is 50%–50% bifurcated into two paths, these two paths will carry 100 units of traffic when $h_{dt} = 200$, and 20 units of traffic when $h_{dt} = 40$.

The difference between oblivious routing and standard non-multihour routing, is that the oblivious routing exploits the knowledge of traffic fluctuations for having a better network performance. Formulation (4.28) shows a flow-path formulation of the multi-hour static routing, optimizing the worst-case unused bandwidth u_u.

$$\max_{\hat{x}, u_u} \; u_u \quad \text{subject to:} \tag{4.28a}$$

$$\sum_{p \in P_d} \hat{x}_p = 1, \quad \forall d \in D \tag{4.28b}$$

$$\sum_{p \in P_e} h_{dt} \hat{x}_p \le u_e - u_u, \quad \forall e \in \mathcal{E}, t = 1, \dots, T \tag{4.28c}$$

$$x_p \ge 0, \quad \forall p \in P \tag{4.28d}$$

Note that the difference between (4.28) and the standard non-multihour formulation is just the existence of one link capacity constraint (4.28c) per link and *time interval t*.

4.6.8.2 Dynamic Routing

In the dynamic routing case, the fractions $\hat{x}_p$ (or $\hat{x}_{de}$ or $\hat{x}_{te}$) can now be different at different time intervals. In the adaptation of (4.28) to this problem, decision variables should be of the form $\hat{x}_{pt}$, denoting the fraction of traffic of demand $d(p)$ being carried in p during time interval t.

Naturally, when the network is able to adapt its routing every time interval, a more efficient resource utilization can be made. However, frequent routing changes are undesirable, and an artificial cost is commonly added to the objective function to avoid excessive routing modifications. This is the case in (4.29).

$$\max_{\hat{x}, u_u} \; u_u - \gamma \sum_{t=1}^{T} \sum_{p} (\hat{x}_{pt} - \hat{x}_{pt+1})^2 \quad \text{subject to:} \tag{4.29a}$$

$$\sum_{p \in P_d} \hat{x}_{pt} = 1, \quad \forall d \in D, t = 1, \dots, T \tag{4.29b}$$

$$\sum_{p \in P_e} h_{dt} \hat{x}_{pt} \le u_e - u_u, \quad \forall e \in \mathcal{E}, t = 1, \dots, T \tag{4.29c}$$

$$x_{pt} \ge 0, \quad \forall p \in P, t = 1, \dots, T \tag{4.29d}$$

Modifications in the problem constraints with respect to static routing are trivial. In the objective function, note that the last term substracts a positive value $\gamma(\hat{x}_{pt} - \hat{x}_{pt+1})^2$ if the routing changes between two consecutive intervals (we assume that time interval $t = T$ is followed by time interval $t = 1$). Parameter γ tunes the penalization given to the routing changes. Note that the objective function is concave, and thus the maximization problem can be computed in polynomial time.

4.7   Notes and Sources

The routing problem as described is an application to communication networks of the classical multicommodity flow problem, appearing in transportation, logistics and many other disciplines. Flow-path, flow-link, and destination-link formulations are common in this context, and can be found in multiple sources. In this book, we partially inherit the notation used in [6] for them. The relation between routing tables (f_{te}) and x_{te} destination-routing variables in Section 4.4.2 is derived from standard results in Markov chains (e.g., see Chapter 1 of [12]).

Candidate Path List Augmentation procedures are an application of standard column-generation techniques of linear programs to flow-path formulations. A background on the topic can be found in [4–5]. [6] describes CPLA techniques in the context of linear routing formulations. The extension to nonlinear convex problems described in Section 4.2.4 is original.

The formulations on the multiple problem variants in Section 4.6 are original, in the sense that were derived from variations of classical network problems, adapted to the book notation, and no effort was made to find them in other sources. The added value is putting them together, and describing them within a common framework and notation, with the aim of growing in the reader the skill of modeling any routing problem. I would like to credit the book in [6] as an encyclopedic compilation of routing problems.

4.8   Exercises

4.1   Let (4.30) be the flow-path formulation of the routing that minimizes the worse case link utilization, and let π^*, λ^*, and v^* be optimum multipliers for the problem, and ρ^* the optimum congestion.

$$\min_{x,\rho} \; \rho \quad \text{subject to:} \tag{4.30a}$$

$$\lambda_d : \sum_{p \in P_d} x_p = h_d, \quad \forall d \in D \tag{4.30b}$$

$$\pi_e : \sum_{p \in P_e} x_p \le \rho u_e, \quad \forall e \in \mathcal{E} \tag{4.30c}$$

$$v_p : x_p \ge 0, \quad \forall p \in P \tag{4.30d}$$

Answer true of false, justify your answer:
1. If a path p does not carry traffic, then $v_p^* > 0$.
2. If a path p carries traffic, then its weight according to π^* equals $\lambda^*_{d(p)}$.
3. If a link e has an utilization $\rho_e < \rho^*$, then it could happen that $\pi_e > 0$.
4. If a link e is a bottleneck ($\rho_e = \rho^*$), then it could happen that $\pi_e = 0$.

4.2   For the problem in formulation (4.30), prove that the expression: $\frac{\sum_d h_d l_d}{\sum_e u_e}$, where l_d is the number of links in the shortest path of demand d, is a lower bound to the optimal congestion ρ^*. *Hint*: Show that this is the solution of the relaxed problem for multipliers $\pi_e = \frac{1}{\sum_e u_e}$.

4.3   For the problem in formulation (4.30), assume that a given path p is constrained to carry at least a traffic $z_p > 0$. Let $(\rho^*, x^*, \pi^*, \lambda^*, v^*)$ be a primal dual optimal solution to the original problem. Use it to compute a lower bound to the network congestion when the new constraint is added.

4.4   For the problem in formulation (4.30), where all the paths are admissible, prove that a necessary optimality condition is that every path in the network should traverse at least a bottleneck link.

4.5   Let Fig. 4.9 show a network where all the links have the same capacity. The offered traffic and routing is unknown, but the carried traffic in each link is written next to the links. Prove that the routing does not optimize the worst-case utilization in the links (all the paths are admissible). *Hint*: Use the results in Exercise 4.4.

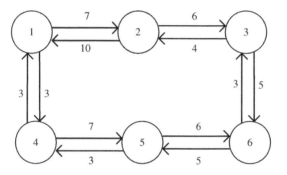

Figure 4.9   Example topology

4.6   Write the flow-path formulation that minimizes the average number of hops in the network. Prove that the CPLA technique described in Section 4.2.4 can be applied to it. Devise a lower bound to the optimum in each iteration and write the stop condition to apply to terminate the algorithm when the average number of hops is guaranteed to be at most 1% worse than the optimum ($\epsilon = 0.01$). Implement previous scheme in a NetPlan algorithm. The CPLA policy should add at least one shortest path for each demand not satisfying the optimality condition. Include ϵ as an input parameter. The optimum solution is returned by updating the traffic routing.

4.7   An optical network is composed of a set $\mathcal{N}$ of nodes and a set $\mathcal{E}$ of fiber links. The length in km of fiber e is denoted as d_e. The offered traffic is composed of a given set D of optical connections, called lightpaths. A lightpath d should be assigned a route among the admissible paths $\mathcal{P}_d$, and a wavelength w in the set $W = \{1, \ldots, W_{max}\}$ of valid wavelengths. Two lightpaths traversing the same fiber cannot be assigned the same wavelength (these are the so-called wavelength clashing constraints). Devise the flow-path formulation which finds the routing and wavelength assignment that minimizes the average number of hops. Adapt the formulation to the multicast case.

Implement a Net2Plan algorithm that solves this formulation with JOM. The maximum length in km to be an acceptable route is set as an input parameter and used to build the candidate path list. The optimum solution is returned by updating the routes of the demands, and using the functions in WDMUtils library of Net2Plan to store the wavelength assignment to each route as route attributes.

4.8   In the same context of Exercise 4.7 (unicast case), adapt the formulation for the case when each lightpath should be assigned a primary and backup route, which are link disjoint and are assigned the same wavelength. Implement a Net2Plan algorithm that solves this formulation with JOM. The optimum solution is returned by updating the routes of the demands and assigning one protection segment to each, with the backup route.

4.9   In the same context of Exercise 4.7 assume that a set $\mathcal{F}$ of SRGs has been defined in the network, and that a lightpath that is affected by a failure can change its route and wavelength. Devise the flow-path formulation which finds the routing and wavelength assignment of each lightpath $d \in D$ in the no-failure state, and in each of the single-SRG failure stages. The optimization target is minimizing the average propagation delay of the lightpaths. Implement a Net2Plan algorithm that solves this formulation with JOM and appropriately returns the optimum solution.

4.10   In the same context of Exercise 4.7 (unicast case), let us assume that each link is a SRG with an unavailability given by U_f (the fraction of time the link is down). We assume that unavailabilities are below 10^{-2}, and link failures are statistically independent. Show that the unavailability of a lightpath can be approximated by the sum of the unavailabilities of the traversed links. Write the formulation of the problem which finds the route and wavelength assignment of the lightpaths that minimizes the average lightpath unavailability. Modify the previous formulation to minimize the worst-case unavailability among the lightpaths. Implement two Net2Plan algorithms that solve previous formulations with JOM and appropriately return the optimum solution.

4.11   Let $\mathcal{G}(\mathcal{N}, \mathcal{E})$ be a network, $\{c_e, e \in \mathcal{E}\}$ be a strictly positive cost assigned to each link, and $\{u_e, e \in \mathcal{E}\}$ the given link capacities. Let d be a multicast demand, from node $a(d)$ to nodes in set $b(d)$. Write a flow-link formulation that finds the multicast tree satisfying d, which minimizes the sum of the costs in the traversed links. Let $\mathcal{P}$ be a set of multicast trees satisfying d. Modify the previous formulation to forbid returning any multicast tree in $\mathcal{P}$ (that is, with exactly the same set of traversed links that a tree in p). Devise an algorithm which uses previous formulation to solve the k-shortest multicast tree problem. Implement such algorithm in a Java function, solving the formulations with JOM.

4.12   Let $\mathcal{G}(\mathcal{N}, \mathcal{E})$ be a network, $\{c_e, e \in \mathcal{E}\}$ be a strictly positive cost assigned to each link, and $\{u_e, e \in \mathcal{E}\}$ the given link capacities. Let $\mathcal{F}$ be a set of SRGs defined in the network. Write an integer flow-link formulation that finds two SRG-disjoint paths for *one* demand d, such that minimizes the sum of the costs of the traversed links. Write a flow-path formulation for the 1+1 SRG-disjoint routing of a set of demands

D. The candidate path list P_d^{l+1} of each demand d is supposed to be populated by SRG-disjoint path pairs (instead of paths). Devise a CPLA scheme where the P_d^{l+1} sets are dynamically populated. Implement such scheme in a Net2Plan algorithm that solves previous formulations with JOM, and appropriately returns the optimum solution.

4.13 Solve Exercise 4.12 for the case of link-disjoint paths (that is, one SRG exists per network link). Can we replace the integer flow-link formulation by a polynomial-time algorithm? Modify the Net2Plan algorithm in Exercise 4.12 to apply the Suurballe's algorithm (built-in in a Net2Plan library) in the case when one SRG is defined per network link.

4.14 Let $G(\mathcal{N}, \mathcal{E})$ be a network and $\{u_e, e \in \mathcal{E}\}$ the given link capacities, and $\{d_e, e \in \mathcal{E}\}$ the link distances in km. Let D be the set of unicast offered demands. Devise a flow-path and a flow-link formulation that minimize the worst-case utilization among the network links, subject to the constraint that no path can have a propagation delay higher than 50 ms, and the routing is non-bifurcated. Modify the flow-path formulation with the constraint that the routing can be bifurcated, but a path cannot carry less than a 10% of the demand traffic.

4.15 Rewrite the formulation (4.26) in Section 4.6.7 using a flow-path approach. Add the constraint that demands unaffected by a failure cannot change its routing.

4.16 Implement a Net2Plan algorithm that solves formulation (4.28) for multihour oblivious routing with JOM. The input traffic is given by a set of T traffic matrices. The routing of each demand is constrained to one out of k loopless shortest paths in number of hops for that demand. Input parameters to the algorithm are the number of time intervals T, the root name of the .n2p files with the traffic matrices (full names are supposed to be root_1.n2p, ..., root_T.n2p), the maximum number k of loopless candidate paths per demand, and a parameter stating whether or not the routing is constrained to be bifurcated.

4.17 Repeat the previous exercise for the case of dynamic routing in formulation (4.29).

References

[1] E. Dijkstra, "A note on two problems in connexion with graphs," *Numerische Mathematik*, vol. 1, no. 1, pp. 269–271, 1959.
[2] D. Eppstein, "Finding the K shortest paths," in *Foundations of Computer Science, 1994 Proceedings., 35th Annual Symposium on.* IEEE, 1994, pp. 154–165.
[3] J. Yen, "Finding the K shortest loopless paths in a network," *Management Science*, pp. 712–716, 1971.
[4] M. Minoux, *Mathematical Programming: Theory and Algorithms*, ser. Wiley-Interscience series in discrete mathematics and optimization. Wiley, 1986.
[5] L. Lasdon, *Optimization Theory for Large Systems*, ser. Dover books on Mathematics. Dover Publications, 2002.
[6] M. Pioro and D. Medhi, *Routing, Flow, and Capacity Design in Communication and Computer Networks.* Morgan Kaufmann Publishers, 2004.
[7] J. Edmonds, *Optimum Branchings.* National Bureau of Standards, 1968.
[8] H. Gabow, Z. Galil, T. Spencer, and R. Tarjan, "Efficient algorithms for finding minimum spanning trees in undirected and directed graphs," *Combinatorica*, vol. 6, no. 2, pp. 109–122, 1986.

[9] B. Fortz and M. Thorup, "Internet traffic engineering by optimizing OSPF weights," in *INFOCOM 2000. Nineteenth Annual Joint Conference of the IEEE Computer and Communications Societies. Proceedings. IEEE*, vol. 2. IEEE, 2000, pp. 519–528.

[10] J. Suurballe, "Disjoint paths in a network," *Networks*, vol. 4, no. 2, pp. 125–145, 1974.

[11] J. W. Suurballe and R. E. Tarjan, "A quick method for finding shortest pairs of disjoint paths," *Networks*, vol. 14, no. 2, pp. 325–336, 1984.

[12] J. R. Norris, *Markov Chains*. Cambridge University Press, 1998, no. 2008.

5

Capacity Assignment Problems

5.1 Introduction

Given a network topology $\mathcal{G}(\mathcal{N}, \mathcal{E})$, with $\mathcal{N}$ the set of nodes and $\mathcal{E}$ the set of links, the capacity assignment problem decides on the capacity u_e allocated to each link $e \in \mathcal{E}$. This problem appears in two main (quite diverse) contexts: (i) as a problem to be solved periodically at long time scales (e.g., every 6 months) to upgrade the capacity of deployed links in a network and (ii) as a control problem to solve at subsecond time scale, to allocate capacity to users in a dynamic environment. We provide two examples:

- *(Slow) Capacity planning*: In Internet Service Provider (ISP) backbone networks, network links are virtual circuits hired to a network carrier, or transport connections established in an own network infrastructure. The link costs depend on the link distance and capacity according to the carrier tariffs or the infrastructure cost structure. ISPs periodically (e.g., every 6 months) execute a so-called *capacity planning* process [1] or *capacity expansion* [2], where the capacity upgrade in the links is planned to match a forecasted traffic demand, at the minimum cost. In this context, the capacity assignment falls into the problems that can be solved without major time constraints, in a centralized form. Chapter 12 is devoted to the design of algorithms for them.
- *(Fast) Capacity allocation*: In many wireless networks technologies, a set of mobile systems (e.g., phones in a cell network) are served by a central base station. To orchestrate the uplink (phone to base station) and downlink (base station to phone) communications, different *medium access protocols* exist. Depending on them, the capacity of a link is established indirectly by deciding on figures such as the link transmission power or the access probability to the channel, which may have a complex relation with the resulting capacities. In particular, the decisions on different links are coupled, mutually affected in many forms. For instance, while increasing the transmission power of a link can augment its capacity, it can also add interferences degrading other links' rates. Also, increasing the access probability to the channel of a link is made at a expense of other link capacities. In this context, when designing an allocation algorithm we should consider that the capacity assignments

Optimization of Computer Networks – Modeling and Algorithms: A Hands-On Approach,
First Edition. Pablo Pavón Mariño.
© 2016 John Wiley & Sons, Ltd. Published 2016 by John Wiley & Sons, Ltd.
Companion Website: www.wiley.com/go/PavonMarinoSol16

should be optimized at a subsecond rate to rapidly adapt to changes in the wireless channel or traffic conditions. Distributed algorithms will be generally preferred. Chapters 9, 10, and 11 provide techniques for the design of such algorithms.

In this chapter, we provide a set of comprehensive examples of capacity assignment problems in both contexts, studying the problem convexity properties and providing an optimality analysis.

5.2 Long-Term Capacity Planning Problem Variants

5.2.1 Capacity Planning for Concave Costs

In this section we address the problem of finding the optimal link capacities that minimize network congestion such that the link costs do not exceed an available budget C. Network congestion is measured as the utilization at the bottleneck link, the one with the highest utilization in the network.

We focus on the case in which link costs are a concave (sublinear) function of link capacities, so that, for example the cost of a link of 10 capacity units is lower than the cost of 10 links of capacity 1. As was shown in Section 3.5, this is the realistic cost structure that reflects the law of economies of scale.

Let $G(\mathcal{N}, \mathcal{E})$ be a network, for which we know (or estimate) the offered traffic and its routing. As a result, the traffic y_e carried by each link $e \in \mathcal{E}$ is known. The cost $c(e)$ of a link e depends on its capacity according to:

$$c(e) = c_e u_e^\alpha$$

where $\alpha \in (0, 1]$ and c_e are the constants that reflect the cost structure. Factor α tunes the intensity of the discount for buying large amounts of capacity in a link. In particular, the ratio between the acquisition cost of (i) k links of U units of traffic and (ii) one link of kU units of traffic, is given by:

$$\frac{kc_e U^\alpha}{c_e(kU)^\alpha} = \frac{k}{k^\alpha} = k^{1-\alpha}$$

When $\alpha = 1$, costs are linear with respect to capacities and there is no large capacity discount. In its turn, the cost of 10 links of one unit of capacity ($k = 10, U = 1$) is more than three times more expensive than buying one link of 10 capacity units when $\alpha = 0.5$ ($10^{0.5} \approx 3.1$), and almost eight times more expensive if $\alpha = 0.1$ ($10^{0.9} \approx 7.9$). Figure 5.1 helps to further illustrate the magnitude of the capacity discounts for different α values.

The decision variables to the optimization problem are:

$$\rho = \{\text{Worst-case link utilization}\}$$

$$u_e = \{\text{Capacity installed in link } e\}, \quad \forall e \in \mathcal{E}$$

The optimization problem to solve is:

$$\min_{\rho, u} \rho \quad \text{subject to:} \tag{5.1a}$$

$$y_e/u_e \le \rho, \quad \forall e \in \mathcal{E} \tag{5.1b}$$

$$u_e \geq y_e, \quad \forall e \in \mathcal{E} \tag{5.1c}$$

$$\sum_e c_e u_e^\alpha \leq C \tag{5.1d}$$

The objective function (5.1a) minimizes network congestion. In (5.1b) we make link utilization y_e/u_e be lower or equal than ρ, the bottleneck utilization. Constraints (5.1c) force that link capacities are greater or equal than carried traffics. Finally, (5.1d) is the budgeting constraint: the total link costs should not exceed the available budget C.

5.2.1.1   Problem Convexification

Objective function of problem (5.1) and constraints (5.1c) are linear, and constraint (5.1b) is satisfied by a convex set of solutions, since function $\frac{y_e}{u_e} - \rho$ is convex with respect to problem variables. However, in constraint (5.1c), the network cost function:

$$c(u) = \sum_e c_e u_e^\alpha \tag{5.2}$$

is concave with respect to link capacities for $\alpha \in (0, 1)$ and the set of points satisfying constraint (5.1c) can be non-convex. For instance, Fig. 5.2a shows the non-convex set of feasible link capacities in a network with two links, 0.5 units of traffic in it, and the cost constraint given by $\sqrt{u_1} + \sqrt{u_2} \leq 3$.

The non-convexity of the feasibility set of (5.1) impedes the application of KKT optimality conditions and hinders the utilization of standard solvers for obtaining a numerical solution to the problem. Fortunately, in this case it is possible to circumvent these issues by a variable

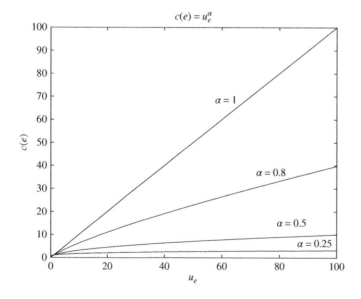

Figure 5.1   Concave cost evolution example, $c(e) = u_e^\alpha$ for different values $\alpha = \{1, 0.8, 0.5, 0.25\}$

change that produces a new equivalent convex formulation. In particular, in (5.3) we rewrite problem (5.1), using the link utilizations $\rho_e = y_e/u_e$ instead of link capacities u_e as decision variables:

$$\min_{\rho_e,\rho} \rho \quad \text{subject to:} \tag{5.3a}$$

$$v_0 : \rho \geq 0 \tag{5.3b}$$

$$v_1 : \rho \leq 1 \tag{5.3c}$$

$$\pi_e : \rho_e \leq \rho, \quad \forall e \in \mathcal{E} \tag{5.3d}$$

$$\lambda : \sum_e c_e y_e^\alpha \rho_e^{-\alpha} \leq C \tag{5.3e}$$

Problems (5.1) and (5.3) are equivalent in the sense that there is a one-to-one translation between a solution in terms of capacities and in terms of utilizations ($\rho_e = y_e/u_e$) and considering this translation both solutions have the same feasibility set and optimum solutions.

Objective function and constraints (5.3b–d) are linear. Linear constraints (5.1b,c) just state that link utilizations should be between zero and one, which means that no link can be over-subscribed and capacities should be non-negative. Also, it is easy to show that the network cost $c(\rho)$ is now a convex function with respect to ρ_e variables, observing its hessian matrix:

$$\frac{\partial^2 c}{\partial \rho_e^2} = c_e y_e^\alpha \frac{\alpha(\alpha+1)}{\rho_e^{\alpha+2}} > 0, \frac{\partial^2 c}{\partial \rho_e \partial \rho_{e'}} = 0, \forall e, e' \in \mathcal{E}, e \neq e'$$

In summary, problem (5.3) is convex and thus amenable to KKT optimality analysis. Figure 5.2b illustrates this, showing the convex feasibility set resulting after the variable change for the example in Fig. 5.2a.

5.2.1.2   Optimality Analysis

We apply KKT conditions to convex problem (5.3). The Lagrangian function is given by:

$$L(\rho_e, \rho, v, \pi, \lambda) = \rho - v_0\rho + v_1(\rho - 1) + \sum_e \pi_e(\rho_e - \rho) + \lambda \left(\sum_e c_e y_e^\alpha \rho_e^{-\alpha} - C \right)$$

The Lagrangian minimization optimality conditions are:

$$\frac{\partial L}{\partial \rho} = 0 \Leftrightarrow 1 - v_0 + v_1 = \sum_e \pi_e \tag{5.4a}$$

$$\frac{\partial L}{\partial \rho_e} = 0 \Leftrightarrow \pi_e = \alpha \lambda c_e y_e^\alpha \rho_e^{-\alpha-1}, \forall e \in \mathcal{E} \tag{5.4b}$$

Multiplier v_0 must be zero, since its respective constraint ($\rho \geq 0$) cannot be tight: a zero link utilization means infinite link capacities. Thus, according to (5.4) at least one π_e multiplier is strictly greater than zero. Then, $\lambda > 0$ and thus $\pi_e > 0$ for all network links, so constraints (5.3d) are *all* tight. That is, in the optimum all the links have the same utilization, equal to ρ. This is intuitively logical: if one or more links had a higher utilization than the rest, we could

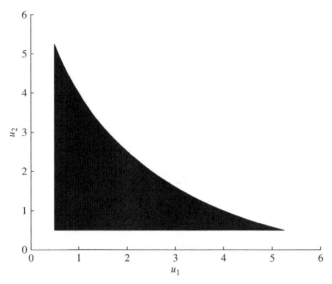

(a) Non-convex feasibility set (u_e variables, (5.1))

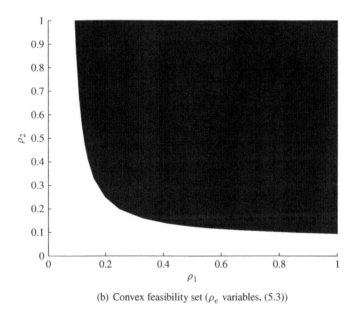

(b) Convex feasibility set (ρ_e variables, (5.3))

Figure 5.2 Example. Feasibility set for problems (5.1) and (5.3) in a network of two links, $y_e = 0.5$ units of traffic in each, $C = 3$, network cost $c(u) = \sqrt{u_1} + \sqrt{u_2} = \sqrt{0.5}\left(\frac{1}{\sqrt{\rho_1}} + \frac{1}{\sqrt{u_2}}\right)$

improve the solution reducing the capacity in the low-utilization links and using the extra budget to increase the capacity of higher utilization links.

Since $\lambda > 0$, constraint (5.3e) is tight, so in the optimum, all the budget C is invested in network links. From this, we can arrive to a closed formula expression to the optimum solution of problem (5.3), given by:

$$C = \frac{\sum_e c_e y_e^{\alpha}}{\rho^{\alpha}} \Rightarrow \rho = \left(\frac{\sum_e c_e y_e^{\alpha}}{C} \right)^{\frac{1}{\alpha}} \tag{5.5}$$

Note that expression $\sum_e c_e y_e^{\alpha}$ is the minimum budget (which we denote C_{min}) needed to have a feasible solution, which installs in each link a capacity equal to the carried traffic. In the next subsections we use formula (5.5) to gain insight on some trade-offs appearing in network design.

5.2.1.3 Diminishing Returns Law With Respect to Available Budget

Differentiating (5.5) with respect to C we obtain the network congestion sensitivity to a variation in budget C.

$$\frac{\partial \rho}{\partial C} = -\frac{\left(\sum_e c_e y_e^{\alpha} \right)^{1/\alpha}}{\alpha C^{1+1/\alpha}} = -\frac{\rho}{\alpha C} \tag{5.6}$$

Logically, higher budgets C always yield improvements in network congestion, since expression (5.6) is negative. Performance improvements are high when C is close to C_{min}, since we are upgrading links that have a capacity close to the carried traffic and congestion can be reduced significantly with small capacity increases. However, the more C grows, the lower are the improvements obtained by each extra budget unit we invest. This is a form of the diminishing returns law in network design. Eventually, the C budget is such that further improvements in congestion are too small to justify investing more resources in the network.

5.2.1.4 Trends in Network Topology Design

In this section, we analyze the optimum capacity assignment (5.5) to show up some trade-offs appearing in network topological design. That is, we are interested in using (5.5) to gain insight on the previous step of capacity optimization: the design phase that decides the set of links to deploy in the network.

Let us assume we want to design a network topology and routing to connect a set $\mathcal{N}$ of nodes with the minimum cost, but satisfying a given network congestion target ρ. Once the links and routing are decided, the cost of the optimal capacity assignment, the only cost considered, would be:

$$C = \frac{\sum_e c_e y_e^{\alpha}}{\rho^{\alpha}} \tag{5.7}$$

If all the links have the same cost factor $c_e = c$ and traffic $y_e = y$, the investment required to meet the congestion target becomes:

$$C = \frac{|\mathcal{E}| c y^{\alpha}}{\rho^{\alpha}} \tag{5.8}$$

We can observe two competing trends:

- Cost C is proportional to the number of links in the network $|\mathcal{E}|$, and thus penalizes the topologies that spread the traffic in multiple links (e.g., full-mesh topologies) and favors those designs that aggregate the traffic of multiple sources in a lower number of links (e.g., tree topologies).
- Cost C is proportional to y^α and thus increases (sublinearly) with the amount of traffic carried in each link. This penalizes topologies with a small number of links that aggregate the traffic of multiple users. In contrast, it favors, for example, full-mesh topologies in which each link just carries the traffic of the users between its end nodes.

Previous analyses show up a trade-off between two opposite trends in network topology design: a trend towards full-mesh topologies with a high number of low loaded links and a trend towards tree-like topologies where a lower number of links carry the aggregated traffic of multiple users. In real designs, the described trade-off is affected by multiple factors like different costs per link, not uniform traffic demands, a fixed cost per link deployment summed to capacity costs (which would favor tree-like topologies), latency, network resiliency, technological constraints, and so on. As we can see in multiple examples along the book, when these factors come into play, the optimum network topologies and routings are usually not regular.

Still, the next example intends to give light on the effect of the α factor on the cost structure: the larger the capacity discounts given by economies of scale (lower α), the more intense the trend toward tree topologies.

Example 5.1 We want to compare three possible topologies for an IP network of $|\mathcal{N}| = 6$ routers, a star, ring and full-mesh topology as shown in Fig. 5.3. The offered traffic matrix is composed of one traffic unit between each node pair. The traffic is routed through the shortest path in number of hops and the ECMP rule is applied for equally bifurcating the traffic if more than one shortest path exists. Then, the number of links and amount of link traffic y in each topology is shown in Table 5.1.

Figure 5.4 plots the cost of the optimum capacity assignment for a congestion target $\rho = 0.5$, ranging values of $\alpha \in [0, 1]$ and assuming that all links have the same cost factor $c_e = 1, \forall e \in \mathcal{E}$. We see that tree and link topologies that aggregate traffic in a small number of links are preferred when the economies of scale are strong (low α values). In turn, when the costs evolve close to linearly ($\alpha \to 1$), the benefits of aggregating the traffic decrease and full-mesh topologies seem a better option.

5.2.2 Capacity Planning with Modular Capacities

No existing network technology permits establishing links of arbitrary capacities eligible in a continuum $u_e \geq 0$. On the contrary, real network technologies restrict link capacities to a discrete set of possibilities, a situation usually called *modular capacities* constraint. For instance, Table 5.2 shows available link rates for some popular layer 1/2 technologies today.

Modular capacities can appear in conjunction with techniques that combine multiple parallel links between two nodes, such that they act as a single link of aggregated capacity. For instance, Link Aggregation Control Protocol (LACP) enables this for Ethernet, and ECMP rule in IP

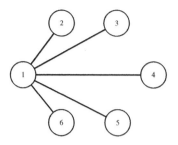

(a) Star topology

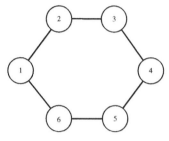

(b) Ring topology

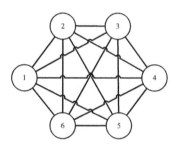

(c) Full-mesh topology

Figure 5.3   Example topologies

Table 5.1   Topology Example 5.4

Topology	$|\mathcal{E}|$	y
Star	10	5
Ring	12	4.5
Full-mesh	30	1

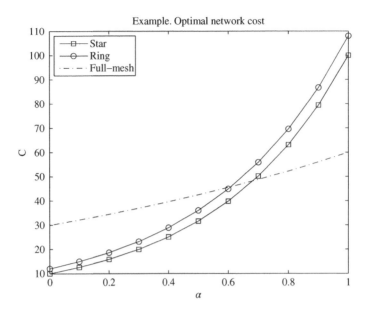

Figure 5.4 Example. Concave cost evolution example, $c(e) = u_e^\alpha$ for different values $\alpha \in [0, 1]$

Table 5.2 Discrete capacities available in different layer 1/2 technologies

Technology	Rates
OTN (Optical Transport Network)	(Gbps) 1.25, 2.5, 10, 40, 100
SONET/SDH	(Gbps) 0.051, 0.155, 0.622, 1.25, 2.5, 10, 40
Ethernet	(Gbps) 0.01, 0.1, 1, 10, 100
802.11 Point-to-point radio link[a]	(Mbps) 15, 30, 45, 60, 90, 120, 135, 150, 180, 200

[a]5 GHz frequency band, 40 MHz bandwidth.

networks can bifurcate the traffic between two nodes among parallel links, producing a similar effect to link aggregation.

Let $G(\mathcal{N}, \mathcal{E})$ be a network and $\mathcal{K}$ the set of available capacity module types eligible for determining the link capacities. We denote $u(k)$ and $c(k)$ the capacity and the cost of a module of type $k \in \mathcal{K}$, respectively. Cost values $c(k)$ usually reflect an economies of scale discount. For instance, if a module has twice the capacity of an other, its cost will be less than twice. Each link e in the network can be composed of an arbitrary number of modules of each type k. We denote this number a_{ek} and they are the decision variables of our assignment problem (5.9):

$$a_{ek} = \{(\text{Integer}) \text{ number of modules of type } k \text{ that make up link } e\}, \quad \forall e \in \mathcal{E}, k \in \mathcal{K}$$

$$\min_a \sum_e \sum_k a_{ek} c(k) \quad \text{subject to:} \qquad (5.9a)$$

$$y_e \leq \sum_k a_{ek} u(k), \quad \forall e \in \mathcal{E} \tag{5.9b}$$

$$a_{ek} \geq 0, \quad \forall e \in \mathcal{E}, k \in \mathcal{K} \tag{5.9c}$$

The objective function minimizes the sum of the costs of the network links. In (5.9b) the capacity of a link, given by the sum of the capacities of its components, is restricted to be higher than or equal to the link traffic. Finally, (5.9c) forbids having a negative amount of modules in a link.

Problem (5.9) can be trivially decomposed into $|\mathcal{E}|$ independent problems, one per network link, which can be fast solved for network scenarios appearing in practice using standard computing facilities. Modular capacity assignment becomes challenging when it appears in CFA problems that jointly optimize traffic routing and link capacities. In the next section, we show a case of joint routing and modular capacity assignment minimizing network cost.

5.2.3 Multi-Period Capacity Planning

Previous models for capacity planning targeted the optimization of the link bandwidth acquisition in a particular moment of time, so that the network meets its performance objectives during the subsequent single period until a new capacity planning process is done, for example 6 months later. In turn, multi-period capacity planning jointly optimizes the link capacity acquisitions to be performed during a sequence of T planning periods, for example, a sequence of four consecutive 6-month periods. Link capacities are usually considered modular and accumulative: that is, the capacity modules assigned to a link at a time period, stay there in the subsequent periods, so the capacity in the links only grows. Multi-period planning should be fed by two estimations:

- *Capacity costs forecast.* The expected evolution of the costs of the different link capacity modules during the next T periods. In general, the cost of the capacity modules for a given bit rate decreases with time and new higher-rate modules can appear. In addition, delaying the acquisitions can reduce the operational expenditures added to the total cost, since new equipment does not incur in maintenance expenses until it is purchased.
- *Traffic growth forecast.* The traffic forecasts can be based on sophisticated own-developed estimations, or use simple compound-annual-growth rate (CAGR) predictions (e.g., traffic growth of 20% per year) that are periodically published by network companies.

The rationale behind multi-period planning is that both previous estimations can permit better planning decisions in the network, which reduce costs in the long term. In particular, a multi-period view can recommend delaying the capacity acquisitions in some links, waiting for price cuts, or anticipating the acquisition of a more cost-efficient large capacity module, not immediately needed, but expected to be needed soon according to the traffic growth forecast. Still, the multi-period optimization should be repeated every single acquisition period, updating the cost and traffic forecasts. In other words, an acquisition plan for, as an example 2 years, produced by the multi-period optimization, is re-optimized and updated every 6 months, for example.

Multi-period planning can be solved as a joint routing and capacity assignment problem. In (5.10) we present a formulation for a multi-period planning, for a sequence of T periods,

$t = 1, \ldots, T$. Link capacities are modular. $\mathcal{K}$ denotes the set of module types, $u(k)$ is the capacity of a module $k \in \mathcal{K}$, and $c(k, t)$ is the estimated cost of a module of type k if acquired during time period t. The offered traffic forecast for a demand $d \in D$ during time period t is given by h_{dt}. Note that in general, h_{dt} volumes are increasing with t: $h_{dt+1} > h_{dt}$. Finally, we assume that the routing can be different in consecutive time periods, opening the door to exploit new trade-offs. For instance, offloading the traffic of some links until a expected cost reduction makes desirable acquiring a large capacity module for it. Decision variables for the problem are:

$a_{ekt} = \{$Number of NEW modules of type k for link e, acquired in time period $t\}$,

$\qquad \forall e \in \mathcal{E}, k \in \mathcal{K}, t = 1, \ldots, T$

$x_{pt} = \{$Amount of traffic of demand $d(p)$ carried by p during time period $t\}$,

$\qquad \forall p \in P, t = 1, \ldots, T$

The formulation of the described multi-period problem is (5.10):

$$\min_{a,x} \sum_t \sum_e \sum_k a_{ekt} c(k,t) \quad \text{subject to:} \tag{5.10a}$$

$$\sum_{p \in P_d} x_{pt} = h_{dt}, \quad \forall d \in D, t = 1, \ldots, T \tag{5.10b}$$

$$\sum_{p \in P_e} x_{pt} \leq \sum_{t'=1}^{t} \sum_k a_{ekt'} u(k), \quad \forall e \in \mathcal{E}, t = 1, \ldots, T \tag{5.10c}$$

$$a_{ek} \geq 0, \quad \forall e \in \mathcal{E}, k \in \mathcal{K} \tag{5.10d}$$

$$x_{pt} \geq 0, \quad \forall p \in P, t = 1, \ldots, T \tag{5.10e}$$

Objective function (5.10a) sums the acquisition costs along the time periods. Constraint (5.10b) means that all the traffic is carried. Constraint (5.10c) makes that in any period, the traffic in each link does not exceed its capacity. Note that the capacity installed in a link at time t is given by the accumulation of all the modules assigned in t and before.

5.3   Fast Capacity Allocation Problem Variants: Wireless Networks

Fast capacity allocation problems appear in network technologies where the link capacities should rapidly adapt to network varying conditions. The paramount example of this are wireless networks, for two reasons. First, the typical random variabilities in the wireless channel, due to a phenomenon called fading, require frequent readjustments of the link capacity, for example, at a subsecond rate. Second, when the wireless links are not point-to-point, but spread the signal energy in a broadcast pattern (e.g., in mobile cellular networks and in wireless LANs), the messages from multiple users can simultaneously arrive to a receiver hindering or even impeding the detection of legitimate communications. Because of this, the spectrum becomes a shared medium and access control or multiplexing mechanisms are needed to coordinate the users communication.

In this section we provide a brief overview of wireless technologies to introduce the technological context, and define some basic concepts appearing in its modeling like the *capacity region*. Sections 5.4 and 5.5 are devoted to model selected capacity assignment problems appearing in realistic case studies on wireless technologies.

5.3.1   The Wireless Channel

Figure 5.5 illustrates a single wireless channel, using as an example the transmission from a base station antenna to a mobile phone. In its travel to the receiving antenna, the base station signal suffers several forms of degradation:

- *Signal attenuation* or *path loss*: The signal power received at a distance d of the transmitter antenna is proportional to $\frac{p}{d^\gamma}$, where p is the signal power irradiated by the transmitter antenna, and γ is a so-called path loss exponent. Path loss exponents range typically from 2 to 5. Larger exponents significantly increase the attenuation, and thus reduce the wireless coverage. As an example, if the distance to the transmitter is multiplied by 2, the received power is divided by 4 when $\gamma = 2$ and divided by 32 when $\gamma = 5$.
- *Fading*: Fading is a significant and relatively fast variation of the received power or signal quality. Fading is caused by the propagation of the signal from the transmitter to the receiver through different paths (multi-path propagation), for example a direct path between both antennas, and several paths where the signal is reflected in surrounding objects (see Fig. 5.5). The signal received from different paths, traverse different distances, and can arrive to the receiver creating *constructive* or *destructive* interferences. A constructive interference increases the signal received power, while destructive interferences reduce it, for example up to 60 dB. The interference patterns depend on the reflecting objects, their shapes, movement, position, channel local conditions, and so on in a form impossible to control. The result is that the observed attenuation can largely and randomly fluctuate in intervals of duration typically between hundreds of milliseconds to several seconds.

  The apparition of fading at a particular frequency band is random, and independent from other bands. Also, the apparition of fading in a reception antenna located in a particular position is also random, and independent from a fading event in an antenna separated a short

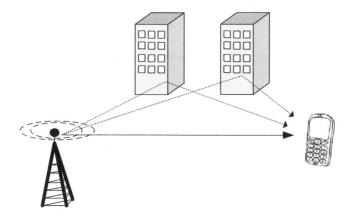

Figure 5.5   Example of multi-path propagation

distance from it[1]. Different diversity strategies can exploit independence of fading events to alleviate their effects. The frequency-hopping technique is an example of this, where the transmission frequency is rapidly switched among many channels in a sequence known by the transmitter and the receiver so that the effects of fading at some frequencies are reduced. Another diversity technique is using multiple transmission and/or reception antennas separated a short distance among them, a technique called MIMO (Multiple Input Multiple Output).

- *Noise*: The noise is an undesired fast and random varying signal (e.g., varying at least two orders of magnitude faster than the fading), coming from the multiple electromagnetic signals produced in nature and by the transmission and reception circuitry, which sum to the legitimate received signals.

A well-known relation coming from information theory, states that the maximum transmission rates u_e attainable at a wireless link e can be written, for a large family of modulations as [3]:

$$u_e = \frac{1}{T} \log(1 + K \cdot \text{SNR}_e) \tag{5.11}$$

Here, constant T is the symbol period, constant $K = (-\phi_1)/(\log(\phi_2 \text{BER}))$, where ϕ_1 and ϕ_2 constants depend on the modulation, and BER is the required bit-error-rate. SNR_e stands for the signal-to-noise ratio of the link, given by:

$$\text{SNR}_e = \frac{P_e^{rx}}{\sigma_e^2}$$

P_e^{rx} is the power received from the legitimate signal, which depends on the transmission power and the total signal attenuation, considering both path loss and fading. σ_e^2 is the noise power, including the thermal noise caused by the electronics at the receiver.

5.3.2 Wireless Networks

In point-to-point wireless links designed to have a *fixed* capacity (e.g., point-to-point radio links between two locations), the link parameters are engineered such that even in some worst-case fading and noise conditions considered, a minimum SNR is obtained to meet the target BER. The link antennas are designed to be directive, sending and receiving most of the energy in the direction towards the other antenna, and thus not affecting or being affected significantly by other surrounding wireless links. These fixed capacity radio links do not add any particular difference from a modeling point of view with respect to any other (wired) link.

In contrast, we are interested here in wireless networks where the signal transmitted by a node is broadcast to the medium and/or shares the spectrum with the transmissions from other nodes. This is the case of mobile cellular networks, and IEEE 802.11 wireless LAN technologies. Depending on the channel conditions (which fluctuate), and the simultaneous transmissions from other users, a legitimate signal received by a node can be mixed with the signals from other transmissions, generally called *interferences*. Interference signals can

---

[1] Usually, it is considered that a separation similar to the signal wavelength or a small multiple of it is enough to consider the apparition of fading in a location statistically independent from the other. Note that a wavelength λ (m) of a signal at frequency f (Hz) is given by $\lambda = c/f$. For instance, at 2.4 GHz the wavelength is 12.5 cm.

hinder, or even prevent the correct reception of the legitimate messages. As a result, the traffic sent by a node can produce a reduction of the capacity of *other nodes* to receive traffic. This is a major difference with respect to wireline networks.

We distinguish two cases:

- *Hard interference scenario*: In some wireless technologies, when two or more simultaneous signals from different sources arrive to a receiver, the detection of any of them is impossible. This situation is commonly referred to as a *collision* between the two (or more) messages. A *Medium Access Control* (MAC) protocol is needed to coordinate network nodes, reducing or eliminating the possibility of such collisions. Wireless communications under IEEE 802.11 protocols (e.g., Wi-Fi) include MAC protocols of this type.
- *Soft interference scenario*: Multiple wireless technologies implement a multiplexing scheme, such that a node can simultaneously receive different communications from different sources. Interfering signals from other users are seen as a noise that penalizes the SNR of the link. If the interfering power is small enough, the messages received from the link can be detected. Code Division Multiplexing (CDM), Orthogonal Frequency Division Multiplexing (OFDM) or Time Division Multiplexing (TDM) are examples of such techniques. In CDM, each information bit is multiplied before transmission by a sequence of bits (called chips) using a particular code. A node receiving simultaneous transmissions is able to demultiplex them using the corresponding transmitter code. User codes are chosen to be *orthogonal*, which means that in perfect channel conditions, each user information can be demultiplexed without corruptions. If not, other users' signals appear as attenuated added noise. CDM is used in the radio interface of mobile phone standards like CDMA2000 or WCDMA (3G). In OFDM the information is transmitted in parallel slower rate signals at different frequencies or subcarriers, where the separation between frequencies is specifically chosen to cancel the interferences between them. Then, users using different subcarriers do not interfere each other. Finally, TDM consists of dividing the time into frames, and the frames into slots. Users are allocated slots, such that two users can transmit in the same frame, as long as they use different slots. TDM is used in old GSM cellular systems. Modern LTE and LTE-advanced 4G use a combination of OFDM and TDM techniques, such that each user is assigned a set of carrier-slot pairs. The higher the number of slot pairs assigned, the larger the capacity allocated to the user.

Both MAC and multiplexed access case studies will be modeled in examples throughout this section.

5.3.3 Modeling Wireless Networks

We reuse the same notation and framework of the book, and denote the set of network nodes with $\mathcal{N}$ and $\mathcal{E}$ is the set of links. Since links in a wireless network can appear and disappear because of channel fluctuations, we will consider that a wireless link $e \in \mathcal{E}$ exists if, in optimum channel conditions, the link source node can send messages to the link end node. This model can include:

- Ad-hoc wireless networks, where all nodes are of the same type and a link between any node pair is possible if channel conditions between them allow it.

- Cellular networks where some nodes are base stations and the rest are mobile systems. In this case, direct communications between mobile systems are not possible. Thus, network links are restricted to (i) links from a base station to and from mobile system, and (ii) base station to base station links.

We define the capacity u_e in a link e as the amount of traffic that can be sent by the origin node, that is *correctly received* by the end node.

5.3.3.1   Capacity Region

To model the coupling between the capacities at different links such that, for example, the increase of a link capacity can imply the reduction of other, we use the concept of *capacity region* $\mathcal{V}$ of a network as the set of vectors $u = \{u_e, e \in \mathcal{E}\}$ that can be attained.

As an example, let us focus on a network like the one in Fig. 5.6a, where two nodes can send information to a common central node. Wireless links have a nominal rate of 1 Mbps, but the wireless technology is such that a node cannot receive messages arriving simultaneously from two incoming links, since a collision would occur destroying both signals. The capacity

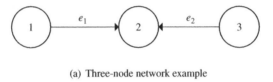

(a) Three-node network example

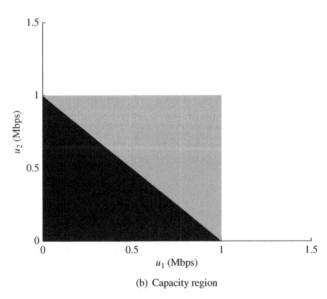

(b) Capacity region

Figure 5.6   Example. Three node network and its capacity region (u_1, u_2), when a node cannot receive simultaneously traffic from two nodes (black) and when this constraint does not exist (black and gray)

region in this network is given by those (u_1, u_2) non-negative vectors, for which $u_1 + u_e \leq 1$ (Fig. 5.6b). Note that the situation $u_1 + u_2 = 1$ would be only achieved if both links can be perfectly coordinated to (i) never use the channel at the same time and (ii) never leave the channel unused. This would be the task of the *Medium Access Control* (MAC) protocol. The concept of capacity region can also be applied to wired networks. In this example, if no external constraint exists to use both wired links at its maximum rate, the capacity region would be the square $\{0 \leq u_1 \leq 1, 0 \leq u_2 \leq 1\}$ (see Fig. 5.6b).

5.3.3.2 Indirect Capacity Allocation

Capacity allocation algorithms in the wireless context, that we will see in Part II of the book, are distributed iterative methods running at network nodes, with the aim of rapidly adapting link capacities to varying network conditions. A characteristic of these algorithms is that the link capacity is usually not directly configured, but comes *indirectly* as a result of controlling other figures. For instance, in some cellular networks, a link capacity is set by deciding on the transmission power of that and other links. Assigning more power to a link could increase its capacity at a cost of augmenting the interference power of surrounding users and thus reducing their capacities. In an other example, MAC protocols can be based on assigning each link a probability to access the channel. Then, increasing the access probability of a link would augment its capacity at a cost of increasing the chances of collisions with other users and thus reducing their capacities.

In this context, a general modeling of the capacity assignment problem maximizing the network utility takes the form:

$$\max_{u,p} \sum_e U_e(u_e) \quad \text{subject to:} \tag{5.12a}$$

$$u = f(p) \tag{5.12b}$$

$$p \in \mathcal{P} \tag{5.12c}$$

where $p = \{p_e, e \in \mathcal{E}\}$ is the set of control variables like transmission power or access probabilities and $\mathcal{P}$ the set of feasible values that these control variables can take (e.g., minimum and maximum powers, probabilities below 1, etc.). U_e are the utility functions of the NUM model and f denotes the *control-to-capacity function*, which determines the capacity of the network links, depending on the decisions taken on p. We remark that usually f is such that a link capacity u_e depends on the p_e decisions on that and other links.

The control-to-capacity function helps us to more formally define the concept of capacity region $\mathcal{U}$ of a network:

$$\mathcal{U} = \{u \text{ such that exists at least an allocation } p \in \mathcal{P}, \text{ for which } u = f(p)\}$$

Given a set of hardware constraints that characterize a technology, we say that a capacity assignment scheme is *throughput optimal* when it is able to find the control variables to produce any capacity vector desired within the capacity region. Naturally, no capacity assignment algorithm can attain a capacity outside the capacity region. But as we will see, not all the capacity allocation schemes are able to cover the full capacity region.

5.4   MAC Design in Hard-Interference Scenarios

In this section we study the design of MAC protocols coordinating the transmissions in a wireless network constrained by hard interference schemes. We consider a network composed of $\mathcal{N}$ wireless nodes and wireless links. A link e exists between origin node $a(e)$ to destination node $b(e)$, if under perfect channel conditions, $a(e)$ can send traffic to $b(e)$ at the link nominal rate $\bar{u}_e$.

We assume that all the nodes share the same wireless channel, such that the transmission on a link can impede the reception in another link/s (hard interference). This can happen for instance if the distance between the receiver of the former node and the transmitter node of the latter is less than some threshold (e.g., d^I km). In general, d^I can be equal to or higher than the coverage distance. In the latter case, a node can cause interferences to more nodes than those to whom it can send traffic.

We use a general framework to model hard interferences, using a Boolean matrix $A_{ee'} \in \{0, 1\}^{|\mathcal{E}| \times |\mathcal{E}|}$, where $A_{ee'} = 1$ if link e interferes link e', and $A_{ee'} = 0$ otherwise. In other words, $A_{ee'} = 1$ means that if e and e' simultaneously transmit traffic, the message in link e' is lost.

The interference map A is a powerful framework to model multiple interference constraints and hardware limitations. Figure 5.7 helps us to illustrate this. We assume a case when $d = d^I$, and thus the coverage range equals to the interference range (drawn in the figure). A link ending in node n is interfered by any other link whose transmitter has n in range. For instance, links e and e' in Fig. 5.7 cannot be simultaneously active, since both end in the same node ($A_{ee'} = 1, A_{e'e} = 1$). Also, $A_{e''e} = 1$ since when the link e'' is active, the signal interferes any link entering in node 2. Note, however, that $A_{ee''} = 0$ and A matrix is in general not symmetric. Finally, the interference map can be also a tool to add other hardware constraints. For instance, in some wireless technologies, nodes cannot transmit and receive simultaneously. In the example of Fig. 5.7, this can be included in A by making $A_{e'''e} = A_{e'''e'} = 1$.

Given an interference map A, we define by $\mathcal{M} \subset \{0, 1\}^{|\mathcal{E}|}$ the set of *valid link schedules*. A valid link schedule $m \in \mathcal{M}$ represents a set of links that do not generate any collision

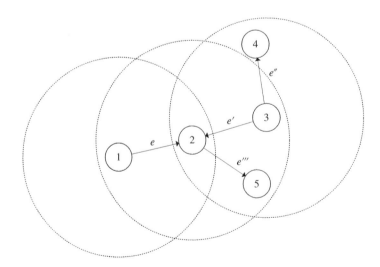

Figure 5.7   Example of interference map

when they are transmitting simultaneously. That is, for any $e \neq e'$ such that $m_e = m_{e'} = 1$, $A_{ee'} = 0$.

We define $\pi_m, m \in \mathcal{M}$, as the proportion of time that the network is operating under a particular scheduling m. Then, the capacity region $\mathcal{U}$ of a network constrained by hard interferences, for any MAC protocol, should be contained in the set $\bar{\mathcal{U}}$ of *feasible capacities*:

$$\bar{\mathcal{U}} = \left\{ u \in \mathbb{R}_+^{|\mathcal{E}|} : \exists \pi \in \mathbb{R}^{|\mathcal{M}|}, \pi_m \geq 0, \sum_{m \in \mathcal{M}} \pi_m = 1, \text{ such that } u \leq \bar{u}_e \sum_{m \in \mathcal{M}, m_e = 1} \pi_m \right\}$$

(5.13)

Note that since $\sum_m \pi_m = 1$, we are implicitly assuming a best-case situation when all the time is spent in *valid* link schedules that are collision free. For this reason we call $\bar{\mathcal{U}}$ the set of *feasible capacities*, since no MAC protocol can attain a capacity vector outside $\bar{\mathcal{U}}$. Interestingly, it is easy to show that the set $\bar{\mathcal{U}}$ is convex, since it is composed of all the convex combinations that can be made with the $|\mathcal{M}|$ capacity vectors associated to valid schedules (Exercise 5.2).

In this context, the target should be the design of MAC protocols that can attain any capacity within set $\bar{\mathcal{U}}$. Unfortunately, this can require complex coordination mechanisms, not amenable to a distributed implementation in practice. In next subsections, we will see a distributed random access MAC scheme whose capacity region is significantly smaller than $\bar{\mathcal{U}}$ and a model of CSMA (Carrier Sense Multiple Access) protocols, which have been shown (under some simplifying assumptions) to attain any capacity within the interior of $\bar{\mathcal{U}}$.

5.4.1 Optimization in Random Access Networks

In this section we model the capacity allocation problem associated to a Aloha-type *random access* MAC protocol[2] for a wireless network composed of a set $\mathcal{N}$ of nodes and a set $\mathcal{E}$ of wireless links. A synchronization scheme exists such that time is divided into consecutive slots of equal duration and slot initial times are aligned for all networks nodes. The access to the wireless medium is based on the so-called *persistence probabilities* as follows:

- In each slot, each node n independently decides to transmit with probability q_n or be idle with probability $1 - q_n$. We call q_n the *persistence probability of node n*.
- When a node n determines to transmit traffic, it chooses one of its outgoing links $e \in \delta^+(n)$ with probability p'_e, such that $\sum_{e \in \delta^+(n)} p'_e = 1$. Then, the probability p_e that a link e transmits traffic is given by: $p_e = p'_e q_{a(e)}$. We call p_e the *persistence probability of link e* and denote $p = \{p_e, e \in \mathcal{E}\}$ to its vector form. When a link e is transmitting traffic during a time slot, it makes it so at its *nominal transmission rate* $\bar{u}_e$, which can be the same for all links, or depend on the link distance, end nodes hardware, and so on. Note that node transmission probabilities q_n are determined by persistence probabilities p_e:

$$q_n = \sum_{e \in \delta^+(n)} p_e, \quad \forall n \in \mathcal{N}$$

---

[2] ALOHAnet was a pioneering wireless network developed in the 1970s demonstrating a wireless packet-based network. Each ALOHA station implemented a random-based access protocol to transmit. This scheme inspired later the access control in Ethernet and other random-access based protocols.

The network is subject to hard interferences and we denote A as the $|\mathcal{E}| \times |\mathcal{E}|$ matrix describing the interference map. From it, we define $\mathcal{N}_{to}^I(e)$ as the set of nodes whose transmissions interfere *to* the receiver of e, excluding the transmitter of e ($a(e)$). Also, $\mathcal{E}_{from}^I(n)$ denotes the set of links that are interfered *from* any transmission of node n, excluding outgoing links of node n. Hence,

- If a link e transmits traffic and any node in $\mathcal{N}_{to}^I(e)$ is transmitting traffic simultaneously, the transmission in e fails.
- If a node n transmits traffic and any link $e \in \mathcal{E}_{from}^I(n)$ is transmitting traffic simultaneously, the transmission in e fails.

Note that it is also possible to model wireless technologies that forbid a node to transmit and receive data at the same time. To represent this situation, sets $\mathcal{N}_{to}^I(e)$ should include the end node of e ($b(e) \in \mathcal{N}_{to}^I(e)$) and sets $\mathcal{E}_{from}^I(n)$ should include all the incoming links to e ($\delta^-(n) \subset \mathcal{E}_{from}^I(n)$).

In this section we are pursuing a model to adjust the persistence probabilities in our time slotted network, with the target of maximizing the network utility and enforcing a fair capacity allocation to the links. The average capacity u_e of a link e is the time average of the traffic that can be *correctly received* at its end. This depends on the persistence probabilities (q,p) in *all the nodes and links*. In particular, if e is active in a time slot, correct reception occurs when all nodes that can interfere with e ($\mathcal{N}_{to}^I(e)$) are silent. Given the statistical independence in transmission decisions in different nodes, the probability of correct reception equals $\prod_{n \in \mathcal{N}_{to}^I(e)}(1 - q_n)$ and thus the relation between persistence probabilities (the control variables) and link capacities is:

$$u_e = \bar{u}_e p_e \prod_{n \in \mathcal{N}_{to}^I(e)} (1 - q_n) \tag{5.14}$$

We apply a NUM (*Network Utility Maximization*) model as described in Section 3.7, where each link e is associated an increasing utility function $U_e(u_e)$. The optimization problem formulation is:

$$\max_{u,q,p} \sum_e U_e(u_e) \quad \text{subject to:} \tag{5.15a}$$

$$u_e \le \bar{u}_e p_e \prod_{n \in \mathcal{N}_{to}^I(e)} (1 - q_n), \quad \forall e \in \mathcal{E} \tag{5.15b}$$

$$u_e^{\min} \le u_e \le u_e^{\max}, \quad \forall e \in \mathcal{E} \tag{5.15c}$$

$$\sum_{e \in \delta^+(n)} p_e = q_n, \quad \forall n \in \mathcal{N} \tag{5.15d}$$

$$0 \le q_n \le 1, \quad \forall n \in \mathcal{N} \tag{5.15e}$$

$$0 \le p_e \le 1, \quad \forall e \in \mathcal{E} \tag{5.15f}$$

The first constraint establishes the capacity in the links. The second constraint sets the limits to u_e capacities between $u_e^{\min}$ and $u_e^{\max}$, input parameters to the problem.

The *capacity region* $\mathcal{U}$ of (5.15) is the set of vectors $u = \{u_e, e \in \mathcal{E}\}$ for which it is possible to find at least one feasible persistent probabilities (q, p) satisfying (5.15def):

$$\mathcal{U} = \{u \text{ satisfying } (5.14) \text{ for at least a } (q, p) \text{ satisfying } (5.15\text{def})\}$$

Note that the capacity region is solely determined by the nominal capacities $\bar{u}$ and the interference model given by the sets $\{\mathcal{N}_{to}^{l}(e), e \in \mathcal{E}\}$. The following example shows us that (i) the capacity region of the problem can be a non-convex set and (ii) since collisions can occur, the capacity region is smaller than the maximal set of capacities $\tilde{\mathcal{U}}$ for this interference map that can be attained by an optimum MAC protocol.

Example 5.2 We focus on a network like the one in Fig. 5.6a, where nodes 1, 3 are the only ones transmitting traffic. The capacity region is composed for those vectors (u_1, u_2) for which there exists at least one valid persistence probability values $(p_1, p_2) \in [0, 1]^2$ such that $u_1 = p_1(1 - p_2), u_2 = p_2(1 - p_1)$. This set is shown in Fig. 5.8a.

The non-convexity of the capacity region (e.g., see Fig. 5.8a as an example), caused by constraint (5.15b), hinders the problem solution by efficient methods. However, under verifiable sufficient conditions on the curvatures of utility functions, we can transform (5.15) into a separable convex optimization problem. To do so, we have to apply the logarithm function to both sides of constraints (5.15b,c):

$$\log u_e \leq \log \left(\bar{u}_e p_e \prod_{n \in \mathcal{N}_{to}^{l}(e)} (1 - q_n) \right), \quad \forall e \in \mathcal{E}$$

$$\log u_e^{min} \leq \log u_e \leq \log u_e^{max}, \quad \forall e \in \mathcal{E}$$

And now rewrite problem (5.15) using $u_e' = \log u_e$ as decision variables. To simplify the writing, we adapt the notation of the constants $u_e'^{min} = \log u_e^{min}$, $u_e'^{max} = \log u_e^{max}$, $\bar{u}_e' = \log \bar{u}_e$, and function $U_e'(u_e') = U_e(e^{u_e'})$:

$$\max_{u', q, p} \sum_e U_e'(u_e') \quad \text{subject to:} \tag{5.16a}$$

$$u_e' \leq \bar{u}_e' + \log p_e + \sum_{n \in \mathcal{N}_{to}^{l}(e)} \log(1 - q_n), \quad \forall e \in \mathcal{E} \tag{5.16b}$$

$$u_e'^{min} \leq u_e' \leq u_e'^{max}, \forall e \in \mathcal{E} \tag{5.16c}$$

$$(q, p) \text{ satisfying } (5.15\text{def}) \tag{5.16d}$$

Problem (5.16) has now a convex feasibility set, since: (i) all constraints but (5.16b) are linear, and (ii) $\log p_e$ and $\log(1 - q_n)$ are concave functions of the decision variables, and thus constraint (5.16b) is satisfied by a convex set of solutions. Another form to state this is that the set of points:

$$\mathcal{U}' = \{\log(u), u \in \mathcal{U}\}$$

composed of the points in the capacity region, where capacities are now measured in a logarithmic scale, is actually a convex set (see Fig. 5.8b as an example). Still, problem (5.16)

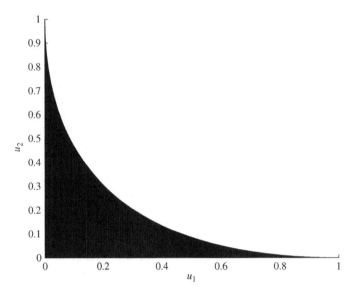

(a) Capacity region in natural units (u_1, u_2).

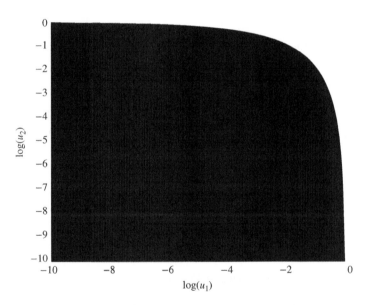

(b) Capacity region in logarithmic units $(\log u_1, \log u_2)$.

Figure 5.8   Example. Capacity region (u_1, u_2) for a network of two nodes, persistence probabilities $p_1, p_2 \in [0, 1]$, and capacities $u_1 \le p_1(1 - p_2), u_2 \le p_2(1 - p_1)$

may not be convex, if its objective function is not concave. Happily, the following proposition shows sufficient conditions that state that if utility functions U are *elastic enough* (concave enough), the transformed utilities U' are still concave functions.

Proposition 5.1 If utility function $U_e(u_e)$ is such that:

$$\frac{d^2 U_e(u_e)}{du_e^2} < -\frac{dU_e(u_e)}{u_e du_e} \qquad (5.17)$$

Then function $U'_e(u'_e) = U_e(e^{u_e})$ is strictly concave with respect to u'_e variables.

Proof. Since $u_e = e^{u'_e}$ we have:

$$\frac{d^2 U'_e(u_e)}{du_e'^2} = \frac{d^2 U'_e(u_e)}{du_e'^2}\left(\frac{du_e}{du'_e}\right)^2 + \frac{dU_e(u_e)}{du_e}\frac{d^2 u_e}{du_e'^2} = u_e^2\left(\frac{d^2 U_e(u_e)}{du_e^2} + \frac{dU_e(u_e)}{u_e du_e}\right)$$

which is below zero if (5.17) holds, and exactly zero (and the function is non-strictly concave) if it holds with equality.

This proposition states that the utility function needs to be not just strictly concave ($\frac{d^2 U_e(u_e)}{du_e^2} < 0$), but with a curvature that is bounded away from zero by as much as $\frac{dU_e(u_e)}{u_e du_e}$. It is easy to show that this holds for α-utility functions like (3.19), for $\alpha \geq 1$.

Proposition 5.2 α-utility functions of the form (3.19) satisfy Proposition 5.1 with a strict inequality for $\alpha > 1$ and equality for $\alpha = 1$.

Then, we can conclude that for α-utilities elastic enough ($\alpha \geq 1$), problem (5.16) is convex (with a unique optimum for $\alpha > 1$). Applying Proposition 3.1, we see that optimum solution of (5.16) in these convex cases is α-proportionally fair.

5.4.2 Optimization in Carrier-Sense Networks

The random access schemes shown in the previous section has the undesirable property of a reduced capacity region with respect to the maximal feasible capacity region $\bar{U}$ (5.13). Its main reason is that no specific hardware is available to reduce the chances of collisions. In this section we model MAC protocols based on CSMA (Carrier Sense Multiple Access), an scheme that has this precise target.

The basic principles behind CSMA is that each node senses the channel (*carrier sense*) before transmitting and refrains from starting a transmission if it is found to be busy. If multiple stations sense the channel is busy and defer their access, they will also simultaneously find when the channel is released and try to seize it. To reduce the chances of the collisions at this point, when a transmitter finds the channel idle, it waits a random time before transmitting, this is called a *back-off time*. If other node seizes the channel before the end of the back-off time, the timer is frozen and restarted when the channel is sensed idle again.

CSMA-based random access algorithms are the most widely used distributed MAC schemes in wireless networks. As an example, IEEE 802.11 is based on a carrier sense phase like the one described. The back-off time is randomly chosen as a multiple between 0 and $W - 1$ times the DIFS (*DCF Interframe Space*). The DIFS parameter varies among different 802.11 flavors between 28 μs and 128 μs and is related to the time needed by the electronics to detect a collision. W is the so-called back-off window. Since each mini-slot (of DIFS duration) is chosen with the probability $1/W$, the larger W the smaller the collision probability. The Distributed Coordination Function (DCF) implementation in 802.11 uses a Binary Exponential Backoff (BEB) to dynamically adjust the window size. BEB works doubling the window size when a collision is detected, unless it already has its possible maximum value, and resetting it to its minimum value after every successful transmission.

The network dynamics under CSMA have been thoroughly studied. We make use in this section of the simplified model described in [4]. It is based on the following assumptions:

- If two links conflict, because their simultaneous transmissions would result in a collision, then the carrier sense of both transmitting nodes can hear when the other transmits. This means that the carrier sense range should be greater than the coverage range and the so-called *hidden node* problem can be avoided[3].

- If a transmitter senses a conflicting link transmission, it keeps silent. If not, it waits for a random period of time with exponential distribution of mean $1/R_e$ and then starts its transmission. If some conflicting link is sensed during the back-off, the timer is frozen and continues when the channel is sensed idle again. We assume that the transmission time of any link has a random duration with exponential distribution, with mean 1^4.

- The carrier sense is instantaneous and the back-off time is chosen in a continuum. Then, the possibility of two nodes choosing the same back-off time is zero and collisions never occur. Note that this assumption is violated in real networks, because of the discretization of the back-off time, the finite speed of light and the time needed to detect a received power.

We call $r_e = \log R_e$ the *transmission aggressiveness* (TA for short) of the link e. The larger the aggressiveness of a link, the smaller the average back-off time when accessing the channel. In window-based schemes where the back-off time is randomly chosen between 0 and W, there is a direct relation between r_e and W given by:

$$\frac{1}{e^{r_e}} = \frac{W}{2} \Rightarrow W = 2e^{-r_e}$$

The TA of each link is the parameter we can control in CSMA protocols to tune the average link capacity. At this point, we are interested in the following questions that will be addressed in subsequent subsections:

- Which is the relation between the TA in the network links and the resulting link capacities?
- Which is the region of attainable capacities in CSMA protocols?

---

[3] The hidden node problem occurs when two nodes cannot communicate each other (are hidden from each other), but have a common neighbor to whom they can send traffic simultaneously, generating collisions. For instance, in Fig. 5.6a, nodes 1 and 3 are hidden between them, and have node 2 as a common neighbor. By assuming that the carrier sense range is large enough, a node can sense the hidden nodes transmissions and avoid collisions. For instance, if carrier sense range in Fig. 5.6a is twice the coverage range, node 1 and 3 could sense each other an avoid collisions in node 2.

[4] It is possible to show [4] that results in this section do not depend on the distributions of the back-off and channel holding times, provided that their mean are the ones stated here.

In Part II of the book, we exploit these results to devise algorithms that optimize the transmission aggressiveness in a distributed form, but unlike, for example BEB schemes, can be tuned to decide the network utility function to maximize.

5.4.2.1 TA-to-Capacity Relation

Given some fixed transmission aggressiveness values $r = \{r_e, e \in \mathcal{E}\}$, the network state randomly transits among conflict-free states, since according to our model collisions cannot occur. Then, the set of possible network states is at most $\mathcal{M}$, the set of valid schedules of the network. For a given network state m, $\mathcal{E}(m)$ denotes its set of active links. Given a link e, $\mathcal{M}(e)$ is the set of states where link e is active.

The system evolution can be described by a continuous time Markov chain. This Markov chain has been studied in [4] and other works, showing that its stationary distribution for particular r values is given by:

$$\pi_m(r) = \frac{e^{\sum_{e \in \mathcal{E}(m)} r_e}}{C(r)}, \quad \forall m \in \mathcal{M} \tag{5.18}$$

$C(r)$ is a normalizing constant chosen to make the sum $\sum_m \pi_m(r) = 1$:

$$C(r) = \sum_m e^{\sum_{e \in \mathcal{E}(m)} r_e}$$

The value $\pi_m(r)$ is the fraction of time that the network is found in state m, when the observation time is large enough. Then, the average traffic that can be transmitted by a link e when transmission aggressiveness is given by r, is:

$$u_e(r) = \bar{u}_e \sum_{m \in \mathcal{M}(e)} \pi_m(r) = \frac{1}{C(r)} \sum_{m \in \mathcal{M}(e)} e^{\sum_{e \in \mathcal{E}(m)} r_e} \tag{5.19}$$

Expression (5.19) is the targeted TA-to-capacity relation, that reflects that the capacity in a network link depends on the transmission aggressiveness in *all* the network links.

5.4.2.2 Capacity Region

The capacity region $\mathcal{V}$ of the CSMA MAC model described is composed of those capacities $u = \{u_e, e \in \mathcal{E}\}$ that can be induced in the network by a TA vector r, where $0 \leq r_e < \infty$:

$$\mathcal{V} = \{u: \text{ there is a TA vector } 0 \leq r < \infty \text{ satisfying (5.19)}\}$$

Naturally, the set $\mathcal{V}$ is contained in the set $\bar{\mathcal{U}}$ of feasible capacities (5.13). It is possible to show (see [5] for details) that the interior of this set **int** $(\bar{\mathcal{U}})$, is given by those capacity vectors u induced by schedules which spend a *non-zero* amount of time in all possible network states:

$$\text{int}\,(\bar{U}) = \{u \in \mathbb{R}_+^{|\mathcal{E}|} : \exists \pi \in \mathbb{R}^{|\mathcal{M}|}, \pi_m > 0, \sum_{m \in \mathcal{M}} \pi_m = 1, \text{ such that } u \leq \bar{u}_e \sum_{m \in \mathcal{M}(e)} \pi_m\} \tag{5.20}$$

we call these capacities *strictly feasible*. Interestingly, Proposition 5.3 states that using finite r values we can attain any *strictly feasible* capacity. This means that if a capacity vector u

is in the interior of the feasible set $\bar{U}$ we can attain it and, if it is in the boundary, we can approximate it as much as we want.

Proposition 5.3 The capacity region U of the CSMA protocol described is the set of strictly feasible capacities.

Proof. Let u be a strictly feasible capacity vector. We focus on the optimization problem that finds π_m distribution that maximizes the entropy among those that provide link capacities equal or higher than u:

$$\max_{\pi} - \sum_m \pi_m \log(\pi_m) \quad \text{subject to:} \tag{5.21a}$$

$$r_e : u_e \leq \bar{u}_e \sum_{m \in \mathcal{M}(e)} \pi_m \quad \forall e \in \mathcal{E} \tag{5.21b}$$

$$w : \sum_m \pi_m = 1 \tag{5.21c}$$

$$v_m : \pi_m \geq 0, \quad \forall m \in \mathcal{M} \tag{5.21d}$$

The objective function to maximize is concave and thus the problem (5.21) is convex. Also, if u is a strictly feasible solution, the problem has a non-empty feasibility set, since at least the π_m values that attain u are feasible. For Slater conditions to hold, the feasibility set should have a non-empty relative interior and (see [5] for further details, and Exercise 5.4 for an example) this is why we need u to be strictly feasible, and not just feasible. Finally, when the problem is feasible and Slater conditions are satisfied, KKT conditions hold for (5.21). Then, the proof is based on showing that the optimum π_m is the distribution enforced by a CSMA protocol when r_e multipliers are the TAs at the nodes, multiplier $w = \log(C(r)) - 1$, and multipliers $v_m = 0$. The complete proof can be consulted in [4].

Note that since $\bar{U}$ is a convex set, its interior is also convex [6]. Then, the following property holds.

Proposition 5.4 The capacity region U (5.20) is a convex set.

A NUM modeling of the capacity allocation problem, applying the capacity-to-TA relation (5.19) is given in (5.22):

$$\max_{u,r} \sum_e U_e(u_e) \quad \text{subject to:} \tag{5.22a}$$

$$u_e \leq \frac{1}{\sum_m e^{\sum_{e \in \mathcal{E}(m)} r_e}} \sum_{m \in \mathcal{M}(e)} e^{\sum_{e \in \mathcal{E}(m)} r_e} \quad \forall e \in \mathcal{E} \tag{5.22b}$$

$$u_e \geq 0, r_e \geq 0, \quad \forall e \in \mathcal{E} \tag{5.22c}$$

where $U_e(u_e)$ are the utility functions. Problem (5.22) is convex as long as the utility functions are concave. Recall that the set of feasible solutions is convex, since the capacity region defined by constraints (5.22b) is a convex set, as stated in Proposition 5.4.

5.4.2.3 CSMA and Maximum Entropy

An intriguing relation derived from the proof of Prop. 5.3 is that when a CSMA protocol with TA values r attains a capacity $u(r)$, the resulting distribution of schedules π_m is the one that maximizes the entropy among those schedules that attain this capacity. We can speculate about this, making an analogy with classical statistical mechanics, where many microscopic behaviors aggregate into macroscopic states and the equilibrium is reached in the macroscopic state that is the *most likely one*.

Let us assume a network whose state evolves jumping among valid schedules $m \in \mathcal{M}$ in discrete steps. We observe the system during T consecutive instants and denote as $\{m(t), t = 1, \dots, T\}$ the sequence of states observed. In our analogy, any of the possible sequences $\{m(t)\}$ is a microscopic state. The number of different microscopic states that can be observed is $|\mathcal{M}|^T$ and we assume that all of them *have the same probability to occur*.

Each sequence $\{m(t)\}$ has associated a number x_m of times that the network was in state m during the observation period. Note that it always holds that $\sum_m x_m = T$. In our analogy, a vector $\{x_m, m \in \mathcal{M}\}$ is our macroscopic state.

Many different microscopic states $\{m(t)\}$ can produce the same macroscopic state $\{x_m\}$. When all the microscopic states are equiprobable, the number of microscopic states that result in the same macroscopic state $\{x_m\}$ is:

$$K_{\{x_m\}} = \frac{T!}{\prod_m x_m!}$$

We want to search for the most likely macroscopic state, that is, the $\{x_m\}$ with the largest K, or equivalently with the maximum $\frac{1}{T} \log K$. Using Stirling's approximation of the factorial for large numbers ($n! \approx n^n e^{-n}$) we have:

$$\max \frac{1}{T} \log \frac{T!}{\prod_m x_m!} \approx \max - \sum_m \frac{x_m}{T} \log \frac{x_m}{T}$$

When the observation period is large x_m/T approximates the probability of finding the system in state m, that is, π_m. This means that the most likely network state distribution, the equilibrium one in the statistical mechanics analogy, is the π_m distribution that maximizes the entropy $-\sum_m \pi_m \log \pi_m$.

5.5 Transmission Power Optimization in Soft Interference Scenarios

In this section we focus on a wireless network where communications apply a multiplexing technique that permits simultaneous communications between nodes. The imperfections of the multiplexing technique mean that some of the power received from a link appears in other links as interference and computes as an added noise.

Let $\mathcal{G}(\mathcal{N}, \mathcal{E})$ be a wireless network where $\mathcal{N}$ is the set of nodes and $\mathcal{E}$ the set of links. Let $G_{ee'}$ denote the gain (in linear units) of the signal propagating from initial node of e to end node of e': the ratio between the power (e.g., in mW) of the transmitted signal at $a(e)$ and the measured signal at $b(e')$. When $e = e'$, G_{ee} values depend on the distance between link e end nodes and wireless channel conditions (which can significantly fluctuate along time). When $e \neq e'$, $G_{ee'}$ factor includes, together with the effect of channel conditions, an extra attenuation added by the ability of the receiver at e' to filter out most of the interfering signal received from link e. For instance,

- In CDM systems this ability is provided by the orthogonality between the codes of different links. Although ideal code orthogonality makes $G_{ee'} = 0$, $G_{ee'} > 0$ in realistic situations.
- In OFDM systems, uplink signals arriving to the base station from different mobile phones are orthogonal and in general we can consider $G_{ee'} = 0$ (where e and e' are the uplinks from different phones to their common base station). However, interferences can still occur if neighbor base stations share the same frequency bands. In this case, the uplinks from a mobile phone can arrive to other base stations aside to the one serving it, causing interferences.

The transmission power at each link e in linear units is denoted by p_e and $p = \{p_e, e \in \mathcal{E}\}$ is the vector form of these transmission powers. The overall Signal-to-Noise Ratio (SNR) at the receiver of e is given by:

$$\text{SNR}_e = \frac{p_e G_{ee}}{\sigma_e^2 + \sum_{e' \neq e} p_{e'} G_{e'e}} \tag{5.23}$$

The numerator in (5.23) is the power transmitted at the initial link node that reaches the receiver. The denominator sums the power (in linear units) at the receiver of those signals that are considered to be undesired noise hindering the correct detection: (i) the power of the thermal noise σ_e^2 at the receiver hardware and (ii) the sum of the interference powers from other links[5].

In those contexts when user interferences dominate in the SNR, the link capacities are limited by them, with a theoretical limit given by [3]:

$$u_e = \frac{1}{T} \log(1 + K \cdot \text{SNR}_e) \tag{5.24}$$

As mentioned in Section 5.3.1, constant T is the symbol period and K is a constant dependent on the modulation and target BER. Without loss of generality, to simplify the notation in the sequel we assume $T = 1$ and absorb K into the G_{ee} factor. Also, applying the assumption that gains $G_{ee'} \ll G_{ee}$, we can approximate $1 + \text{SNR}_e \approx \text{SNR}_e$. Note that this approximation requires working in sufficiently high SNR regions, which may not happen if a significant amount of interfering transmitters are much closer to the receiver than the legitimate transmitter.

With the previous assumptions, we can model the problem of assigning transmission power p_e to each link e to maximize the utility of the resulting capacity allocation in the network ($u = \{u_e, e \in \mathcal{E}\}$) as follows (5.25):

$$\max_{u,p} \sum_e U_e(u_e) \quad \text{subject to:} \tag{5.25a}$$

$$u_e \leq \log\left(\frac{p_e G_{ee}}{\sigma_e^2 + \sum_{e' \neq e} p_{e'} G_{e'e}}\right), \quad \forall e \in \mathcal{E} \tag{5.25b}$$

$$u_e^{min} \leq u_e \leq u_e^{max}, \quad \forall e \in \mathcal{E} \tag{5.25c}$$

$$p_e \geq 0, \quad \forall e \in \mathcal{E} \tag{5.25d}$$

[5] Some systems set a limit to the amount of interference power that can handle, as a multiple with respect to the receiver thermal power. This is called the *Rise Over Thermal* (ROT), typically between 3 dB and 10 dB.

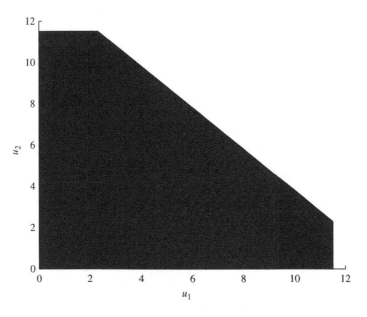

Figure 5.9 Capacity region of a network composed of two links $\{e_1, e_2\}$ with a common receiver node and $G_{e_1e_1} = G_{e_2e_2} = 1$, $G_{e_1e_2} = G_{e_2e_1} = 0.001$, $\sigma_{e_1}^2 = \sigma_{e_2}^2 = 10^{-10}$, and transmission powers from 0 to 100 mW

The *capacity region* of the system is the set of rates $u = \{u_e, e \in \mathcal{E}\}$ for which at least a feasible power allocation p exists such that (5.25b,d) hold. It can be shown (see Exercise 5.5) that the capacity region is a convex set (see Fig. 5.9 for an example).

Although the feasibility set of problem (5.25) is convex, the right-hand side of constraint (5.25b) is a non-concave function. This can create difficulties when devising solution algorithms for the problem. Interestingly, it is possible to construct an equivalent convex formulation solving this issue, by using a convenient set of decision variables $\tilde{p}_e$ that are simply p_e variables expressed now in a logarithmic scale:

$$\tilde{p}_e = \log p_e \Rightarrow p_e = e^{\tilde{p}_e}, \quad \forall e \in \mathcal{E} \tag{5.26}$$

Problem (5.25) becomes:

$$\max_{u,\tilde{p}} \sum_e U_e(u_e) \quad \text{subject to:} \tag{5.27a}$$

$$u_e \leq \log\left(\frac{e^{\tilde{p}_e} G_{ee}}{\sigma_e^2 + \sum_{e' \neq e} e^{\tilde{p}_{e'}} G_{e'e}}\right), \quad \forall e \in \mathcal{E} \tag{5.27b}$$

$$u_e^{min} \leq u_e \leq u_e^{max}, \quad \forall e \in \mathcal{E} \tag{5.27c}$$

The next proposition, adapted from [7], proves the validity of the transformation.

Proposition 5.5 The function:

$$f(\tilde{p}) = \log\left(\frac{e^{\tilde{p}_e}G_{ee}}{\sigma_e^2 + \sum_{e' \neq e} e^{\tilde{p}_{e'}}G_{e'e}}\right)$$

is a concave function with respect to $\tilde{P}$ variables, for all $e \in \mathcal{E}$.

Proof. Let $\tilde{p}$ and $\tilde{q}$ be two different power assignments in the network. We need to show that for all $\tilde{p}$ and $\tilde{q}$ assignments and all $\alpha \in [0, 1]$:

$$\alpha f(\tilde{p}) + (1 - \alpha)f(\tilde{q}) \leq f(\alpha\tilde{p}_e + (1 - \alpha)\tilde{q}_e)$$

The left- and right-hand sides of previous expressions can be rewritten as:

$$\alpha f(\tilde{p}) + (1 - \alpha)f(\tilde{q}) = \log\left(\frac{e^{\alpha\tilde{p}_e}e^{(1-\alpha)\tilde{q}_e}G_{ee}^{\alpha+1-\alpha}}{(\sigma_e^2 + \sum_{e' \neq e} e^{\tilde{p}_{e'}}G_{e'e})^\alpha(\sigma_e^2 + \sum_{e' \neq e} e^{\tilde{q}_{e'}}G_{e'e})^{1-\alpha}}\right)$$

$$f(\alpha\tilde{P}_e + (1 - \alpha)\tilde{q}_e) = \log\left(\frac{e^{\alpha\tilde{p}_e+(1-\alpha)\tilde{q}_e}G_{ee}}{\sigma_e^2 + \sum_{e' \neq e} e^{\alpha\tilde{p}_{e'}+(1-\alpha)q_{e'}}G_{e'e}}\right)$$

Then, the proposition holds if:

$$\log\left(\frac{\sigma_e^2 + \sum_{e' \neq e} e^{\alpha\tilde{p}_{e'}+(1-\alpha)q_{e'}}G_{e'e}}{(\sigma_e^2 + \sum_{e' \neq e} e^{\tilde{p}_{e'}}G_{e'e})^\alpha(\sigma_e^2 + \sum_{e' \neq e} e^{\tilde{q}_{e'}}G_{e'e})^{1-\alpha}}\right) \leq 0 \tag{5.28}$$

Previous condition is guaranteed by the celebrated Holder's inequality, which states that if $x_i, y_i \geq 0, p > 1$ and $1/p + 1/q = 1$, then:

$$\left(\sum_i x_i^p\right)^{\frac{1}{p}}\left(\sum_i y_i^q\right)^{\frac{1}{q}} \geq \sum_i x_i y_i$$

By making $p = \frac{1}{\alpha}, q = \frac{1}{1-\alpha}$, and:

$$x = \{\sigma_e^{2\alpha}, e^{\alpha\tilde{p}_{e'}}G_{e'e}^\alpha \forall e' \neq e\}$$

$$y = \{\sigma_e^{2(1-\alpha)}, e^{(1-\alpha)\tilde{q}_{e'}}G_{e'e}^{1-\alpha} \forall e' \neq e\}$$

Then the numerator in (5.28) equals $\sum_i x_i y_i$ and the denominator equals $\left(\sum_i x_i^p\right)^{\frac{1}{p}}\left(\sum_i y_i^q\right)^{\frac{1}{q}}$ and thus Holder's inequality proves the proposition.

5.6 Notes and Sources

The inclusion in this chapter of long-term capacity planning problems and transmission power or random access allocation problems in wireless networks is consistent with the aim of the book of putting together related problems, even if they have been traditionally studied separately.

The problem convexification described in Section 5.2.1 is original, together with the case study illustrating the trade-offs in network design coming from the optimal capacity allocation

formula. However, this case study is inspired by the minimum average delay capacity allocation problem studied in [8]. In particular, in Chapter 5 of [8] a closed formula was obtained
just for the linear cost case, which made it difficult to connect the topology trade-offs with the
concave cost structure.

The models for capacity planning with modular capacities and multi-period capacity planning are original, but similar to other models found in the literature (e.g., [2]).

The modeling of the persistence probability optimization in random access protocols,
together with Prop. 5.1 and Prop. 5.2 that characterize its convexity, come from [9].

The modeling of CSMA protocols presented is adapted from [4, 5, 10, 11]. A detailed model
of 802.11 protocol can be found in [12].

The optimization of transmission power in wireless networks has been investigated in multiple works from different perspectives. The problem of optimizing the transmitter powers that
match a *known* set of data rates (and thus minimum SNRs) has been extensively investigated,
and distributed schemes like the Foschini and Miljanic algorithm [13] are widely used by the
industry in such transmission power adaptations. Section 5.5 targets the problem of finding the
SNRs or link rates that make an efficient utilization of the shared channel and is based mostly
on the works in [7, 14, 15], and [16].

5.7   Exercises

5.1   (Chapter 5, [8]) We focus on a modified version of problem (5.1), where the objective function is replaced by the average network estimation with M/M/1 link delays
and constraint (5.1b) is removed. Show that the problem is convex using the variable
change $\rho_e = y_e/u_e$. Apply KKT conditions to obtain a closed formula for the optimum
capacities in the linear cost case ($\alpha = 1$).

5.2   Prove that the set of feasible capacities (5.13) in a hard-interference scenario is always
convex.

5.3   Prove Prop. 5.2 applying Prop. 5.1.

5.4   For the network of Fig. 5.6a, show that the capacity vector $u_1 = 0.5$ and $u_e = 0.5$ is
feasible, but not strictly feasible. Also, show that it cannot be attained using a CSMA
protocol with finite transmission aggressiveness values. *Hint*: Use (5.19) and note that
the fraction of time when no link is active should be zero.

5.5   [15] Show that the capacity region in problem (5.25) is convex.

5.6   Implement a Net2Plan algorithm that finds the optimum routing and capacity assignment in a multi-period context for a given network, solving formulation (5.10) with
JOM. Input parameters are the set of capacity modules $\mathcal{K}$, their capacities $u(k)$ and current costs $c(k)$, the number of planning periods T, and c_f and h_f parameters: a capacity
module cost at period $t + 1$ is supposed to be a fraction c_f of its cost at time t, and
the traffic of a demand at $t + 1$ increases in a fraction h_f with respect to the traffic in
period t. The optimum solution for the first time interval is returned by updating the
link capacities and routing.

5.7   Implement a Net2Plan algorithm that finds the optimum persistence probabilities for a given network, by solving with JOM a modified version of (5.16), where decision variables u'_e and constraints (5.16bc) are eliminated, and u'_e in the objective function is replaced by $\bar{u}'_e + \log p_e + +\sum_{n\in\mathcal{N}^l_{to}(e)} \log(1 - q_n)$. Input parameters to the problem are the α factor, the minimum link persistence probability and maximum node persistence probabilities (different to 0 and 1, respectively, to avoid numerical instabilities in the solver), the nominal link capacities $\bar{u}_e$ (the same for all links), and a Boolean argument stating if a node can or cannot transmit and receive simultaneously. The optimum solution is returned by updating the link capacities, setting the p_e values as link attributes, and q_n values as node attributes. *Hint*: In the case $\alpha \neq 1$, use the relation $e^{\log x} = x$ in JOM formulation.

5.8   Implement a Net2Plan algorithm that finds the optimum transmission aggressiveness in a CSMA network with JOM solving a modified version of (5.21), where the objective function is replaced by $-\sum_m \pi_m \log \pi_m + \beta\sum_e U_e(u_e)$, where U_e is an α-utility function, and $\beta \geq 0$ a parameter controlling the importance of utility maximization in the objective function. Input parameters to the problem are the α-fairness factor, β, the link nominal capacities $\bar{u}_e$ (the same for all links), and the minimum value accepted for π_m. The optimum solution is returned by updating the link capacities and setting the r_e values as link attributes.

5.9   Implement a Net2Plan algorithm that finds the optimum transmission power in a wireless network with JOM solving the formulation:

$$\max_p \sum_e U_e\left(\log\left(\frac{p_e G_{ee}}{\sigma_e^2 + \sum_{e'\neq e} p_{e'} G_{e'e}} \right) \right), \text{ subject to: } p^{min} \leq p_e \leq p^{max}, \quad \forall e \in \mathcal{E}$$

p_e are the transmission power in logarithmic units, U_e are α-utility functions. G_{ee} values are given by $G_{ee} = d_e^{-\gamma}$, where d_e is the link distance and γ the path loss exponent. When $e \neq e'$, $G_{ee'} = d(a_e, b_{e'})^{-\gamma} A$, where $d(n_1, n_2)$ is the euclidean distance between nodes n_1 and n_2 and A an attenuation factor given by the receiver ability to filter out some of the interfering signals. The thermal noise σ_e^2 is set as $\sigma_e^2 = i_{wc}/\text{ROT}$, i_{wc} is the worse case interference received if all the links transmit at the maximum power, and ROT is the Rise over Thermal parameter (e.g., ROT=10). Input parameters to the algorithm are $\alpha, p^{min}, p^{max}, \gamma, A$, and ROT. The optimum solution is returned by updating the link capacities and setting the transmission power values as link attributes.

References

[1] A. Nucci and K. Papagiannaki, *Design, Measurement and Management of Large-Scale IP Networks: Bridging the Gap Between Theory and Practice*. Cambridge University Press, 2009.

[2] M. Pioro and D. Medhi, *Routing, Flow, and Capacity Design in Communication and Computer Networks*. Morgan Kaufmann Publishers, 2004.

[3] A. Goldsmith, *Wireless Communications*. Cambridge University Press, 2005.

[4] L. Jiang and J. Walrand, "A distributed csma algorithm for throughput and utility maximization in wireless networks," *IEEE/ACM Transactions on Networking (TON)*, vol. 18, no. 3, pp. 960–972, 2010.

[5] L. Jiang and J. Walrand, "A distributed algorithm for maximal throughput and optimal fairness in wireless networks with a general interference model," *EECS Department, University of California, Berkeley, Tech. Rep*, 2008.

[6] S. Boyd and L. Vandenberghe, *Convex Optimization*. New York, NY, USA: Cambridge University Press, 2004.

[7] C. W. Sung, "Log-convexity property of the feasible sir region in power-controlled cellular systems," *Communications Letters, IEEE*, vol. 6, no. 6, pp. 248–249, 2002.

[8] L. Kleinrock, *Queueing Systems*. New York, NY, USA: John Wiley & Sons, Inc., Wiley Interscience, 1976, vol. II: Computer Applications.

[9] J.-W. Lee, M. Chiang, and A. R. Calderbank, "Utility-optimal random-access control," *Wireless Communications, IEEE Transactions on*, vol. 6, no. 7, pp. 2741–2751, 2007.

[10] B. Nardelli, J. Lee, K. Lee, Y. Yi, S. Chong, E. W. Knightly, and M. Chiang, "Experimental evaluation of optimal CSMA," in *INFOCOM, 2011 Proceedings IEEE*. IEEE, 2011, pp. 1188–1196.

[11] J. Liu, Y. Yi, A. Proutiere, M. Chiang, and H. V. Poor, "Towards utility-optimal random access without message passing," *Wireless Communications and Mobile Computing*, vol. 10, no. 1, pp. 115–128, 2010.

[12] G. Bianchi, "Performance analysis of the ieee 802.11 distributed coordination function," *Selected Areas in Communications, IEEE Journal on*, vol. 18, no. 3, pp. 535–547, 2000.

[13] G. J. Foschini and Z. Miljanic, "A simple distributed autonomous power control algorithm and its convergence," *Vehicular Technology, IEEE Transactions on*, vol. 42, no. 4, pp. 641–646, 1993.

[14] M. Chiang, S. H. Low, J. C. Doyle *et al.*, "Layering as optimization decomposition: A mathematical theory of network architectures," *Proceedings of the IEEE*, vol. 95, no. 1, pp. 255–312, 2007.

[15] D. ONeill, D. Julian, and S. Boyd, "Seeking Foschini's genie: optimal rates and powers in wireless networks," *IEEE Transactions on Vehicular Technology*, 2003.

[16] D. C. ONeill, D. Julian, and S. Boyd, "Adaptive management of network resources," in *Vehicular Technology Conference, 2003. VTC 2003-Fall. 2003 IEEE 58th*, vol. 3. IEEE, 2003, pp. 1929–1933.

6

Congestion Control Problems

6.1 Introduction

Communication networks are intended to carry the traffic offered by the sources. When traffic volumes are too high, network links saturate, and there is a situation called congestion. To prevent it, a mechanism is needed to perform a fair and effective sharing of the link bandwidth by assigning the traffic volume each source is allowed to inject.

The network elements taking the bandwidth sharing decisions traditionally receive different names in different contexts. When traffic demands are virtual circuits or traffic connections at a fixed rate, the bandwidth sharing is governed by an *admission control* logic that decides for each connection request whether to carry it or block it. In turn, if the target is adjusting periodically (e.g., at a subsecond rate) the traffic injected by a demand, such that network resources are not saturated, the process is called *congestion control*. This chapter is mostly devoted to congestion control in packet switched networks.

Congestion control algorithms are code running in the traffic sources that computes an upper limit to the allowed rate to inject in the network. Some sources do not have much traffic to transmit and the congestion control upper limit is actually not restrictive for them. These are typically called *mouse* connections and correspond to occasional traffic between standard users. In contrast, other sources are always willing to transmit as much traffic as they are allowed. They are connections for large file downloads as an example, mirroring between servers, bulk data retrievals, and so on. In the congestion control field these sources are referred to as *elephants*.

Commonly, the number of mouse connections in a network is much higher than the number of elephants. However, a small number of elephant connections can solely deplete the bandwidth in the traversed links if allowed, unfairly leaving little for the rest. The role of a congestion control mechanism is precisely to avoid this, enforcing two goals: (i) a *fair* bandwidth assignment among the connections and (ii) utilizing the network links as much as possible, but still keeping traffic losses and packet delays at acceptable levels.

The most challenging example of congestion control context is the Internet: a network of millions of users and links, with a multitude of technologies and link bandwidths ranging from Kbps to Gbps and loosely organized as an interconnection of thousands of subnetworks

Optimization of Computer Networks – Modeling and Algorithms: A Hands-On Approach,
First Edition. Pablo Pavón Mariño.
© 2016 John Wiley & Sons, Ltd. Published 2016 by John Wiley & Sons, Ltd.
Companion Website: www.wiley.com/go/PavonMarinoSol16

administered by different institutions. The majority of the traffic in the Internet is exchanged through TCP (Transmission Control Protocol) connections between end users. TCP congestion control is the paramount example of how it is possible to successfully adjust connection rates in a *fair* and scalable *decentralized* form. Because of its special interest, we will use it is a modeling case study in Section 6.3.

6.2 NUM Model

The congestion control can be modeled as a Network Utility Maximization (NUM) problem. The success of NUM model is its flexibility to explain the behavior of already existing congestion control schemes and give insights to build new enhanced variants. We will see several examples of this in this chapter.

Let $\mathcal{G}(\mathcal{N}, \mathcal{E})$ be a network, where $\mathcal{N}$ is the set of nodes and $\mathcal{E}$ the set of links. The sources of traffic are the demands in a demand set $\mathcal{D}$. The traffic h_d injected by a demand d is routed through a path (or multicast tree if the demand is multicast) p_d. Each traffic source d has an associated *utility function* $U_d(h_d)$, which returns the utility in the sense of profit or abstract benefit that the source perceives when it transmits at rate h_d.

The basic NUM congestion control model assigns the rates $h = \{h_d, d \in \mathcal{D}\}$ that solve (6.1):

$$\max_h \sum_d U_d(h_d) \quad \text{subject to:} \tag{6.1a}$$

$$\pi_e : \sum_{d:e \in p_d} h_d \leq u_e, \quad \forall e \in \mathcal{E} \tag{6.1b}$$

$$\upsilon_d : h_d \geq 0, \quad \forall d \in \mathcal{D} \tag{6.1c}$$

The objective function (6.1a) maximizes the so-called *network utility*, defined as the sum of the utilities of all the demands in the network. As explained in Section 3.7, this concept is the application to network problems of the global welfare maximization target in economics. Constraint (6.1b) represents how the rates assigned are coupled among them: the sum of the rates of the sources traversing a particular link (that is, the total traffic in the link), cannot exceed its capacity. Finally, (6.1c) shows the standard non-negativity constraints.

6.2.1 Utility Functions for Elastic and Inelastic Traffic

Different demands $d \in \mathcal{D}$ can have different utility functions U_d, which are always non-decreasing: a demand perceives a higher bandwidth assignment as better (or "not worse"). If U_d is differentiable, this means that $\frac{\partial U_d}{\partial h_d} \geq 0$.

The particular non-decreasing shape of utility functions characterizes the elasticity of the demands:

- *Elastic sources*: As described in Chapter 2, pure elastic sources are always willing to transmit traffic and can fill any extra bandwidth assigned. This is represented by a strictly increasing utility function $U'_d(h_d) > 0$. In addition, a law of diminishing returns applies, meaning that the extra utility coming from increasing the bandwidth from $h_d + 1$ to $h_d + 2$, is lower than the increase occurred from h_d to $h_d + 1$. That is, the increases in the utilities ($U'(h_d)$)

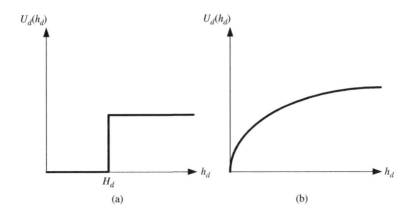

Figure 6.1   Examples of utility functions, (a) inelastic source and (b) elastic source

decrease with h_d and thus U_d functions are strictly concave $((U'(h_d))' < 0)$. Note that the behavior of elastic demands described coincides with elephant elastic connections.

- *Inelastic sources*: A pure inelastic or rigid traffic demand d requires a fixed bandwidth (H_d) without significant variations. Being assigned a bandwidth lower than H_d, is not useful for it $(U_d(h_d) = 0, h_d < H_d)$. If it is assigned exactly H_d units, the utility is maximum (e.g., $U_d(H_d) = \gamma > 0$). Any extra bandwidth assigned over H_d does not mean any extra utility $(U_d(h_d) = \gamma, h_d > H_d)$. This corresponds, for instance, with a multimedia source that requires a more or less constant bit rate to transmit its content. A lower bit rate means that the multimedia flow cannot be satisfactorily received. An excess rate above its request is not used.

Figure 6.1 shows an utility function example for inelastic and elastic traffic sources. Note that NUM problem (6.1) becomes a convex optimization problem with a unique optimal solution in the case of pure elastic utility functions and a non-convex problem with potentially multiple local optimums in the case of pure inelastic sources.

6.2.2   Fair Congestion Control

Section 3.7 exposed the relation between the shape of the utility functions and fairness in the resource allocation resulting of solving the NUM problem with such functions. Applying Prop. 3.1 to the NUM problem (6.1) shows that utility functions of the form:

$$U_d(h_d) = \begin{cases} w_d h_d & \text{if } \alpha = 0 \\ w_d \log h_d & \text{if } \alpha = 1 \\ w_d \dfrac{h_d^{1-\alpha}}{1-\alpha} & \text{if } \alpha > 0, \alpha \neq 1 \end{cases} \qquad (6.2)$$

enforce congestion controls where rate allocation is (w, α)-fair. This means allocations $h^* = \{h_d^*, d \in D\}$ such that, for any other different feasible allocation $h = \{h_d, d \in D\}$ it holds that:

$$\sum_d w_d \frac{h_d - h_d^*}{h_d^{*\alpha}} \leq 0, \quad \forall h \text{ feasible} \qquad (6.3)$$

As was discussed in Section 3.7 (see example in Fig. 3.10 and the associated plot in Fig. 3.11), utility functions associated to different α values enforce different types of fairness, and there is no consensus on which particular value of α is "fair enough".

- $\alpha = 0$ corresponds to NUM problems that intend to maximize the total network throughput $\sum_d h_d$. These allocations can be arbitrarily unfair. For instance, demands with one-hop routes deplete all the bandwidth of their traversing links, leaving zero bandwidth for any other demand sharing the link, but with a longer route.
- $\alpha = 1$ corresponds to allocations which are *proportionally fair*. In this context, a proportionally fair allocation h^* is such that in any other allocation h, some demands can receive a higher percentage (proportion) of traffic and some others a lower. However, the sum of the percentages with respect to h^* will always be strictly negative.
- $\alpha = 2$ corresponds to network utilities $\sum_d w_d/h_d$. If we interpret w_d values as the size of a file to be transmitted by demand d, then w_d/h_d is the delay incurred in fully transmitting it. For this reason, this type of fairness is commonly called *min-delay*.
- $\alpha \to \infty$ approximates the max-min fairness. In such case, the allocation h^* is such that no demand d can receive a higher rate ($h_d > h_d^*$) without reducing the rate of a "poorer" demand d' ($h_{d'} < h_{d'}^*$, for d' for which $h_{d'}^* \leq h_d^*$). This maximizes the rate of the demand with the minimum rate allocated.

In summary, lower α-fairness values tend to produce solutions where the network throughput $\sum_d h_d$ is higher, but with larger differences between the rates h_d of different sources (more "unfair"). In its turn, higher α values tend to reduce the difference between sources, commonly at a cost of a lower network throughput. Finally, as we will see, any $\alpha > 0$ fairness forbids allocating zero bandwidth to a demand in the optimum.

6.2.3 Optimality Conditions

For concave utility functions (e.g., elastic demands), problem (6.1) is a convex optimization problem with linear constraints, for which KKT optimality conditions hold. The Lagrange function of (6.1) is given by:

$$L(h, \pi, v) = \sum_d U_d(h_d) + \sum_e \pi_e \left(\sum_{d:e\in p_d} h_d - u_e \right) - \sum_d v_d h_d$$

When $\alpha > 0$, all demands receive a non-zero bandwidth in the optimum, since otherwise the utility increase $\partial U_d/\partial h_d$ in the optimality conditions becomes infinite. Thus, constraints (6.1c) are loose and multipliers $v_d = 0$. Lagrange minimization optimality conditions become:

$$\frac{\partial U_d(h_d)}{\partial h_d} = \sum_{e\in p_d} \pi_e, \quad \forall d \in D \tag{6.4}$$

Where $\pi_e \geq 0$ multipliers satisfy the complementary slackness conditions: $\pi_e = 0$ for non-saturated links for which the carried traffic is strictly below the capacity $\left(\sum_{d:e\in p_d} h_d < u_e \right)$.

We can interpret π_e as the price per bandwidth unit that any demand traversing link e has to pay. The reading of (6.4) is that, in the optimum, the variation of the benefit per bandwidth unit obtained modifying the rate $\left(\frac{\partial U_d(h_d)}{\partial h_d} \right)$, equals the variation of the cost per bandwidth unit incurred if this modification is done $\left(\sum_{e \in p_d} \pi_e \right)$.

For α-fair utility functions (6.2), we have that:

$$\frac{\partial U_d(h_d)}{\partial h_d} = \sum_{e \in p_d} \pi_e \Rightarrow h_d = \left(\frac{w_d}{\sum_{e \in p_d} \pi_e} \right)^{1/\alpha}, \quad \forall d \in D$$

Then optimality conditions expose that:

- A single non-negative number per link π_e is enough information for determining the optimum congestion control for all the network demands.
- To compute its optimum rate, each demand just needs to know its *own* utility function, and the *sum* of the optimum π_e values in their *traversing* links. No specific information is needed from other demands.

Previous observations are exploited for the design of decentralized congestion control algorithms, as will be shown later in this chapter and Part II of this book. We can classify these schemes according to how the links notify congestion information (link multipliers) to the sources, distinguishing between *explicit* and *implicit* feedback controls.

- *Explicit feedback* means that a proper signaling message is sent from the links to the sources. This is possible in technologies like ATM, which reserve space in Resource Management (RM) cells to let the links feedback information to the traversing flows ingress nodes.
- *Implicit feedback* means that the source is able to estimate the congestion situation in the network (link multipliers in the NUM model), without receiving any specific message from the links. The congestion conditions is derived primarily from (i) detection of packet losses or (ii) increments in the end-to-end delays of the connection. Both implicit approaches correspond to the operation of TCP protocol in the Internet, which for its special interest is analyzed in deeper detail in the next section.

6.3 Case Study: TCP

The Transmission Control Protocol (TCP) is the largely dominating protocol used by applications in the Internet for exchanging data. A TCP connection established between two applications: a_1 in node n_1 and a_2 in node n_2, acts as a bidirectional pipe, permitting a_1 sending data to a_2 and vice versa. Nodes n_1 and n_2 are usually end-user computers identified by their IP address and an IP network (typically Internet) is in charge of forwarding the datagrams of the TCP connection.

In the following analysis, we focus on the data flow from a_1 to a_2. An analogous scheme occurs for the data in the opposite direction. The transmission buffer in a_1 side stores the data that a_1 is willing to send to a_2. TCP chops the buffered data into individual segments of maximum size MSS (Maximum Segment Size)[1], a parameter negotiated during connection establishment. Each segment has enough information in its TCP header to identify the position of the data carried. It is encapsulated in (typically one) IP datagram targeted to n_2. If it arrives successfully to the destination, the data is placed in the TCP reception buffer for a_2 and an acknowledgement segment[2] (ACK) is sent from n_2 to n_1. These ACKs may be segments carrying or not data of the opposite direction $a_2 \rightarrow a_1$. The ACK segment notifies the identifier of the next segment that n_2 is willing to receive, implicitly acknowledging all the segments of lower identifiers.

6.3.1   Window-Based Flow Control

TCP uses a window-based flow control to pace the transmission of packets. A transmission window is defined within the transmission buffer. The window starts in the first byte of the first segment of the buffer (the oldest segment not acknowledged) and has a size of W bytes. All the data placed in the transmission window is immediately chopped into segments and sent to the other end. A retransmission timer is dedicated to each segment. During normal operation, the segment ACK is received before the timer expires. Then, the acknowledged bytes are removed from the buffer and the window *slides*, moving its initial byte to the first byte not acknowledged. However, if a timer expires before the segment ACK was received, it is retransmitted[3].

Previous operation means that by controlling the transmission window size W, it is possible to effectively modulate the amount of data segments injected in the network. This is illustrated in Fig. 6.2. We show the case of an elephant TCP connection and an ideal IP network in which

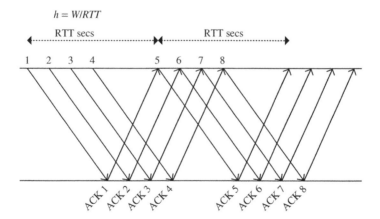

Figure 6.2   Window-based congestion control. The average rate of a connection h is limited by $\frac{W}{RTT}$

---

[1] Ideal MSS is the largest possible segment size that can reach the destination in one single IP packet without fragmentation. This depends on the MTU (Maximum Transfer Units) of the intermediate links, an information that is in general unknown. A default value for the MSS is 536 bytes.

[2] Under some conditions, an ACK can be sent every two received packets instead of one, a situation called *delayed ACKs*. Delayed ACKs add no significant insight to the picture and will not be included in our analysis.

[3] If the segment to retransmit is not in the transmission window (which can happen if the window size W was reduced), the retransmission is delayed until the segment re-enters into the transmission window.

segments have a constant end-to-end delay and are not dropped. To simplify the notation, we consider that all the segments are of the MSS size, and identify them by a correlative number[4]. In Fig. 6.2, the window size is $W = 4$ segments, and assumes that transmission and reception buffers at each connection end are also at least this size. The round-trip-time (RTT) of the connection is the time between a segment is transmitted and its associated ACK is received. Initially, the transmission buffer is empty and all the segments are immediately transmitted at the bit rate of the network card of the node (which can be assumed as infinite, or any large value). However, a second round of transmissions cannot start until the ACKs arrive, RTT seconds later[5]. Thus, the source sends W bytes every RTT seconds and has an average rate h:

$$h = \frac{W}{RTT} \tag{6.5}$$

Note that the relation $W = hRTT$ means that larger windows are needed for high throughput connections, between ends separated a long distance. These networks, for which the bandwidth-delay product can be high, are popularly called *Long Fat Networks* (LFN). As an example, consider two Internet users that experience a RTT of 200 ms (a reasonable value for, e.g., users at different continents). A TCP connection between them of average rate 100 Mbps would require a transmission window (and thus transmission and reception buffers) of size $W = 2.5$ MBytes.

Congestion control is implemented in TCP by adjusting the transmission window size W according to the signs observed that may suggest a congestion condition in the network. This approach was first proposed by Van Jacobson in 1988 [1] and the associated TCP variant was called TCP Tahoe. Several refinements to the TCP congestion control have appeared since then and TCP is generally recognized as a successful protocol that has performed remarkably well in the last decades, while the Internet scaled up several orders of magnitude in size and bit rates.

In the next subsection we will investigate, for didactic purposes, two main TCP variants that capture the two main approaches for TCP congestion control:

- *TCP Reno* that uses segment losses as a sign of network congestion.
- *TCP Vegas* that uses increments in the round-trip-time as a sign of network congestion.

TCP implementations in modern operative systems like Compound TCP (mainly Microsoft systems) and PRR (RFC 6937, in Linux kernels), apply the ideas of one or both of the approaches.

6.3.2 TCP Reno

The congestion control in TCP Reno was originally presented in 1990 [2] and has gone through several enhancements after it such as TCP New Reno (RFC 6582) and TCP SACK (RFC 2018

---

[4] Note that TCP actually identifies bytes, not segments. The identifier of a segment is the identifier of its first byte. During connection establishment, the identifier of the first byte is randomly chosen and subsequent bytes are numbered consecutively.

[5] Subsequent window transmission rounds are paced by the arrival of ACKs. If the window size is kept, each ACK slides the window, a new segment enters in it, and is transmitted. For this reason TCP is termed a *self-clocking* protocol. The time separation between consecutive ACK receptions depends on the bandwidth and traffic conditions of the traversed links. Still, the relation $W = hRTT$ remains an average.

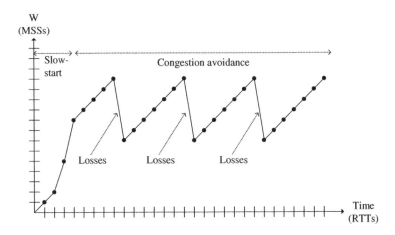

W
(MSSs)

Slow-start

Congestion avoidance

Losses   Losses   Losses

Time
(RTTs)

Figure 6.3 TCP Reno slot window evolution

and 2883). Figure 6.3 helps us to illustrate the baseline Reno congestion control version that we will study. It shows the time evolution of the window size W of an elephant TCP connection. For the sake of simplicity, window size W is measured in number of segments, and time t in number of RTTs after connection establishment, assuming that all RTTs have the same duration.

TCP Reno connections start with a phase called *slow-start*. Initial window size is $W = 1$, and during slow-start the transmitter increases the window size in one unit for every ACK received. Since during an RTT[6] the sender receives as many ACKs as the number of segments sent (W), the window size at time $t + 1$ is twice the window size at time t ($W(t + 1) = 2W(t)$). See that this results in an (everything but slow) exponential growth of the window during slow-start.

Slow-start ends when window size reaches a configured threshold. Then, the so-called congestion avoidance phase begins. During congestion avoidance the window size W evolves as follows:

$$W(t + 1) = \begin{cases} W + 1, & \text{if no segment losses detected during previous RTT} \\ \frac{W}{2}, & \text{otherwise (one or more losses detected)} \end{cases} \quad (6.6)$$

This is the core of the method, which enforces the generally called AIMD operation (*Additive Increase Multiplicative Decrease*). There are two manners in which a TCP sender can detect segment losses:

- *Retransmission timeouts*. If the retransmission timer of a segment expires, TCP assumes that the segment is lost because of a severe congestion situation in the network, resets the transmission window $W = 1$, and enters into slow-start phase again.
- *Duplicate ACKs*. Duplicate ACKs are consecutive ACK segments received that request the same segment identifier. They suggest that the repeatedly requested segment was lost, while

---

[6] Each TCP connection is permanently updating a RTT estimation, using information of the elapsed time between data segment transmission and ACK reception.

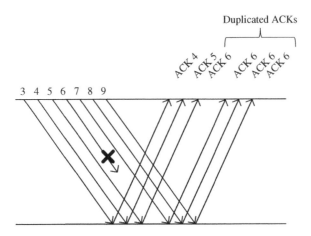

Figure 6.4 Duplicated ACKs example

some or all of the next segments in the window were correctly received[7]. This is illustrated in Fig. 6.4. When a number of duplicated ACKs are detected (typically four) in an RTT, TCP Reno halves the congestion window and performs the so-called *fast recovery* where the requested segment is immediately retransmitted. Fast recovery is repeated until all the window segments are acknowledged. If this is acquired, TCP continues in congestion avoidance mode. If not, it reverts to a slow-start starting from $W = 1$.

In ideal situations, TCP Reno remains in the AIMD congestion avoidance phase permanently: it increases linearly its window size every RTT until eventually losses occur, then halves its window, attempts a fast recovery, and starts again. The resulting sawtooth shape in Fig. 6.3 is the typical stationary behavior of elephant connections under TCP Reno, that we model in the next subsection.

6.3.2.1 Losses and Throughput Relation in Macroscopic Equilibrium

We model a TCP connection as two unidirectional demands $d, d' \in D$, one for each direction. We focus in the sequel on a particular direction represented by demand d. We denote p_d the set of links traversed by the demand data segments (from transmitter to receiver). T_d denotes the RTT of the connection d, which we assume to be constant.

We can reasonably estimate TCP's average performance by making some simplifications. We assume that only data segments can be lost and ACK segments carried in the opposite direction do not suffer congestion. We denote q_e the segment loss probability in link e and q_d denotes the end-to-end loss probability, which we approximate as follows:

$$q_d \approx 1 - \prod_{e \in p_d}(1 - q_e) \approx \sum_{e \in p_d} q_e \tag{6.7}$$

---

[7] They can be also a sign that a packet out of sequence occurred, without segment losses. For this reason TCP is sensitive to packet misorders, since the resulting duplicated ACKs are interpreted as a sign of packet losses.

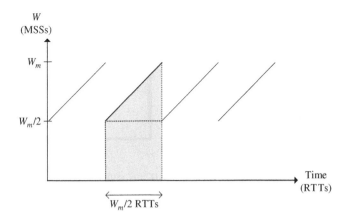

Figure 6.5   Simplified AIMD evolution

To simplify the analysis, we consider that losses do not occur randomly, but happen period-ically. In the equilibrium, an ideal periodic sawtooth is created as shown in Fig. 6.5.

Let the maximum value of the window be W_m segments and $W_m/2$ the minimum, occurring when losses are detected and the window is halved. The duration of each cycle is exactly $W_m/2$ RTTs, since during congestion avoidance the window grows in one unit each RTT. The total number of delivered segments during a cycle is the shaded area under the sawtooth, which is:

$$\left(\frac{W_m}{2}\right)^2 + \frac{1}{2}\left(\frac{W_m}{2}\right)^2 = \frac{3}{8}W_m^2$$

The average bandwidth h_d is thus given by the ratio between the traffic delivered in a cycle and the cycle duration:

$$h_d = \frac{MSS\frac{3}{8}W_m^2}{T_d\frac{W_m}{2}} = \frac{3}{4}\frac{MSS}{T_d}W_m \tag{6.8}$$

In equilibrium, exactly one segment loss occurs within a cycle. Then, neglecting the retrans-mission during the fast recovery process of the segment lost, we have that:

$$q_d = \frac{1}{\frac{3}{8}W_m^2} \Rightarrow W_m = \sqrt{\frac{8}{3q_d}} \tag{6.9}$$

Substituting (6.9) into (6.8) we have that:

$$h_d = \frac{MSS}{T_d}\sqrt{\frac{3}{2q_d}} \Leftrightarrow q_d = \frac{3}{2}\left(\frac{MSS}{T_d h_d}\right)^2 \tag{6.10}$$

Equilibrium relation (6.10), derived in [3], has been shown to be significantly precise. Other approaches considering, for example, random losses (instead of periodic) or the presence of delayed ACKs lead to similar h_d estimations, where constant factor $\sqrt{3/2} \approx 1.22$ in (6.10) is replaced by others between 1.31 and 0.87.

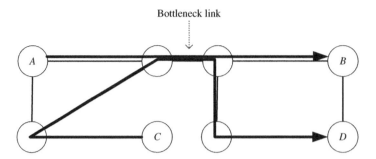

Figure 6.6 Example. Connection d ($A \rightarrow B$), RTT of 100 ms, connection d' ($C \rightarrow D$), RTT of 200 ms. Losses only occur in the (shared) bottleneck link. According to (6.10), $h_d = 2h_{d'}$

Expression (6.10) shows an inverse proportional dependence of the connection rate with respect to the RTT. This means that a connection d with twice the RTT of other connection d' and the same loss probability would receive half bandwidth. For instance, in Fig. 6.6 two connections d, d' share a link. If RTTs of both connections are related by $T_d = \frac{1}{2}T_{d'}$ and the shared link is the only dropping packets, then both connections would observe the same segment loss probability and their average rates would be such that $h_d = 2h_{d'}$.

6.3.2.2   NUM Modeling of the Macroscopic Equilibrium

Let $G(\mathcal{N}, \mathcal{E})$ be a network and D a set of demands on it, where each demand represents one direction of a TCP Reno elephant-type connection. In macroscopic equilibrium, the average rate of each demand is given by (6.10).

We assume that the equilibrium of TCP Reno described in (6.10) is an optimum assignment of a NUM problem with utility functions U_d and link multipliers π_e. Then, the NUM optimality conditions hold:

$$\frac{\partial U_d}{\partial h_d} = \sum_{e \in p_d} \pi_e \tag{6.11}$$

In our NUM modeling, we interpret the Lagrange multiplier π_e as the probability of dropping a segment traversing link e, which we denoted q_e in our TCP model. Then, assuming that loss probabilities q_e are small enough for approximation (6.7) to apply, we have that $\sum_{e \in p_d} \pi_e$ is the probability or fraction of segments lost:

$$\sum_{e \in p_d} \pi_e = q_d \tag{6.12}$$

Putting together (6.11), (6.12), and TCP equilibrium (6.10), setting $MSS = 1$ without loss of generality, we have:

$$\frac{\partial U_d}{\partial h_d} = \sum_{e \in p_d} \pi_e = q_d = \frac{3}{2}\left(\frac{1}{T_d h_d}\right)^2 \Rightarrow \frac{\partial U_d}{\partial h_d} = \frac{3}{2}\left(\frac{1}{T_d h_d}\right)^2$$

Integrating the previous expression we see that the utility function of the NUM problem TCP Reno solves $U_d(h_d)$:

$$U_d(h_d) = -\frac{3}{2T_d^2}\frac{1}{h_d} \qquad (6.13)$$

We make the following considerations:

- *Fairness.* Utility function (6.13) corresponds to a (w, α) utility function like (6.2) with fairness parameter $\alpha = 2$ and weight $w_d = \frac{3}{2T_d^2}$. Then, we can conclude that Reno congestion control enforces a (w, α)-fair solution in the network with the previous values.
- *Implicit signaling.* Lagrange multipliers π_e are signaled to the sources *implicitly*, without the need of the links to actually compute any π_e value, nor communicate it with any explicit message to the traversing sources. This is because the sources can deduce the end-to-end segment loss information indirectly from the reception of duplicated ACKs. Then, TCP Reno is a purely decentralized system in which each connection autonomously decides its rate and yet converges to a global fair allocation. This is done in a robust form that can adapt automatically to fluctuations of congestion conditions of the links, number of competing TCP connections, and so on.

6.3.3 TCP Vegas

TCP Vegas was introduced in 1994 [4] as an alternative to TCP Reno. The Vegas congestion control modifies TCP Reno congestion avoidance phase with the intention of correcting the oscillatory sawtooth-like behavior of TCP Reno. In contrast to the Reno scheme that increases permanently the window until queues are saturated and congestion is produced, a Vegas source anticipates the onset of congestion by estimating, from RTT observations, the end-to-end *queueing delay* of the connection segments. To do so, if no losses occur, congestion avoidance updates the window size every RTT according to:

$$W_d(t+1) = \begin{cases} W_d(t) + 1, & \text{if } \frac{W_d(t)}{T_d^{min}} - \frac{W_d(t)}{T_d(t)} < \beta_d \\ W_d(t) - 1, & \text{if } \frac{W_d(t)}{T_d^{min}} - \frac{W_d(t)}{T_d(t)} > \gamma_d \end{cases} \qquad (6.14)$$

Window size is measured in number of MSS segments. $T_d(t)$ is the current RTT estimation, and T_d^{min} is the minimum value observed in the historical RTT monitoring, and intends to approximate the end-to-end propagation part of the RTT if all the queues were empty. That is, the queuing delay would be given by $T_d(t) - T_d^{min}$.

Interpreting (6.14) we have that:

- $h_d^{max}(t) = \frac{W_d(t)}{T_d^{min}}$ in (6.14) is the best possible *expected rate*, occurring if all the queues in the path of the connection are empty.
- $h_d(t) = \frac{W_d(t)}{T_d(t)}$ is the *actual rate*.
- Control (6.14) intends to keep the difference between the expected and actual rate within the interval $[\beta_d, \gamma_d]$, typically $\beta_d = 1/T_d^{min}$ and $\gamma_d = 3\beta_d$.

We can have a second interpretation of Vegas if we multiply expression (6.14) by T_d^{min}:

$$W_d(t+1) = \begin{cases} W_d(t) + 1, & \text{if } W_d(t) - h_d(t)T_d^{min} < 1 \\ W_d(t) - 1, & \text{if } W_d(t) - h_d(t)T_d^{min} > 3 \end{cases} \qquad (6.15)$$

- $W_d(t)$ is the volume of traffic injected in a RTT.
- $h_d(t)T_d^{min}$ is the average number of packets that are "on the fly", not waiting in the queues. This is the direct application of the Little's law that states if a conservative system (everything that gets in, gets out) receives elements at a rate h, the average number of elements inside the system N and the average time each element stays in the system T are related by: $h = \frac{N}{T}$.
- Then, Vegas control intends to keep the sum of the number of packets in the queues (typically called *backlog*) of the traversed nodes at a very low value: between 1 and 3. So, in contrast with Reno, Vegas intends to adjust rates to have approximately empty queues in the routers.

6.3.3.1 NUM Modeling of the Macroscopic Equilibrium

To simplify the analysis, we make $\beta_d = \gamma_d$, denote T_d to be the constant RTT, and W_d as the equilibrium window size. Then, in macroscopic equilibrium, it holds that:

$$\frac{W_d}{T_d^{min}} - \frac{W_d}{T_d} = \beta_d, \quad \forall d \in D$$

Rewriting the previous expression we have the equilibrium condition:

$$\beta_d = \frac{1}{T_d^{min}}(W_d - h_d T_d^{min}) \qquad (6.16)$$

Where $(W_d - h_d T_d^{min})$ is the total number of packets queued along the source path p_d. We consider that, in equilibrium, TCP Vegas is assigning rates that optimize a particular NUM model with utility functions U_d and link multipliers π_e. Then, in equilibrium, optimality conditions should hold:

$$\frac{\partial U_d}{\partial h_d} = \sum_{e \in p_d} \pi_e, \quad \forall d \in D \qquad (6.17)$$

We now consider that π_e multiplier for Vegas is the queueing delay in link e. This is given by the backlog b_e at queue e (number of packets in the queue) divided by link capacity u_e:

$$\pi_e = \frac{b_e}{u_e}, \quad \forall e \in \mathcal{E}$$

These multipliers satisfy the complementary slackness condition of NUM problem (6.1), since $\pi_e = 0$ (no packets in the queue) if the aggregated rate in e is below the link capacity $\left(\sum_{d:e \in p_d} h_d < u_e\right)$.

If b_e is the backlog at link e, under the first-in-first-out queueing discipline we can assume that the fraction of queued packets that belong to a source d is given by h_d/u_e. Thus, the total number of packets of source d in the backlog of link $e \in p_d$ equals $b_e h_d / u_e$. Summing this expression along all the source traversed links we have the sum of the queued packets of source d:

$$W_d - h_d T_d^{min} = \sum_{e \in p_d} b_e \frac{h_d}{u_e}, \quad \forall d \in \mathcal{D}$$

Introducing the previous expression into the equilibrium conditions (6.16) we have:

$$\beta_d = \frac{1}{T_d^{min}} \sum_{e \in p_d} b_e \frac{h_d}{u_e}, \quad \forall d \in \mathcal{D}$$

If we compare previous expression with optimality condition (6.17) we have by simple inspection that:

$$\frac{\partial U_d}{\partial h_d} = \sum_{e \in p_d} \pi_e \Rightarrow \frac{\partial U_d}{\partial h_d} = \frac{\beta_d T_d^{min}}{h_d}$$

Thus, the utility function Vegas is optimizing is given by:

$$U(h_d) = \beta_d T_d^{min} \log h_d$$

We make the following considerations:

- *Fairness.* Since Vegas sets β_d as a multiple of the inverse of T_d^{min} (typically $\beta_d = 1/T_d^{min}$), we have that Vegas applies an utility function $U(h_d) = \log h_d$, thus enforcing so-called proportional fairness ($\alpha = 1$ in α-fairness functions (6.2)). Note that this is a somewhat "less fair" α factor than Reno ($\alpha = 2$), which would suggest that Vegas penalizes the longer connections more than Reno. However, this is compensated for by the factor $1/T_d^2$ in (6.13) that Reno applies penalizing the utility of connections with a larger RTT.
- *Implicit signaling.* As in Reno, Vegas Lagrange multipliers π_e are signaled to the sources *implicitly*, without the need of the links to actually compute any π_e value, nor communicate with it any explicit message to the traversing sources. This is because the sources can deduce the end-to-end queueing delay by monitoring the connection RTT, which is done observing the elapsed time between the transmission of a data segment and the reception of its associated ACK. Then, TCP Vegas is also a purely decentralized system in which each connection autonomously decides its rate and converges to a global fair allocation. Note that it also adapts automatically to fluctuations of congestion conditions of the links, number of TCP connections, and so on. without modifying its behavior.
- *Vegas and segment losses.* TCP Vegas is quite effective at sharing the bandwidth in the links among Vegas sources, avoiding oscillations (sawtooths) and reducing the queueing delays. However, it requires the absence of losses to reach this equilibrium. If a TCP Vegas connection shares a bottleneck link with Reno connections, the Reno source will steadily grow the window until losses occur, according to its standard AIMD operation while Vegas reduces its window to reduce the queueing delay. For this reason, initial versions of Vegas received an unfairly low bandwidth when competing with Reno sources. This has been somewhat addressed in subsequent versions of the protocol, where Vegas reverts to a Reno-like operation in the presence of losses.

6.4   Active Queue Management (AQM)

Congestion control in the Internet was and is dominated by loss-based Reno-like AIMD behavior in which TCP elephant connections steadily increase their rate until packet drops are eventually produced. The interplay of AIMD sources and the *tail drop* standard behavior of queues in the routers[8], brings two main issues:

- *Excessive buffer delays*: Loss-based congestion control keeps the queues highly occupied in every link that is a bottleneck, since traffic is increased until buffers saturate and losses are produced. Queues are persistently full and each queue can add up to tens of milliseconds of buffering delay. TCP connections traversing multiple bottlenecks, even mouse connections not causing congestion, can experience a large end-to-end latency and RTT.
- *Global synchronization*: When a queue is full and the router receives a burst of packets, all of them are dropped and all the affected connections will at the same time reduce the window and ramp up again. If connection RTTs are similar, they may get synchronized, increasing and reducing their windows simultaneously. This phenomenon is called global synchronization, and impairs network performance, since it makes the resulting traffic pattern more bursty.

Active Queue Management (AQM) is the name given to a set of techniques implemented in the routers that replace the standard tail drop queue policy by making some *preventive* packet drops before the queues fill up, with the aim of addressing previous problems. One of the most popular AQM techniques available in the routers is the Random Early Discard (RED) queues (or some of its variants, see [5] for details). Upon packet reception, a RED queue on a link e observes its backlog length b_e (averaged with its recent past). Then, RED will make a preventive drop of the packet with a probability p given by a piece-wise linear distribution like the one in Fig. 6.7, where b_e^{min}, b_e^{max} and m_e are RED configuration parameters.

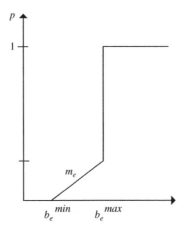

Figure 6.7   RED marking probability example. m_e is the slope of the line

---

[8] Tail drop operation means that packets are dropped only when they arrive at a router with full queues.

AQM techniques can be combined or not with Explicit Congestion Notification (ECN). Under ECN mechanism, when a link detects incipient congestion, it activates the ECN bit reserved in the IP header of the affected packet, which is not dropped but forwarded. When it arrives to the receiver TCP side, the congestion notification is conveyed back in the TCP header of the ACK, to warn the transmitter. Then, TCP reduces its rate as if the acknowledged segments were lost, but without the need for retransmissions.

6.4.1 A Simplified Model of the TCP-AQM Interplay

What follows is a discussion to give some insights of the interplay of TCP and AQM techniques *in the macroscopic equilibrium*. We make the following assumptions:

1. We model an AQM scheme for a link e as a function $G_e(b_e(t), y_e(t), v_e(t))$, which returns the probability $p_e(t)$ of dropping or marking a packet arriving at instant t, depending on the current length of the queue backlog $b_e(t)$, the amount of arriving traffic $y_e(t)$, and an internal link state $v_e(t)$. The tail drop policy for a buffer of B packets can be modeled as one more AQM with G_e given by:

$$p_e(t) = \begin{cases} 0, & \text{if } b_e(t) < B \\ 1, & \text{if } b_e(t) = B \end{cases}$$

2. Since the dynamics of retransmissions are not captured in our model, we do not further distinguish between packet drops and ECN markings as congestion notifications and results in this section are applicable to both. We use the term "marking" to refer to both packets drops and ECN-bit settings.
3. We do not focus here on the transient time evolution and short time scale dynamics of the congestion loop engaging the source and AQM links. We assume that the network reaches a *macroscopic* equilibrium and consider that, in this equilibrium, the application of the AQM technique does not change the equilibrium relation of AIMD sources:

$$h_d = \frac{MSS}{T_d} \sqrt{\frac{3}{2q_d}} \tag{6.18}$$

where q_d is estimated as the sum of the marking probabilities in the path links and RTT T_d *is considered constant*, as is customary in the literature (e.g., [3]).

4. As we did in Section 6.3.2, we interpret marking probabilities as Lagrange multipliers of the NUM problem (6.1), when AIMD Reno-like sources are being used ($p_e = \pi_e$).
5. The AQM behavior is consistent with the complementary slackness optimality conditions of the multipliers: $\pi_e > 0 \Rightarrow \sum_{d:e \in p_d} h_d = u_e$. This means that a link drops packets ($\pi_e > 0$) only when its queues in the equilibrium are non-empty ($b_e > 0$), since this means that the link is saturated ($y_e = u_e$). Note that any AQM that does not drop packets when the queues are empty ($b_e = 0$) meets this condition and all satisfy this in practice.

According to previous assumptions, the application of an AQM does not modify the utility function (6.13) that describes the behavior of the TCP source. This means that the same NUM model (6.1) with the same utility function applies whatever AQM (or tail drop) is used. For them, the NUM model predicts the same average macroscopic rate allocation h_d and the same loss probabilities in the links $\pi_e = p_e$. The only difference *partially* captured by the model would be the average queueing delay in the links b_e, which would depend on the AQM. In particular, after solving the model, the resulting link loss probabilities can be used to derive the average queueing delay as follows:

$$b_e = \begin{cases} 0, & \text{if } y_e < u_e \\ (G_e)^{-1}(\pi_e), & \text{if } y_e = u_e \end{cases}, \quad \forall e \in \mathcal{E}$$

As an example, in a RED queue with marking probability given by Fig. 6.7, any estimated loss probability p_e coming from the NUM model that falls in the range $(0, m_e(b_e^{max} - b_e^{in}))$ is unambiguosly associated with an average backlog b_e in the range (b_e^{min}, b_e^{max}). In turn, any loss probability $p_e > 0$ in a tail drop link, is associated with an average queue length equal to the maximum queue size $b_e = B$.

Interestingly, the model presented predicts that (i) the loss probability and average rate allocated in the network *does not* depend on the AQM nor the buffer sizes, (ii) queueing delays *do* depend on them, but (iii) so that larger buffers only degrade the performance bringing higher queuenig delays! In other words, the model suggests that cutting down the buffers would reduce the RTTs (considered constant in the model), actually improving the TCP responsiveness and performance.

Naturally, this result should be taken with care, since the simplified model supporting it is not considering many of the microscopic dynamics where buffers are really needed. In particular, buffers are required to absorb normal short-scale traffic fluctuations in packet switching networks without discarding datagrams. For that, they are *good queues*. However, this model suggests that too big buffers do not produce an improvement in the congestion loop: any buffer put would be filled up during congestion and larger buffers just mean longer delays for the traversing traffic, without a payoff of more link utilization. Actually, the increased RTT just degrades TCP performance. Thus, they are *bad queues*.

Previous discussion is actually a hot topic today, and TCP and AQM foundations are being revisited under a new light. This is stemmed from the path-breaking works [6, 7], where the distinction between *good queues* and *bad queues* (too large queues) is set. The rationale behind this approach is that current IP routers are equipped with *bad queues*: buffers of hundreds of MBs per link, which normal congestion control loops keep persistently full. This results in hundreds of milliseconds of extra RTT caused by queueing delays in long-distance connections and a chaotic and laggy performance observed in the network. To address this situation, a redefinition of the AQM schemes, queuing dimensioning, and other processes are being conducted. The interested reader is referred to the set of projects under the keyword *buffer-bloat* [8] and similar initiatives.

6.5   Notes and Sources

Congestion control is the first network topic where the utility maximization framework was applied [9]. In this context, [9] introduced proportional fairness, min-delay fairness was presented in [10], and α-utilities in [11].

Multiple works have successfully explored the NUM model, producing insights for reverse engineering existing congestion control schemes, inspiring enhancements to existing algorithms as well as devising new ones [12–17]. Excellent books covering the topic are [18, 19], and [20].

TCP Tahoe was presented in [1] and the first reference to TCP Reno is usually credited to [21]. TCP Reno has gone through several enhancements in RFCs like RFC 1323, RFC 2001, RFC 2018, RFC 2581, and RFC 2582. AIMD behavior was previously analyzed in [22]. TCP Vegas was presented in [4], and analyzed in other sources like [23]. The macroscopic equilibrium relation presented for Reno is the one in [3]. Other approaches leading to similar estimations can be found in [24, 25]. A comprehensive description of the internals of multiple TCP variants can be found in various sources like [26].

Some other selected TCP models not based on utility functions are [27, 28], and [29]. Information on congestion control schemes in ATM networks can be found in [30]. A seminal reference on RED-like AQM techniques is [31]. [5] is an organized index on RED proposals. Other marking proposal based on an optimization model is [32]. The model of the TCP-AQM interplay is extracted from [33]. Seminal papers on buffer bloat discussions are [6, 7], while [8] provides updated references.

6.6   Exercises

6.1   Let $G(\mathcal{N}, \mathcal{E})$ be a network, u_e and d_e the capacity and length of link e, and $\mathcal{D}$ a set of unicast demands. Each traffic demand $d \in \mathcal{D}$ consists of sending a file of size W_d bits through a known path p_d. Each demand has an end-to-end delay estimated using the M/M/1 queue model limited to T. The maximum difference between the maximum and minimum bandwidth assigned to a demand is bounded by F. Write the formulation of the program that finds the bandwidth assignment to the demands that minimize the average file delivery times, under the mentioned constraints.

6.2   Figure 6.8 shows a network with $k + 1$ nodes, k links, k one-hop demands and one demand (d_0) traversing all the links. All the links have one unit of capacity. Find the expressions for the optimum carried traffic for each demand, the network throughput, and the network utility for the NUM problem using α utility functions. Study the effect of α factor and the number of links k in the trade-off network throughput versus traffic assigned to the long connection d_0.

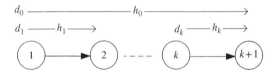

Figure 6.8

6.3   Let $G(\mathcal{N}, \mathcal{E})$ be a network and $\{u_e, e \in \mathcal{E}\}$ the link capacities in bps. Demand set $\mathcal{D}$ is composed of one traffic demand between each node pair. Each demand d is served through the shortest path p_d between the demand end nodes. Each demand can be served using a high quality or a low quality connection, requiring R_h and R_l bps, respectively, where $R_l < R_h$. The network operator has a benefit of 1 unit per each demand carried in a low quality connection, 1.5 units for each served with a high quality connection and has a penalization of 4 units if the demand is not served. Formulate the problem that optimizes the bandwidth allocation to the sources. Implement a Net2Plan algorithm that solves this formulation with JOM, using R_l, R_h and the per-demand benefits/penalization quantities as input parameters. The optimum solution is returned by updating the demand routes and offered traffic.

6.4 The multi-path congestion control problem consists of optimizing the demand rates, in the case when the traffic of each demand traffic $d \in D$ can be bifurcated in several paths P_d. We use P to denote the set of all paths in the network, P_e the paths traversing link e. Devise a formulation for the NUM problem that maximizes the sum of the α-utilities of the demands.

6.5 Implement a Net2Plan algorithm that finds the optimum α-fair bandwidth allocation for a network with JOM. The algorithm creates a number K of demands between each node pair, all of them carried through a shortest path between demand end nodes. If more than one shortest path exists, one is selected out of them and used for all the demands. Every demand has a minimum injected traffic h_{min}. Input parameters to the algorithm are α, K, and h_{min}. The optimum solution is returned by updating the demand routes and offered traffic. Use previous algorithm to explore the network throughput versus fairness trade-off for $\alpha = \{0, 0.5, 1, 2, 3, 5\}$ in several chosen non-regular topologies. As a measure of fairness use the Jain index: $F = \frac{\left(\sum_d h_d\right)^2}{|D| \sum_d h_d^2}$.

6.6 Implement a Net2Plan algorithm that finds the macroscopic bandwidth allocation of TCP Reno or TCP Vegas connections, using the NUM model in Section 6.3.2 and Section 6.3.3, respectively. The algorithm creates a number K of demands between each node pair, all of them carried through a shortest path between demand end nodes. If more than one shortest path exists, one is selected out of them and used for all the demands. Demand RTTs are computed from the link distances assuming a propagation speed v. Input parameters to the algorithm are the type of TCP connections (Reno or Vegas), K and v. The optimum solution is returned by updating the demand routes and offered traffic.

6.7 Use a modification of previous algorithm to study the behavior of Reno sources with different RTTs that share a bottleneck link. For this, use a topology like the one in Fig. 6.8, with K demands from node i to node $k + 1$, $i = 1, \ldots, k$. All links have a capacity of two, but the link between nodes k and $k + 1$ with a capacity of one (the bottleneck). Observe the traffic assigned to demands of different RTTs depending on the TCP type and factor K.

6.8 A set of TCP Reno connections D traverse one single shared bottleneck link with a capacity of one. T_d denotes the RTT of demand $d \in D$. Deduce the expression of the throughput h_d of each demand and of the packet drop probability in the bottleneck as a function of the RTTs of all the demands. Do the traffics h_d change if all the demands multiply their RTT by any common factor? And the drop probability? Assume that K TCP connections exist for each demand instead of just one. Answer the previous questions and comment on the effect of K in the throughputs and link packet drop probability.

References

[1] V. Jacobson, "Congestion avoidance and control," in *ACM SIGCOMM Computer Communication Review*, vol. 18, no. 4. ACM, 1988, pp. 314–329.

[2] V. Jacobson, "Berkeley TCP evolution from 4.3-tahoe to 4.3-Reno," in Proceedings of the 18th Internet Engineering Task Force, University of British Colombia, Vancouver, BC, 1990, p. 365.

[3] M. Mathis, J. Semke, J. Mahdavi, and T. Ott, "The macroscopic behavior of the tcp congestion avoidance algorithm," *ACM SIGCOMM Computer Communication Review*, vol. 27, no. 3, pp. 67–82, 1997.

[4] L. S. Brakmo and L. L. Peterson, "Tcp vegas: End to end congestion avoidance on a global internet," *Selected Areas in Communications, IEEE Journal on*, vol. 13, no. 8, pp. 1465–1480, 1995.

[5] References on red (random early detection) queue management. Available online at: http://www.icir.org/floyd/red.html

[6] J. Gettys and K. Nichols, "Bufferbloat: Dark buffers in the internet," *Queue*, vol. 9, no. 11, p. 40, 2011.

[7] K. Nichols and V. Jacobson, "Controlling queue delay," *Communications of the ACM*, vol. 55, no. 7, pp. 42–50, 2012.

[8] Bufferbloat projects Available online at www.bufferbloat.net.

[9] F. Kelly, "Charging and rate control for elastic traffic," *European transactions on Telecommunications*, vol. 8, no. 1, pp. 33–37, 1997.

[10] L. Massoulié and J. Roberts, "Bandwidth sharing: objectives and algorithms," in *INFOCOM'99. Eighteenth Annual Joint Conference of the IEEE Computer and Communications Societies. Proceedings. IEEE*, vol. 3. IEEE, 1999, pp. 1395–1403.

[11] J. Mo and J. Walrand, "Fair end-to-end window-based congestion control," *IEEE/ACM Transactions on Networking (ToN)*, vol. 8, no. 5, pp. 556–567, 2000.

[12] F. P. Kelly, A. K. Maulloo, and D. K. Tan, "Rate control for communication networks: shadow prices, proportional fairness and stability," *Journal of the Operational Research society*, pp. 237–252, 1998.

[13] S. H. Low and D. E. Lapsley, "Optimization flow control: basic algorithm and convergence," *IEEE/ACM Transactions on Networking (TON)*, vol. 7, no. 6, pp. 861–874, 1999.

[14] H. Yaïche, R. R. Mazumdar, and C. Rosenberg, "A game theoretic framework for bandwidth allocation and pricing in broadband networks," *IEEE/ACM Transactions on Networking (TON)*, vol. 8, no. 5, pp. 667–678, 2000.

[15] S. Kunniyur and R. Srikant, "End-to-end congestion control schemes: Utility functions, random losses and ecn marks," *Networking, IEEE/ACM Transactions on*, vol. 11, no. 5, pp. 689–702, 2003.

[16] T. Alpcan and T. Başar, "A utility-based congestion control scheme for internet-style networks with delay," in INFOCOM 2003. *Twenty-Second Annual Joint Conference of the IEEE Computer and Communications. IEEE Societies*, vol. 3. IEEE, 2003, pp. 2039–2048.

[17] D. X. Wei, C. Jin, S. H. Low, and S. Hegde, "Fast TCP: motivation, architecture, algorithms, performance," *IEEE/ACM Transactions on Networking (ToN)*, vol. 14, no. 6, pp. 1246–1259, 2006.

[18] R. Srikant, *The Mathematics of Internet Congestion Control*. Springer Science & Business Media, 2012.

[19] S. Shakkottai, S. G. Shakkottai, and R. Srikant, *Network Optimization and Control*. Now Publishers Inc, 2008.

[20] R. Srikant and L. Ying, *Communication Networks: An Optimization, Control, and Stochastic Networks Perspective*. Cambridge University Press, 2013.

[21] V. Jacobson, "Modified TCP congestion avoidance algorithm," *End2end-Interest Mailing List*, 1990.

[22] D.-M. Chiu and R. Jain, "Analysis of the increase and decrease algorithms for congestion avoidance in computer networks," *Computer Networks and ISDN Systems*, vol. 17, no. 1, pp. 1–14, 1989.

[23] S. H. Low, L. L. Peterson, and L. Wang, "Understanding TCP Vegas: a duality model," *Journal of the ACM (JACM)*, vol. 49, no. 2, pp. 207–235, 2002.

[24] S. Floyd, "Connections with multiple congested gateways in packet-switched networks part 1: One-way traffic," *ACM SIGCOMM Computer Communication Review*, vol. 21, no. 5, pp. 30–47, 1991.

[25] T. Lakshman and U. Madhow, "The performance of tcp/ip for networks with high bandwidth-delay products and random loss," *Networking, IEEE/ACM Transactions on*, vol. 5, no. 3, pp. 336–350, 1997.

[26] K. R. Fall and W. R. Stevens, *TCP/IP illustrated, Volume 1: The protocols*. Addison-Wesley, 2011.

[27] E. Altman, K. Avrachenkov, and C. Barakat, "A stochastic model of tcp/ip with stationary random losses," *ACM SIGCOMM Computer Communication Review*, vol. 30, no. 4, pp. 231–242, 2000.

[28] F. Baccelli and D. Hong, "AIMD, fairness and fractal scaling of TCP traffic," in INFOCOM 2002. Twenty-First Annual Joint Conference of the IEEE Computer and Communications Societies. Proceedings. IEEE, vol. 1. IEEE, 2002, pp. 229–238.

[29] J. Padhye, V. Firoiu, D. Towsley, and J. Kurose, "Modeling TCP throughput: A simple model and its empirical validation," *ACM SIGCOMM Computer Communication Review*, vol. 28, no. 4, pp. 303–314, 1998.

[30] E. J. Hernandez-Valencia, L. Benmohamed, R. Nagarajan, and S. Chong, "Rate control algorithms for the ATM ABR service," *European Transactions on Telecommunications*, vol. 8, no. 1, pp. 7–20, 1997.

[31] S. Floyd and V. Jacobson, "Random early detection gateways for congestion avoidance," *IEEE/ACM Transactions on Networking*, vol. 1, no. 4, pp. 397–413, 1993.

[32] S. Athuraliya, S. H. Low, V. H. Li, and Q. Yin, "Rem: active queue management," *Network, IEEE*, vol. 15, no. 3, pp. 48–53, 2001.

[33] S. H. Low, "A duality model of TCP and queue management algorithms," *Networking, IEEE/ACM Transactions on*, vol. 11, no. 4, pp. 525–536, 2003.

7

Topology Design Problems

7.1 Introduction

Topology design refers to the network problems that optimize the set of nodes and/or links to deploy in a network. These problems typically appear in the elaboration of long-term network planning projects for network deployments or upgrades. When these planning studies correspond to a network to build from scratch, they are referred to as *greenfield planning*. When they are recommendations for adding or upgrading links/nodes in an existing network, they are commonly called *brownfield planning*.

Topology planning tasks are typically conducted offline by consultant companies or planning departments within the network operator. Since the deployment of nodes and links involves a high cost and impacts on the performance merits achievable later, an accurate modeling of the problem and a good method to find numerical solutions are critical points. In this context, topology design problems can be solved without major time constraints in a centralized form. However, the problem variants of practical interest are typically $\mathcal{NPO}$-complete, formulated using integer programs and thus it is $\mathcal{NP}$-hard finding even approximate solutions, and there are no algorithms that guarantee obtaining them in worse case polynomial time. Then, we should get along with just suboptimal solutions for medium to large networks. This can be achieved solving the integer formulations like the ones in this chapter using numerical solvers (e.g., interfaced through JOM), configured with a maximum running time so that the best solution found so far is returned if the time limit is reached before the optimal is hit. As we will see in Chapter 12, another customary option relies on ad-hoc developed heuristics for that.

There are two classical types of topological design problems:

- *Node location problems*, where the number and location of the network nodes is optimized. For instance, given a set of access node locations, find the optimum placement of the network core nodes and connect each access node to at least one core node, such that the deployment cost is minimum.

Optimization of Computer Networks – Modeling and Algorithms: A Hands-On Approach,
First Edition. Pablo Pavón Mariño.
© 2016 John Wiley & Sons, Ltd. Published 2016 by John Wiley & Sons, Ltd.
Companion Website: www.wiley.com/go/PavonMarinoSol16

- *Full topology design* that receive as an input a set of sources and destination nodes of traffic and the forecasts of traffic volumes between them. From this, the problem should find (i) the set of links connecting these nodes, potentially adding new transit nodes to the network (that just forward traffic), (ii) the capacity of these links, and (iii) the routing of the traffic, so that network cost is minimized.

Node location and full topology design problems appear in all the network technologies in one or other form. In this chapter we provide example models for both, together with some illustrative variations. In addition, we will present results obtained solving small-scale versions of the described problems to optimality using Net2Plan. These results will expose how the cost structure of the problem (e.g., link to node cost ratio or fixed to variable link cost ratio) affect the resulting topology.

7.2 Node Location Problems

Let $\mathcal{N}_a$ be a set of known locations, each one hosting an access node in the network, or any entity that generates traffic. Let $\mathcal{N}_c$ be the set of known locations where core nodes can be potentially placed. Each core location $n_j \in \mathcal{N}_c$ can host at most one core node. Each access node must be connected by a direct link to one and only one core node. For technological limitations, each core node can be connected to at most K access nodes.

We use index $i = 1, \ldots, |\mathcal{N}_a|$ to refer to access nodes and index $j = 1, \ldots, |\mathcal{N}_c|$ for potential core node locations. We denote c_{ij} as the cost of connecting access node n_i to the core node placed at location n_j and denote c_j as the cost of placing a core node at location n_j (including core node cost).

We are interested in determining the placement of core nodes and how the access nodes are connected to them, such that the total network cost is minimum. We formulate the problem as an integer program with the following decision variables:

$z_j = \{1$ if a core node is placed at location $n_j \in \mathcal{N}_c$, 0 otherwise$\}$, $\quad \forall j = 1, \ldots, |\mathcal{N}_c|$

$e_{ij} = \{1$ if access node $n_i \in \mathcal{N}_a$ is connected to nore location $n_j \in \mathcal{N}_c$, 0 otherwise$\}$,

$\quad \forall i = 1, \ldots, |\mathcal{N}_a|, j = 1, \ldots, |\mathcal{N}_c|$

The problem formulation is given by (7.1):

$$\min_{z,e} \sum_{ij} c_{ij} e_{ij} + \sum_j c_j z_j \quad \text{subject to:} \tag{7.1a}$$

$$\sum_j e_{ij} = 1, \quad i = 1, \ldots, |\mathcal{N}_a| \tag{7.1b}$$

$$\sum_i e_{ij} \leq K z_j, \quad j = 1, \ldots, |\mathcal{N}_c| \tag{7.1c}$$

Objective function (7.1a) sums the total network cost, including the link costs ($\sum_{ij} c_{ij} e_{ij}$) and node costs ($\sum_j c_j z_j$). Constraint (7.1b) means that each access location should be connected to exactly one core node. Finally, constraint (7.1c) means that:

- If a location j has no core node ($z_j = 0$), then it cannot be connected to any access node ($\sum_i e_{ij} = 0$).
- If a location j has a core node ($z_j = 1$), then it can be connected to at most K access nodes ($\sum_i e_{ij} \leq K$).

7.2.1 Problem Variants

Multiple problem variants exist for the baseline node location problem described. Below, we provide some of them:

- *Multiple core nodes per location*: By allowing variables z_j to take any non-negative integer value (not just 0 or 1), without any other changes in (7.1), we model the case when more than one core node can be placed in the same location.
- *Maximum traffic per core node*: Let us assume that each access node generates an amount of traffic h_i and core nodes, because of technological constraints, cannot handle more than H traffic units. These limitations can be incorporated into the problem by adding the constraints:

$$\sum_i h_i e_{ij} \leq H z_j, \quad j = 1, \dots, |\mathcal{N}_c|$$

- *Maximum number of core nodes*: Inventory limitations that settle a maximum number M of available core nodes to place can be modeled by constraint:

$$\sum_j z_j \leq M$$

- *Brownfield design*: In brownfield design, some of the core nodes $\mathcal{N}_c' \subset \mathcal{N}_c$ are already placed and the decision to make is how to upgrade the network with new core nodes. This can be easily modeled by adding the constraints:

$$z_j = 1, \quad \forall j : n_j \in \mathcal{N}_c'$$

Then, the formulation searches for the minimum cost solution among those that include existing nodes in the design.

- *Full-mesh core nodes connection costs*. Let us assume that core nodes must be fully connected between them in a bidirectional full mesh. We denote $\xi_{jj'}$ as the cost of a bidirectional connection between core node in location j and core node in location j'. We are interested in adding these costs into the total network cost. We add new decision variables to problem (7.1):

$$\tilde{e}_{jj'} = \{1 \text{ if core location } n_j \text{ is connected to core location } n_{j'}, \ 0 \text{ otherwise}\},$$

$$\forall j, j' = 1, \dots, |\mathcal{N}_c|, j > j'$$

Limiting the cases $j > j'$ is a form of not considering twice each bidirectional link between two locations. For instance, a connection between location 10 and 5 includes both $5 \rightarrow 10$

and $10 \rightarrow 5$. Then, we can update the objective function of problem (7.1) by adding the core link costs:

$$\min_{z,e,\tilde{e}} \sum_{ij} c_{ij} e_{ij} + \sum_{j} c_j z_j + \sum_{j>j'} \tilde{e}_{jj'} \xi_{jj'}$$

And add the following constraints to make $\tilde{e}_{jj'}$ variables reflect a full mesh between the core nodes placed:

$$\tilde{e}_{jj'} \geq z_j + z_{j'} - 1, \quad \forall j, j' = 1, \ldots, |\mathcal{N}_c|, j > j'$$

Previous constraint makes that if both locations j and j' have a node, then a core link should be considered ($\tilde{e}_{jj'} \geq 1$). If only one location or none location has it, then previous constraint converts into $\tilde{e}_{jj'} \geq 0$ or $\tilde{e}_{jj'} \geq -1$, which in the optimum means that $\tilde{e}_{jj'} = 0$ and no core link is computed.

7.2.2   Results

This section is devoted to illustrating some trade-offs appearing in node location design. We expose selected tests conducted in a topology of $|\mathcal{N}_a| = 30$ access site locations plotted using a random uniform distribution. Access sites are also the candidate locations for placing core nodes; that is, $\mathcal{N}_a = \mathcal{N}_c$. To simplify our tests, the cost of each link is considered equal to its euclidean distance and costs c_{ij} are normalized such that its maximum value is one ($\max_{ij} c_{ij} = 1$). We denote as C to the cost of a core node, independently from where it is placed.

We use Net2Plan to solve to optimality several problem instances of (7.1), ranging different K and C values. Tests were completed interfacing to a CPLEX solver using the built-in JOM library, meaning a total computation of less than 1 min. Figure 7.1 graphically illustrates two solutions for core node costs $C = \{5, 1\}$. Maximum core node capacity is set to $K = 30$ and thus one single core node could serve all access nodes if needed. When the node cost is sufficiently high (e.g., $C = 5$ in Fig. 7.1a), the optimum solution places one single core node in a central position. Reducing C incentives placing more core nodes to reduce the link costs (e.g., Fig. 7.1b, for $C = 1$).

In Fig. 7.2 we show the number of core nodes in the optimal solutions for a set of parameters $K = \{3, 5, 10, 20, 30\}$ and $C = \{0.05, 0.1, 0.25, 0.5, 0.75, 1, 2, 5\}$. This helps us to explore the trade-off between link costs and node costs in node location problems. For a given K value, the optimal number of core nodes is a non-increasing function of C. When core nodes are inexpensive with respect to link costs, the optimal solution places one in each possible location (30 in our case). When they are expensive, the minimum possible amount of core nodes is placed (Z_{min}). This is given by:

$$Z_{min} = \lceil \frac{|\mathcal{N}_a|}{K} \rceil$$

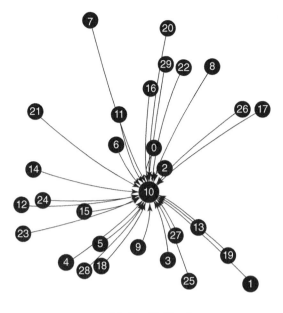

(a) $K = 30, C = 5$

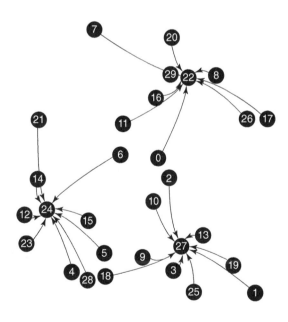

(b) $K = 30, C = 1$

Figure 7.1 Example: node location plots

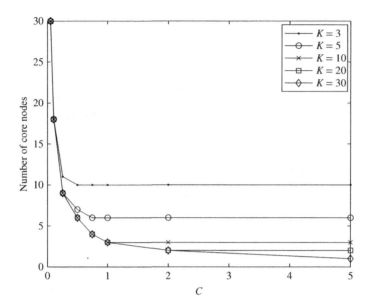

Figure 7.2   Node location trends: number of core nodes for different values of maximum node connectivity (K) and core node costs (C)

where $\lceil x \rceil$ is x rounded up to the closest integer. Previous expression just means that if each core node can serve a maximum of K access nodes, we will need at least $\frac{|\mathcal{N}_a|}{K}$ of them (rounded up) to serve all access nodes. Finally, for a given cost factor C, higher values of K mean less restrictions in the placement process that results in a lower or equal cost and a lower equal number of core nodes used.

7.3   Full Topology Design Problems

Let $\mathcal{N}$ denote a given set of already located core nodes, each concentrating the traffic coming from and going to a set of users in its area of coverage, not included in the model. The traffic offered to the network is given by a set D of unicast demands between core nodes. The traffic volume h_d of each demand d is known, coming from a traffic forecast study.

The basic full-topology design problem we address in this section intends to find (i) the links between core nodes to be deployed in the network, (ii) the capacity in each link, and (iii) how the traffic is routed in the resulting network. The optimization target is the minimization of the total network cost C, given by the sum of the costs of the links. The cost $c(e)$ of deploying a link e with capacity u_e is given by:

$$c(e) = c_f(e) + c_v(e)u_e$$

$c_f(e)$ is a fixed opening cost, applicable if the link is deployed, whatever capacity is installed in it. In its turn, $c_v(e)$ is the cost factor multiplying the link capacity, such that the increments of link capacity have a linear cost impact, $c_v(e)$ being the increase slope.

To simplify the problem modeling we use the concept of *candidate link set*. A candidate link set $\mathcal{E}'$ of a topology design problem is the set of links that *could be* included in the network design. If no specific constraints exist, $\mathcal{E}'$ contains a full-mesh with one candidate link per each node pair:

$$\mathcal{E}' = \{(n_1, n_2), \quad \forall n_1, n_2 \in \mathcal{N}, n_1 \neq n_2\}$$

But if the problem context forbids the deployment of links between some particular nodes, they are just removed from the candidate link set.

We formulate in (7.2) the topology design problem, using a flow-link formulation for the routing. Decision variables are:

$$z_e = \{1 \text{ if candidate link } e \text{ is deployed, } 0 \text{ if it is not}\}, \quad e \in \mathcal{E}'$$

$$u_e = \{\text{Capacity of link } e\}, \quad \forall d \in \mathcal{D}, e \in \mathcal{E}'$$

$$x_{de} = \{\text{Traffic of demand } d \text{ that traverses link } e\}, \quad \forall d \in \mathcal{D}, e \in \mathcal{E}'$$

$$\min_{z,x,u} \sum_e c_f(e) z_e + c_v(e) u_e \quad \text{subject to:} \tag{7.2a}$$

$$u_e \leq U z_e, \quad \forall e \in \mathcal{E}' \tag{7.2b}$$

$$\sum_{e \in \delta^+(n)} x_{de} - \sum_{e \in \delta^-(n)} x_{de} = \begin{cases} h_d, & \text{if } n = a(d) \\ -h_d, & \text{if } n = b(d), \quad \forall d \in \mathcal{D}, n \in \mathcal{N} \\ 0, & \text{otherwise} \end{cases} \tag{7.2c}$$

$$\sum_d x_{de} \leq u_e, \quad \forall e \in \mathcal{E}' \tag{7.2d}$$

$$u_e \geq 0, x_{de} \geq 0, \quad \forall d \in \mathcal{D}, e \in \mathcal{E}' \tag{7.2e}$$

Objective function sums for each link (i) its fixed cost $c_f(e)$ only if the it is deployed ($z_e = 1$), and (ii) its variable cost $c_v(e) u_e$, where we assume that if a link is not deployed, then its capacity should be zero ($z_e = 0 \Rightarrow u_e = 0$). This is enforced by constraint (7.2b): if $z_e = 0$ then, the capacity is constrained to be zero ($u_e \leq 0$), and if $z_e = 1$, the capacity should be below U ($u_e \leq U$), where U is any constant larger than any reasonable value that a link capacity can take (e.g., $U = \sum_d h_d$).

Constraints (7.2c–e), are the standard flow conservation, link capacity and non-negativity constraints present in flow-link problems (e.g., see Section 4.3 for details), where link capacities u_e are also decision variables. The important difference is that these constraints are defined as if all the candidate links $\mathcal{E}'$ are accepted in the network. For instance, sets of outgoing and incoming links of a node n, ($\delta^+(n)$ and $\delta^-(n)$), include all such links in the candidate set.

In summary, topology design problems can be modeled as any routing and capacity assignment problem over a candidate link set $\mathcal{E}'$ including all the possible network links. Then, z_e binary decision variables and associated constraints (7.2b) are the basic modifications needed to introduce the link deployment decision into the problem, as is done in the objective function (7.2a).

7.3.1 Problem Variants

Below, we sketch some of the multitude of problem variants to the baseline topology design problem (7.2):

- *Different routing types*: It is possible to easily produce versions of the problem (7.2) in flow-path and destination based formulations. In the former, the candidate path list $\mathcal{P}$ should be computed using the candidate link set $\mathcal{E}'$ as the baseline topology. Then, if x_p represents the traffic carried in path $p \in \mathcal{P}$, (7.2c,d) constraints should be replaced by:

$$\sum_{d \in \mathcal{P}_d} x_p = h_d, \quad \forall d \in D$$

$$\sum_{e \in \mathcal{P}_e} x_p \le u_e, \quad \forall e \in \mathcal{E}'$$

A destination-based routing version of (7.2), would replace (7.2c,d) with:

$$\sum_{e \in \delta^+(n)} x_{te} - \sum_{e \in \delta^-(n)} x_{te} = \begin{cases} h_{nt}, & \text{if } n \neq t \\ -\sum_s h_{st}, & \text{if } n = t \end{cases}, \quad \forall t, n \in \mathcal{N}$$

$$\sum_t x_{te} \le u_e, \quad \forall e \in \mathcal{E}'$$

where x_{te} is the traffic targeted to node t carried by link $e \in \mathcal{E}'$ and h_{st} the traffic offered from node s to node t.

- *Brownfield design*: In brownfield design, some of the links are already installed and we should optimize the extra links to deploy. Let $\mathcal{E}^0 \in \mathcal{E}'$ denote the set of already installed links. By adding constraints:

$$z_e = 1, \quad \forall e \in \mathcal{E}^0$$

we restrict the search to those solutions that include the already deployed links.

- *Bidirectional topology constraints*. In a bidirectional topology, if a link $n \to n'$ is part of the topology, then the opposite link $n' \to n$ also does. We can enforce bidirectional topologies adding the constraints:

$$z_e = z_{e'}, \quad \forall n \neq n' \in \mathcal{N}, e = (n, n'), e' = (n', n) \tag{7.3}$$

- *Ring-topology constraints*. A unidirectional ring over the set of nodes $\mathcal{N}$ is a connected topology of as many links as nodes, which passes through each node exactly once. Let us assume that the demand set D is such that the offered traffic can only be routed through connected topologies (e.g., one demand exists with non-zero offered traffic for each node pair). Then, adding the constraint:

$$\sum_{e \in \delta^+(n)} z_e = 1, \quad \sum_{e \in \delta^-(n)} z_e = 1, \quad \forall n \in \mathcal{N}$$

means that each node has exactly one incoming link and exactly one outgoing link. Since the resulting topology should be connected, previous constraint forces it to be a ring.

We can easily restrict the topology to be a bidirectional ring by making:

$$\sum_{e \in \delta^+(n)} z_e = 2, \quad \sum_{e \in \delta^-(n)} z_e = 2, \quad \forall n \in \mathcal{N}$$

$$z_e = z_{e'}, \quad \forall n \neq n' \in \mathcal{N}, e = (n, n'), e' = (n', n)$$

where each node has two incoming and outgoing links and the topology is bidirectional.

- *Tree-topology constraints.* A bidirectional tree in a topology over the set $\mathcal{N}$ of nodes is a connected bidirectional topology with the minimum possible number of (unidirectional) links: $2(|\mathcal{N}| - 1)$. Similar to the ring case, if the demand set D can only be routed through connected topologies, then any feasible solution of (7.2) is a connected topology. By adding the constraint:

$$\sum_e z_e = 2(|\mathcal{N}| - 1)$$

$$z_e = z_{e'}, \quad \forall n \neq n' \in \mathcal{N}, e = (n, n'), e' = (n', n)$$

we restrict the topology to be a bidirectional tree.

- *Joint link and node location.* In this type of problems, the designer should find the topology, link capacities, and routing that connects a given set $\mathcal{N}$ of core nodes, carrying the set D of demands between them. The difference with the basic topology design problem (7.2) is that now there is also the possibility to add *transit nodes* to the network, which do not generate nor consume traffic and just forward it. The set of candidate locations for the transit nodes is given by $\mathcal{N}_t$, where each location can host one or none transit nodes. We denote as $c_t(n), n \in \mathcal{N}_t$ the cost of placing a transit node in location n, and G_n the maximum number of incoming or outgoing links that n can have, because of any applicable technological constraints. The modeling of the joint link and node location problem can be easily accomplished by extending the candidate link set $\mathcal{E}'$ to all the links between the set of nodes $\mathcal{N}' = \mathcal{N} \bigcup \mathcal{N}_t$:

$$\mathcal{E}' = \{(n_1, n_2), \forall n_1, n_2 \in \mathcal{N}', n_1 \neq n_2\}$$

That is, the topology can contain not only links between core nodes, but also transit-transit and core-transit links. We use the same notation $c_f(e)$ and $c_v(e)$ to reflect the costs of any candidate links, including the new link candidates involving transit nodes. If technological restrictions forbid the use of some of these links, they can just be removed from $\mathcal{E}'$ set. Formulation (7.2) just needs three simple variations to include the possibility of placing transit nodes. First, the following decision variables should be added to the problem, to capture the transit node placement:

$$z_n = \{1 \text{ if a transit node is place in location } n, 0 \text{ otherwise}\}, \quad n \in \mathcal{N}_t$$

Then, the objective function (7.2a) should be extended to include the transit node placement costs:

$$\sum_e c_f(e)z_e + c_v(e)u_e + \sum_{n \in \mathcal{N}_t} c_t(n)z_n$$

In addition, the following constraints should be added to baseline problem (7.2):

$$\sum_{e \in \delta^+(n)} z_e \leq G_n z_n, \quad \forall n \in \mathcal{N}_t$$

$$\sum_{e \in \delta^-(n)} z_e \leq G_n z_n, \quad \forall n \in \mathcal{N}_t$$

Previous constraints mean that if a location n does not have a transit node ($z_n = 0$), then there can be no incoming nor outgoing links to it ($\sum_{e \in \delta^+(n)} \leq 0, \sum_{e \in \delta^-(n)} \leq 0$). If a location n has a transit node, the number of incoming or outgoing links is limited to G_n ($\sum_{e \in \delta^+(n)} \leq G_n, \sum_{e \in \delta^-(n)} \leq G_n$).

- *Other problem variants*: The candidate link set approach eases the modeling of the topology-design version for multiple problem variants, like non-bifurcated routing, multicast routing, modular capacities, network design for protection and restoration schemes, maximum end-to-end delay for a demand, and so on. The exercises at the end of this chapter provide illustrative examples.

7.3.2 Results

In this section we illustrate with test examples how the relative values of fixed and variable link costs (c_f and c_v factors), influences the design making it range from tree-like topologies (when fixed costs dominate) to full-mesh like topologies (when variable costs dominate).

We solve problem (7.2) with the extra constraint of obtaining a bidirectional topology (7.3). Optimal solutions are obtained with Net2Plan, interfacing to a CPLEX solver using the JOM built-in library. We consider three node locations of $|\mathcal{N}| = \{7, 12, 14\}$ nodes. They correspond to the nodes of the example topologies `example7nodes.n2p`, `abilene_N12_E30.n2p` and `NSFNet_N14_E42.n2p` in Net2Plan tool. The offered traffic is uniform, meaning that all the node pairs (s, t) offer the same amount of traffic, normalized such that the total offered traffic in the network sums one unit. Thus:

$$h_{st} = \frac{1}{|\mathcal{N}|(|\mathcal{N}| - 1)}, \quad \forall s \neq t \in \mathcal{N}$$

We assume that $c_v(e) = d_e$ and $c_f(e) = d_e C$, where d_e is the euclidean distance between end nodes of candidate link e, and C is the fixed cost per km of link e. The fixed C_f and variable C_v part of the network cost is:

$$C_f = \sum_e C d_e$$

$$C_v = \sum_e u_e d_e$$

Higher values of C mean that deploying links is the expensive resource, while putting more capacity to an existing link is comparatively inexpensive. This penalizes intensely connected topologies. In its turn, tree-like topologies would be preferred, since (i) they concentrate the traffic in a small number of links, the expensive resource, (ii) and augment the average number

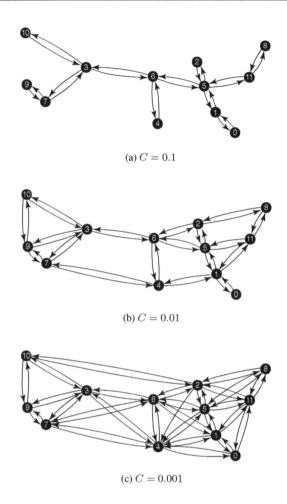

(a) $C = 0.1$

(b) $C = 0.01$

(c) $C = 0.001$

Figure 7.3 Example of topologies in an Abilene node set. $c_f(e) = Cd_e$, $c_v(e) = d_e$, $h_{st} = \frac{1}{12 \times 11}$

of hops in the network $\bar{n}$ and thus the total amount of capacity installed $\bar{n} \sum_d h_d$, which is comparatively cheap.

Conversely, smaller values of C would make inexpensive the addition of new links. Then, topologies tend to be a full-mesh, where a lot of traffic is routed by a direct link (reduced $\bar{n}$), and the total amount of capacity installed (the expensive resource) is reduced. Figure 7.3a–c illustrates the different topologies obtained for different C values in Abilene node set.

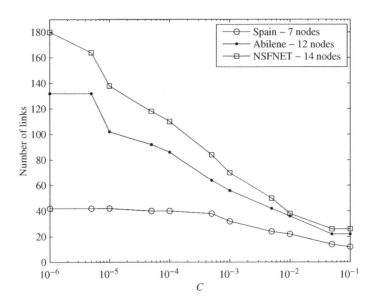

Figure 7.4   Topology trends: number of links in the topology for different values of link cost per km (C)

The impact of the relative cost per km C in the optimal number of links to deploy, is illustrated in Fig. 7.4 for the three node topologies of our case study. The fixed cost factors C ranged are in a logarithmmic scale:

$$C = \{10^{-6}, 5 \times 10^{-6}, \ldots, 10^{-2}, 5 \times 10^{-2}, 10^{-1}\}$$

Higher fixed costs produce tree-like bidirectional topologies with a number of links equal to $2(|\mathcal{N}| - 1)$, while lower link costs result in full-mesh topologies with $|\mathcal{N}|(|\mathcal{N}| - 1)$ links. However, a large range of intermediate cost factors C produce topologies that are far from being trees nor full-meshs. In other words, topology design is needed to optimize the complex trade-offs appearing in these scenarios, and the results cannot be predicted in advance.

7.4   Multilayer Network Design

Communication networks are organized into layers, governed by different protocols, and potentially managed by different companies or institutions, such that the links in an upper layer appear as traffic demands carried by the lower layer in an underlying topology. For instance, in IP over WDM optical networks:

- The upper layer is composed of a set of IP routers, connected through optical connections of fixed capacity (e.g., 10 Gbps, 40 Gbps, 100 Gbps) called lightpaths. The IP routers see each lightpath as a link and the traffic is routed on top of the lightpaths according to the IP nodes routing tables.

- In the lower layer, each lightpath is a demand to carry on top of the underlying topology of optical fibers. Each lightpath is assigned a wavelength that cannot be changed along its route, unless wavelength conversion devices are available. The optical switching nodes forwarding the lightpaths are called Optical Add/Drop Multiplexers (OADMs).

Thus, in the previous example, a lightpath appears to the IP layer as a direct link between two routers of a fixed capacity, irrespective of the actual route of the lightpath across the fibers. The topology of IP links (each corresponding to a lightpath) is usually referred to as the *virtual topology*, since each link is not backed by an actual wire but by a lightpath that follows an arbitrary optical path across the fiber topology.

Many other examples of multilayer networks exist. For instance, a common three-layer structure is that of IP routers connected through a topology of MPLS virtual circuits, which are routed on top of a topology of lightpaths that are routed on top of a topology of optical fibers. What follows is a general illustrative example of a multilayer design with two layers.

Let $\mathcal{N}$ be a set of sites, each one hosting a lower layer switching node and an upper layer switching node. Lower layer nodes in the sites are connected through a given set of transport links $\mathcal{E}_{lo}$ and lower layer connections (called circuits) are forwarded through them. Each transport link can carry a maximum of U_{lo} circuits. In the upper layer, a traffic matrix $h = \{h_{st}, s, t \in \mathcal{N}, s \neq t\}$ is routed by the upper layer nodes over the circuits using destination-based routing. Each circuit appears as a link in the upper layer with a capacity of U_{hi} traffic units.

Formulation (7.4) models the problem that finds the network design satisfying all the upper layer traffic at minimum cost, given by the number of circuits needed. To model the possible virtual topology of circuits, we define $\mathcal{E}_{up}$ as a full-mesh of *virtual links* between all nodes in $\mathcal{N}$. Decision variables are:

$z_c = \{$Number $0, 1, 2, \ldots$ of circuits in virtual link $c\}, \quad c \in \mathcal{E}_{up}$

$x_{tc} = \{$Upper layer traffic targeted to t, in virtual link $c\}, \quad \forall t \in \mathcal{N}, c \in \mathcal{E}_{up}$

$x_{ce} = \{$Number of circuits of virtual link c that traverse lower layer link $e\}, \forall c \in \mathcal{E}_{up}, e \in \mathcal{E}_{lo}$

$$\min \sum_c z_c \quad \text{subject to:} \tag{7.4a}$$

$$\sum_{c \in \delta^+(n) \cap \mathcal{E}_{up}} x_{tc} - \sum_{c \in \delta^-(n) \cap \mathcal{E}_{up}} x_{tc} = \begin{cases} h_{nt}, & \text{if } n \neq t \\ -\sum_s h_{st}, & \text{if } n = t \end{cases}, \quad \forall t, n \in \mathcal{N} \tag{7.4b}$$

$$\sum_t x_{tc} \leq U_{hi} z_c, \quad \forall c \in \mathcal{E}_{up} \tag{7.4c}$$

$$\sum_{e \in \delta^+(n) \cap \mathcal{E}_{lo}} x_{ce} - \sum_{e \in \delta^-(n) \cap \mathcal{E}_{lo}} x_{ce} = \begin{cases} z_c, & \text{if } n = a(c) \\ -z_c, & \text{if } n = b(c), \\ 0, & \text{otherwise} \end{cases} \quad \forall c \in \mathcal{E}_{up}, n \in \mathcal{N} \tag{7.4d}$$

$$\sum_c x_{ce} \leq U_{lo}, \quad \forall e \in \mathcal{E}_{lo} \tag{7.4e}$$

The objective function counts the number of circuits. Equations (7.4b,c) are the destination-link and link capacity constraints for the routing of the upper layer traffic over the virtual topology. Equations (7.4d,e) are the flow-link and link capacity constraints for the routing of the circuits over the lower layer topology.

The problem variations seen in single-layer problems like multicast/anycast traffic, bifurcated/non-bifurcated routing, protection, restoration, and so on can appear in one or more layers of a multilayer design. The modeling of the virtual topology using a full-mesh of candidate virtual links simplifies how these variations can be applied in multilayer problems.

7.5 Notes and Sources

Node location and general topology design are classical network problems, appearing in one form or another, with multiple variants, in all the network technologies. Multilayer network design is motivated by the advantages of the joint optimization of several layers of the network and has become popular in IP-over-any networks.

Some reference books covering the general node location and topology design models in this chapter are [1] [2] [3], and [4]. In particular, [4] collects and studies multiple variants of multilayer design problems. Readers interested in topology problems applied to particular network technologies should refer to its specialized literature.

7.6 Exercises

7.1   Implement a Net2Plan algorithm that solves a variation of formulation (7.1) with JOM where the access nodes are constrained to be connected to exactly two different core nodes. This technique is called dual-homing and appears in networks like SS7 (signaling networks in telephone systems) for improving the fault tolerance. The nodes of the input topology are the access node locations, equal to the core node locations. The cost of link (i,j) is given by its distance d_{ij} normalized so that the maximum link cost equals 1. The core node cost is C, an input parameter to the algorithm, together with K in (7.1). The optimum solution is returned by updating the network links, and adding an attribute hasCoreNode to each node, with the value true or false accordingly.

7.2   Modify Exercise 7.1 to permit an arbitrary number of core nodes per location, but limiting the total number of core nodes to M an algorithm input parameter.

7.3   Modify Exercise 7.1 to optimize the case when all core nodes are fully connected between them. The cost of inter-core nodes links is proportional to its distance, but normalized so that a core link between the two most distant locations in the input network had a cost of C'.

7.4   In Exercise 7.1, assume that each possible access to core link e_{ij} is assigned an availability value assuming 0.05 failures per year and link km, and a mean repair time of 12 h. Modify Exercise 7.1 to optimize the case when the probability of an access node to be disconnected from the core node is not higher than 10^{-3}.

7.5 Modify Exercise 7.4 for the case when each access node is constrained to connect to two core nodes and that an access node is considered disconnected when both of its links fail.

7.6 A set $\mathcal{N}_a$ of users of known location should be covered placing base stations (BS) in a set $\mathcal{N}_b$ of potential sites. Each BS is able to cover a maximum of K users, among those located at a distance of d_{max} or less. In addition, each BS should be assigned a frequency, so that those BSs whose coverage region overlap (located at less that $2d_{max}$ km), cannot have the same frequency. The cost of each BS is one. The total cost sums the BSs cost and a cost of C multiplied by the number of *different* frequencies needed. Write the formulation that finds the BS placement and frequency assignment that minimizes the total cost. Implement a Net2Plan algorithm that solves this problem with JOM. The nodes of the input topology are the user locations, equal to the potential BS locations. C and d_{max} are input parameters to the algorithm. The optimum solution is returned by updating the network links and adding an attribute hasBS to each node, with the value true or false accordingly, and an attribute bsFrequency with its frequency (e.g., 0, 1, 2, ...).

7.7 Let $\mathcal{N}$ be a set of node locations where to place SDH digital cross-connects. Let $\mathcal{D}$ be a set of virtual circuit demands to carry, and h_d the virtual circuit volume in Mbps. The routing of the virtual circuit cannot be bifurcated. Model the problem that finds the bidirectional ring of minimum cost (summing the cost of the links) which carries all the traffic. Ring links are all of the same capacity u, chosen among a set of available capacities (in Mbps) $\{155, 622, 2488, 9953, 39813\}$, and associated cost of the capacity module per bidirectional link $\{c_1, \dots, c_5\}$, respectively. The total link cost is C times its length in km, plus the cost of its capacity modules. Implement a Net2Plan algorithm that solves previous formulation with JOM. Input parameters to the algorithm are the cost factors C and $\{c_1, \dots, c_5\}$. The optimum solution is returned by updating the network links, link capacities, and routing accordingly.

7.8 Modify Exercise 7.7 assuming that the offered traffic is composed of a set $\mathcal{D}$ of multicast demands.

7.9 Modify Exercise 7.7 assuming that each virtual circuit is protected through a link-disjoint path. Backup paths are returned in Net2Plan using protection segments with the appropriate reserved capacity.

7.10 Modify Exercise 7.7 assuming that in the case of link failure, the network reacts rerouting all the affected virtual circuits, keeping the unaffected circuits unchanged. Then, the design is constrained to carry the 100% of the traffic under any failure of a single bidirectional link. The optimum solution is returned by updating the network links, link capacities, and routing in the no-failure state accordingly.

7.11 Implement a Net2Plan algorithm that solves formulation (7.2) with JOM and repeat the results in Section 7.3.2.

7.12 Let $\mathcal{N}$ be a set of node locations. Model the problem that finds the minimum cost bidirectional tree spanning all the nodes, such that the maximum propagation time between most distant nodes in the tree is limited to T_{max} and maximum number of links outgoing from a node is limited to δ_{max}. The link cost is proportional to the sum of the distances of the links. Implement a Net2Plan algorithm that solves previous formulation with JOM. Input parameters to the algorithm are T_{max} and δ_{max}. The optimum solution is returned by updating the network links accordingly.

7.13 In Exercise 7.12, we have a set $\mathcal{D}$ of offered traffic demands. All links have the same capacity U, an input parameter. Modify Exercise 7.12 such that the design is constrained to carry all the traffic.

7.14 Implement a Net2Plan algorithm that solves the problem of joint link and node location in Section 7.3.1. Transit node candidate locations are those nodes in the input design with no ingress nor egress demands. Link capacities are constrained to be an integer multiple of U. Link costs are proportional to its distance, normalized such that the cost of a link between most distant nodes is one. The topology is constrained to be bidirectional and the number of outgoing links of a transient node is limited to G. The cost of a module of capacity U in a link is c_u and the cost of a transit node is c_t. Input parameters to the problem are U, G, c_u, and c_t. The optimum solution is returned by updating the network links, link capacities, and routing accordingly.

7.15 Implement a Net2Plan algorithm that solves formulation (7.4) with JOM. Input parameters to the algorithm are U_{lo} and U_{hi}. The optimum solution is returned by creating a two layer design using standard multilayer Net2Plan functions and updating the network links, capacities, and routes accordingly.

7.16 Modify Exercise 7.15 in the case that each circuit is assigned in the lower layer two link disjoint routes.

7.17 Modify Exercise 7.15 in the case that upper layer offered traffic is composed of a set $\mathcal{D}$ of unicast demands and a flow-path formulation in the upper layer is used with the constraint that each upper layer demand can traverse a maximum of three circuits.

7.18 Modify Exercise 7.15 in the case that circuits are lightpaths in an optical network, which are assigned a wavelength $w \in \{1, \ldots, U_{lo}\}$, and such that two lightpaths using the same wavelength cannot traverse the same lower layer link (optical fiber).

7.19 Modify Exercise 7.18 in the case that OADMs have an internal blocking constraints, such that a maximum C of lightpaths can be originated or terminated in a node with the same wavelength.

7.20 Modify Exercise 7.18 in the case that upper layer traffic has a multihour profile and is given by a set $\mathcal{D}$ of demands, with estimated traffic volume during time interval t of $h_{dt}, d \in \mathcal{D}, t = 1, \ldots, T$. The routing in the upper layer is oblivious and the lightpaths established do not change along time intervals. Compare with the case when the upper layer routing is dynamic.

7.21 Let $\mathcal{N}$ be a set of nodes, and $\mathcal{D}$ a set of demands between them. Let us assume that all network links are constrained to have a capacity of U units. Show that the following lb_{in} and lb_{out} are two lower bounds lb to the number of links in the network, if all the traffic is to be carried:

$$lb_{in} = \sum_n \lceil \frac{\sum_{d:n=a(d)} h_d}{U} \rceil, \quad lb_{out} = \sum_n \lceil \frac{\sum_{d:n=b(d)} h_d}{U} \rceil,$$

References

[1] L. Kleinrock, *Queueing Systems*. New York, NY, USA: John Wiley & Sons, Inc., 1976, vol. II: Computer Applications.

[2] D. Bertsekas and R. Gallager, *Data Networks*. Englewood Cliffs, NJ: Prentice Hall, 1992.

[3] R. S. Cahn, *Wide Area Network Design: Concepts and Tools for Optimization*. Morgan Kaufmann, 1998.

[4] M. Pioro and D. Medhi, *Routing, Flow, and Capacity Design in Communication and Computer Networks*. Morgan Kaufmann Publishers, 2004.

Part Two

Algorithms

8

Gradient Algorithms in Network Design

8.1 Introduction

In this chapter, we introduce the application of gradient methods to network design problems. Most of these problems, like congestion control or fast capacity allocation in wireless networks, involve a potentially large amount of loosely coupled elements that operate more or less independently. Then, there is no possibility of having a central decision unit that receives all the problem inputs, implements a sophisticated algorithm and returns the outputs to the network nodes that have to apply them. For these important cases, we pursue algorithms where each element can proceed to make decisions on its own, coordinated by a small amount of signaling information. As we will see, gradient algorithms are a strong theoretical tool that permits creating such schemes, with convergence guarantees.

We concentrate on constrained *convex* problems of the form:

$$\min_x f(x), \text{ subject to: } x \in \mathcal{X} \tag{8.1}$$

where x is a vector in $\mathbb{R}^n$, f is convex in $\mathcal{X} \subset \mathbb{R}^n$, and $\mathcal{X}$ is an arbitrary closed convex set. The unconstrained problem when $\mathcal{X} = \mathbb{R}^n$ is dealt with as a particular case of (8.1).

There are multiple alternatives for solving this type of problems, the reader is referred to excellent sources such as [1–4] for a survey. Among them, we will concentrate on some variants of the so called *gradient projection* iterative methods, because of its theoretical simplicity and its successful applicability in multiple network optimization problems.

A gradient projection algorithm is an iteration of the form:

$$x(t+1) = P_{\mathcal{X}}(x(t) - \gamma(t)B(t)^{-1}\nabla f(x(t))), \quad t = 0, 1, \dots \tag{8.2}$$

In the first iteration $t = 0$, the method starts in a feasible point $x(0) \in \mathcal{X}$. Given a current point $x(t)$, the next iteration moves in a direction given by minus the gradient of f at point $x(t)$,

Optimization of Computer Networks – Modeling and Algorithms: A Hands-On Approach,
First Edition. Pablo Pavón Mariño.

multiplied by the inverse of a symmetric positive definite scaling matrix $B(t)$ and a strictly positive factor $\gamma(t) > 0$. Note that:

- $-\nabla f(x(t))$ is the direction of steepest decrease of f, so if $\nabla f(x(t)) \neq 0$ and we move from $x(t)$ infinitesimally in that direction, function f decreases with the maximum slope compared to any other direction.
- Since $B(t)$ is positive definite, $B(t)^{-1}$ also is, and the direction $-B(t)^{-1}\nabla f(x(t))$ is also a descent direction, which makes an angle of less than $90°$ with $-\nabla f(x(t))$.
- Depending on, for example $\gamma(t)$, the resulting point:

$$y = x(t) - \gamma(t)B(t)^{-1}\nabla f(x(t))$$

can be outside the feasibility set. If so, the projection operation $P_\mathcal{X}(y)$ returns the closest feasible point to y.

There are multiple variants of gradient methods within (8.2). For instance, Newton methods are based on computing at each iteration a matrix $B(t) = \nabla^2 f(x(t))$ and quasi-Newton methods an approximation to it. These schemes have fast convergence properties. However, the computation of the hessian matrix in each iteration makes them computationally expensive and not amenable to a distributed implementation. Therefore, we will in general restrict to methods where: (i) factor $\gamma(t)$ is a known constant for all the iterations $\gamma(t) = \gamma$, (ii) $B(t)$ matrices are constant (e.g., $B(t) = I$) or its structure is such that it does not hinder the distributed implementation of the method (e.g., $B(t)$ is a diagonal matrix).

Previous limitations will result in slower convergence rates than, for example, sophisticated Newton approaches. We will briefly review the theoretical basis quantifying this to provide practical recommendations in this aspect. Still, slow convergence rates are considered an acceptable sacrifice in our contexts of interest. For instance, in congestion control, the algorithms are supposed to be running permanently, adapting to the fluctuations of network conditions. Algorithm *robustness* and algorithm *stability* are in this scenario as relevant features as algorithm convergence rate:

- *Algorithm robustness* considers the ability of an algorithm to converge even in the presence of faults like the loss or out of sequence delivery of signaling messages.
- *Algorithm stability* requires an algorithm to converge to a solution in a finite amount of time, without eternal oscillations around the optimum, independent of the initial starting solution. Similarly, it means that even if the network conditions change abruptly, the algorithm converges to the new equilibrium situation in a finite amount of time.

Convergence guarantees will be provided for some versions of projected gradient methods. In particular, sufficient conditions are shown in the important cases when the implementation is asynchronous and gradient information used in an iteration can be outdated, signaling messages are lost or delivered out of order, and the case when the gradient information handled has random measurement errors. Both are frequent engineering situations, for which gradient algorithms can be robust.

As we will see, convergence guarantees coming from the analysis usually take the form of maximum step sizes γ. In practice, these conditions are very restrictive and produce such small step sizes γ that the algorithm would be extremely slow to converge. This issue should not be taken as a strong inconvenient. Our interest in gradient algorithms is providing a theoretical insight and guidelines on how new network design algorithms should be engineered, or why some other existing algorithms behave so well. Actually, in real implementations, network algorithms often deviate from a pure application of the gradient iteration. For example internal parameters are dimensioned such that convergence is not guaranteed in theory, but is observed in practice at a reasonable pace according to a battery of tests performed based on the cases of interest. Also, some sophisticated techniques can be added to an algorithm, for example to fast cut down oscillations, even if the algorithm remains in equilibrium in a slightly suboptimal solution. They are a form to address the typical tension between convergence guarantees and convergence speed in control problems. These techniques are problem dependent, mostly heuristics, and are outside the scope of this book. We will contempt with providing some suggestions in the last section of the chapter.

8.2 Convergence Rates

The convergence rate of an algorithm quantifies how fast the algorithm approaches the optimum. The classical methodology to study this is called *local analysis*, which focuses on the behavior of the algorithm in the neighborhood of an optimal solution[1]. Let x^* denote an optimum solution to a minimization problem, $f(x^*)$ its optimal cost, and $x(t)$ a sequence of iterates generated by the algorithm. Convergence rate analysis evaluates the asymptotic evolution of an error function $e(t) = \|x(t) - x^*\|$ or $e(t) = |f(x(t)) - f(x^*)|$ that measures how far we are from an optimum solution. We distinguish the following cases:

- *Sublinear convergence rate*. The error drops according to a power function of the iteration counter. For example:

$$e(t) \leq \frac{c}{t}$$

  where $c > 0$ is a constant. In this case, to find a solution with a maximum error ϵ, we need at least c/ϵ iterations and the approximation is in the order $\mathcal{O}(1/\epsilon)$. Sublinear rates are considered slow. This is the rate we will find (in the worst-case) in our basic and diagonally scaled algorithms in the next sections, when the objective function is non-smooth or it is not strongly convex (second derivative is not bounded away from zero in every direction).
- *Linear or geometric convergence rate*. The error drops according to an exponential (geometric) function of the iteration counter. For example,

$$e(t) \leq c\beta^t \Rightarrow e(t+1) \leq e(t)\beta$$

  where $\beta \in (0, 1), c > 0$. Then, the number of iterations t to reach an error ϵ is related by: $t \leq \frac{-\log 1/\epsilon - \log c}{\log \beta}$, which is in the order of $\log 1/\epsilon$. The convergence can be considered fast, since improving the accuracy of the answer in one right digit, takes a constant amount of computations. This is the worst-case rate we can observe in basic or diagonally scaled gradient projection algorithms, when the objective function is differentiable and strongly convex.

---

[1] Local analysis does not account for the rate of progress of the algorithm in its initial iterations, far from the optimum, for which there is little theoretical support. Still, it is generally considered a good insight to compare the quality of algorithms.

- *Superlinear convergence rate.* The error drops faster than the geometric progression, in a sort of double exponential form. In particular, there exists $p > 1, \beta(0, 1), c > 0$ such that:

$$e(t) \le c(\beta)^{p^t} \Rightarrow e(t+1) \le c^{1-p} e(t)^p$$

The case $p = 2$ is called quadratic convergence, $p = 3$ cubic convergence, etc. This rate is extremely fast, since we can double (in $p = 2$) the number of precision digits in the answer in a constant time. Superlinear convergence is provided, for example, by Newton methods that include full second-derivative information in the scaling. As discussed previously, these methods do not yield in general to a distributed implementation when applied to our network optimization problems of interest and thus are not further considered in this chapter. Multiple software packages (e.g., IPOPT solver, accessible with JOM from Net2Plan) include this type of method, which are of practical use when the problem allows a centralized implementation.

8.3   Projected Gradient Methods

A projected gradient scheme is an iteration of the form:

$$x(t+1) = P_{\mathcal{X}}(x(t) - \gamma(t)B(t)^{-1}\nabla f(x(t))) \tag{8.3}$$

where $P_{\mathcal{X}}(y)$ is the projection of vector y on set $\mathcal{X}$. That is, the point $x \in \mathcal{X}$ closest to y according to a particular distance:

$$P_{\mathcal{X}}(y) = \{x' \in \arg\min_{x \in \mathcal{X}} \|y - x\|\}$$

Different norms in the projection can yield to different vectors x. If unspecified, the Euclidean norm is assumed. When matrices $B(t) = I$, the iteration (8.3) is called the *basic gradient projection* method. If not, the method is called the *scaled gradient projection* method.

Projecting a vector y in a set $\mathcal{X}$ can be a problem as difficult to solve as the original one. Thus, the interest of projection methods lays on those cases when the projection operation is easy to compute. In particular, this occurs when $\mathcal{X}$ is a box-like set:

$$\mathcal{X} = \{x = (x_1, \dots, x_n) : l_i \le x_i \le u_i\} \tag{8.4}$$

That is, each coordinate is limited to belong to an interval $[l_i, u_i]$, where l_i can be $-\infty$ and/or u_i can be ∞. In this case, the projection of a vector $y \in \mathbb{R}^n$ is computed independently for each coordinate i:

$$P_{\mathcal{X}}(y) = (x_1, \dots, x_n), x_i = [y]_{l_i}^{u_i} = \begin{cases} l_i, \text{ if } y_i \le l_i \\ y_i, \text{ if } l_i \le y_i \le u_i , \\ u_i, \text{ if } y_i \ge u_i \end{cases} \quad \forall i = 1, \dots, n \tag{8.5}$$

Note that if an independent element exists handling each coordinate of a vector y, it can compute the i-th coordinate of its projection without knowing the values in other coordinates: leaving y_i as it is, if it belongs to the interval $[l_i, u_i]$, or taking l_i (u_i) when y_i is lower than (greater than) the interval limits. This occurs since the feasibility set is *separable*; that is, it can be expressed as the Cartesian product of other sets, in this case one set per coordinate:

$$\mathcal{X} = \{\mathcal{X}_1 \times \dots \times X_n, \mathcal{X}_i = [l_i, u_i]\}$$

8.3.1   Basic Gradient Projection Algorithm

We focus on an iteration:

$$x(t + 1) = P_{\mathcal{X}}(x(t) - \gamma \nabla f(x(t))) \tag{8.6}$$

for solving problem (8.1), where f is a differentiable function and feasibility set $\mathcal{X}$ is non-empty. In the unconstrained case, $\mathcal{X} = \mathbb{R}^n$.

The convergence of basic projected gradient algorithms is affected by the so-called Lipschitz constant of the gradient of the objective function, which measures how fast $\nabla f(x)$ can change when we move from $x(t)$ to $x(t + 1)$. Given a continuously differentiable function $f : \mathcal{X} \subset \mathbb{R}^n \to \mathbb{R}$, we say that it has a Lipschitz continuous gradient with constant K, if it holds:

$$\|\nabla f(x) - \nabla f(y)\| \leq K \|x - y\|, \quad \forall x, y \in \mathcal{X}$$

The larger constant K is, the steeper that the change of the gradient of f can be. If f is twice differentiable, then constant K is given by the modulus of the largest eigenvalue of the hessian matrix $\nabla^2 f(x)$ in $\mathcal{X}$.

Example 8.1 The Lipschitz constant of function $f(x) = \log x, x > 0$ or $g(x) = \sqrt{x}, x \geq 0$ is infinite, since when $x \to 0$, the second derivative is given by: $f''(x) = -\frac{1}{x^2}$ and $g''(x) = -\frac{1}{4x^{3/2}}$, which tends to infinite when $x \to 0$. Restricted to the domain $x \geq 1$, it happens that $\max_{x \geq 1} |f''(x)| = 1$ and $\max_{x \geq 1} |g''(x)| = 1$, so $K = 1$ for both.

Proposition 8.1 In problem (8.1), we assume that $\mathcal{X}$ is a non-empty, closed and a convex set, f is convex, bounded below in $\mathcal{X}$ and differentiable, and ∇f is Lipschitz continuous with constant K. Then, if $0 < \gamma < \frac{2}{K}$:

- the basic gradient projection scheme (8.6) converges to a global optimum, for any initial point $x(0)$.
- if f is twice differentiable, and m and M are the minimum and maximum modulus of the eigenvalues of the hessian matrix in $\mathcal{X}$, the convergence rate is linear and the best progression factor is achieved for $\gamma = \frac{2}{M+m}$, and is given by:

$$\beta = \frac{\|x(t + 1) - x^*\|}{\|x(t) - x^*\|} = \frac{M - m}{M + m}$$

The proof of this property (e.g., see [5], p. 214, [6], p. 207) is based on showing that the step γ is small enough to guarantee that the objective function is improved in every iteration.

Example 8.2 In the application of gradient algorithm to problem $\min f(x)$, for $f(x) = x^2$, we have that since $f''(x) = 2$, convergence is guaranteed for $\gamma < 1$. Figure 8.1a shows how the algorithm converges starting in $x(0) = 1$ when $\gamma = 0.9$ and Fig. 8.1b shows how it diverges for $\gamma = 1.1$, since in each iteration the optimum is overstepped.

8.3.2   Scaled Projected Gradient Method

Scaled projection methods are a sophistication of the basic gradient step (8.6) that intends to improve its convergence rate. The scaled projection iteration is given by:

$$x(t + 1) = P_{\mathcal{X}}(x(t) - \gamma(t)B(t)^{-1}\nabla f(x(t))) \tag{8.7}$$

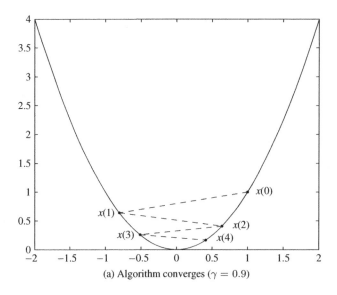

(a) Algorithm converges ($\gamma = 0.9$)

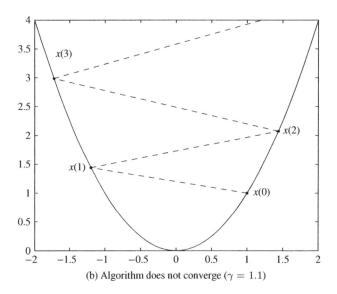

(b) Algorithm does not converge ($\gamma = 1.1$)

Figure 8.1 Gradient algorithm application to problem $\min x^2$. Lipschitz constant for the gradient of the objective function is $(x^2)'' = 2$. Convergence is guaranteed for $\gamma < \frac{2}{K} = 1$

where $B(t)$ is a definite positive matrix. Its role is that the new jumping direction in each iteration given by $B(t)^{-1}\nabla f(x(t))$, makes a better progress to the optimum than the basic gradient approach. Many scaling techniques have been studied and proposed. When $B(t) = \nabla^2 f(x(t))$, we have a form of the celebrated Newton method with superlinear convergence, but seldom amenable to distributed implementations.

The scaled gradient projection methods converge for a γ sufficiently small in unconstrained problems. Proposition 8.2 provides some convergence guarantees.

Proposition 8.2 In the unconstrained problem $\min f(x)$, let f be a continuously differentiable and bounded below function, and ∇f be Lipschitz continuous with constant K. $B(t)$ matrices are bounded above, and the minimum eigenvalue of $B(t)$ along all t is K_2. Then, the scaled gradient scheme:

$$x(t+1) = x(t) - \gamma B(t)^{-1} \nabla f(x(t)) \tag{8.8}$$

converges for $0 < \gamma < 2\frac{K_2}{K}$ (see [5], p. 205). In particular, if f is twice differentiable and strictly convex, the method converges for the scaling matrices:

$$B(t)_{ii} = \frac{\partial^2 f}{\partial x_i^2}(x(t)), \quad B(t)_{ij} = 0, i \neq j \tag{8.9}$$

and now K_2 is the lowest diagonal value of $B(t)$, along t.

Unfortunately, the convergence in constrained problems is not straightforward. The reason is that convergence under general feasibility sets requires applying a norm different to the Euclidean. We illustrate this in a counterexample. Let us assume that at any iteration t, the method (8.7) reaches the optimum $x(t) = x^*$, which is in the boundary of the feasibility set, with one active constraint. In this situation:

- A basic gradient method would have a searching direction $-\nabla f(x^*)$, which is orthogonal to the active constraint, according to KKT conditions (see Fig. B.9 in Appendix B). Thanks to this, if we project back the point $x^* - \gamma \nabla f(x^*)$ into the feasibility set, $x(t+1) = x^*$ again. That is, when the algorithm reaches the optimum, stays in it.
- A scaled gradient method would have a searching direction $-B(t)^{-1} \nabla f(x^*)$, which now is *not* orthogonal to the active constraint. Then, if we project back the point $x^* - \gamma \nabla f(x^*)$ into the feasibility set, $x(t+1) \neq x^*$. Thus, the algorithm does not stay in the optimum when it reaches it.

To rule out convergence issues in constrained problems, we have to modify the projection operation, so that now it is performed using a new norm. If $B(t)$ matrices are symmetric and positive definite, we can use them to induce a norm:

$$\|x\|_{B(t)} = (x^T B(t)x)^{1/2}$$

This norm is applied to redefine the projection of a point y into the feasibility set $\mathcal{X}$, $P_{\mathcal{X}}^{B(t)}(y)$ as the vector $x \in \mathcal{X}$ such that $\|y - x\|_{B(t)}$ is minimum[2] over all $x \in \mathcal{X}$. Now, it can be shown that the scaled gradient projection scheme:

$$x(t+1) = P_{\mathcal{X}}^{B(t)}(x(t) - \gamma(t)B(t)^{-1} \nabla f(x(t))) \tag{8.10}$$

has restated convergence guarantees for $\gamma > 0$ sufficiently small (see [5], p. 217).

---

[2] Note that if $B(t) = I$, the projection $P_{\mathcal{X}}^{B(t)}(y)$ coincides with the standard projection.

Previous assert helps us in the particular case of diagonal scaling, when (i) $B(t)$ matrices are diagonal and also (ii) the feasibility set $\mathcal{X}$ is box-like. In box-like sets, the projection $P_{\mathcal{X}}^{B(t)}(y)$ is the same for any diagonal positive definite matrix $B(t)$, for example the same as with $B(t) = I$ and given by (8.5). In this case, the following proposition, twin of Prop. 8.2 for box-like constraints and diagonal scaling matrices, provides sufficient conditions for convergence.

Proposition 8.3 In problem $\min_{x \in \mathcal{X}} f(x)$, we assume that $\mathcal{X}$ is a non-empty, box-like set (8.4). Function f is convex, continuously differentiable and bounded below, and ∇f is Lipschitz continuous with constant K. $B(t)$ are diagonal matrices bounded above, and the minimum diagonal term of $B(t)$ along all t is $K_2 > 0$. Then, the scaled gradient scheme:

$$x(t+1) = P_{\mathcal{X}}(x(t) - \gamma(t)B(t)^{-1}\nabla f(x(t)))$$

converges for $0 < \gamma < 2\frac{K_2}{K}$ (see [5], p. 217). In particular, if f is twice differentiable and strongly convex, the method converges for the scaling matrices (8.9).

As established, for example, in [2], the diagonal scaling method with second derivatives in the diagonal values is not guaranteed to improve the convergence rate of the basic gradient algorithm, but it is simple and often surprisingly effective in practice.

8.3.3   Singular and Ill-Conditioned Problems

Gradient projection algorithms described yield to linear convergence rates provided that the objective function is twice differentiable and strongly convex. Then, according to Prop. 8.1, the error drops geometrically in the worse case with a progress factor β:

$$\beta = \frac{\|x(t+1) - x^*\|}{\|x(t) - x^*\|} = \frac{M - m}{M + m} = \frac{\kappa - 1}{\kappa + 1}$$

where M and m are the modulus of the largest and smallest eigenvalue of the hessian matrix in the proximity of the optimum, and $\kappa = \frac{M}{m}$ is its so-called *condition number*. Then:

- The best convergence guarantees are obtained for smaller $\beta \to 0$, which happens when $M \approx m$, or equivalently $\kappa \approx 1$. These are called *well-conditioned problems*. A condition number $\kappa = 1$ corresponds to functions that, in the proximity of the optimum, have a symmetry in the curvature in all the dimensions: whatever direction we move from x^*, the gradient modulus increases at a similar rate. As a result, the gradient at any point x near x^*, has an angle close to $0°$ with the line between x and x^*. That is, the gradient vector heads to the optimum.
- When $M \gg m$ (or $\kappa \to \infty$), we have $\beta \to 1$ and the progress towards the optimum can be very slow. These are called, *ill-conditioned problems*. Large κ values mean that in the proximity of the optimum, f is almost flat in some directions (eigenvectors associated to m), and curves very steeply in other directions (eigenvectors associated to M). Graphically, the contours of the objective function are quite asymmetrically elongated ellipsoids. As a result, depending on the point x, the gradient $\nabla f(x)$ can make an angle of almost $90°$ with the line connecting x and x^*, making little progresses in the direction towards the optimum. Visually,

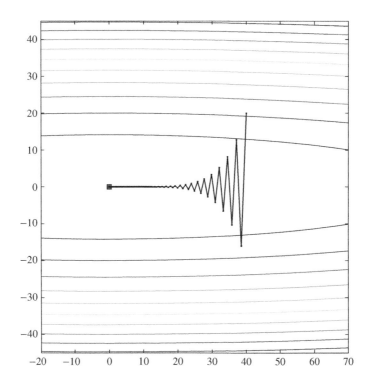

Figure 8.2 Basic gradient algorithm iterations for $\min f(x)$, with $f(x) = x_1^2 + 50x_2^2$, $x(0) = (40, 20)$. Lipschitz constant $K = 100$, $\gamma = 0.9\frac{2}{K} = 0.018$. Condition number of $\nabla^2 f(x)$ is $\kappa = 50$. The method needs 290 iterations to arrive to a distance of 10^{-3} units of the optimum $x^* = (0, 0)$

the method jumps making unproductive zigzags in successive iterations, as illustrated in Fig. 8.2.

- In the limit, an infinite condition number $\kappa = \infty$ is obtained when:
  - Objective function f is smooth ($M < \infty$) and convex, but not strongly convex, so it is flat in some directions ($m = 0$). In these cases, linear convergence rates are no longer attained. As an example, the constant γ gradient method provides a sublinear convergence rate, such that the number of iterations needed to make $f(x(t)) - f(x^*) \le \epsilon$ is $\mathcal{O}(1/\epsilon)$ [2].
  - The objective function is not differentiable ($M = \infty$), as occurs in the dual function of linear programs, or any program for which strong duality does not hold. In these cases, gradients must be replaced by subgradients and convergence is guaranteed only *to the proximity of the optimum* for constant γ, as will be shown in Section 8.5.

8.4 Asynchronous and Distributed Algorithm Implementations

In this section we concentrate on the case when the gradient projection algorithm implementation is *distributed*, in the sense that it is executed in parallel by a set of n *agents*, each agent handling a coordinate of the vector x. Also, we assume that the feasibility set is box-like (8.4) and thus separable for each coordinate[3]. Then, the update process for each agent i handling the i-th coordinate, in the basic gradient method is:

$$x_i(t+1) = \left[x_i(t) - \gamma \frac{\partial f}{\partial x_i}(x(t)) \right]_{l_i}^{u_i}, \quad t = 0, 1, \dots, i = 1, \dots, n \tag{8.11}$$

where $[x]_{l_i}^{u_i}$ is the projection of x in the interval $[l_i, u_i]$ (8.5). The gradient scheme (8.11) is synchronous, in the sense that:

- The time evolution is discrete $t = 0, 1, \dots$ In an iteration, all the agents operate at the same time, simultaneously computing their coordinate $x_i(t+1)$.
- The result of each agent i, have to be signaled to the rest of the agents before the next iteration starts. This is because each agent needs to know the full $x(t)$ vector to compute $\frac{\partial f}{\partial x_i}(x(t))$ in (8.11).

We now describe a distributed partially asynchronous version of the method (8.11), applying the partially asynchronous parallel implementation described in [5]. We assume that there is a set of times $\mathcal{T} = \{0, 1, \dots\}$ at which one or more components x_i are updated, and define:

$$\mathcal{T}^i = \{\text{Set of times when } x_i \text{ is updated}\} \subset \mathcal{T}$$

The agent updating x_i may not have access to the most recent values of the components of x. Thus, we assume that agent i, at time t has a view of vector x given by $x_i^{old}(t)$, composed of potentially outdated components from different past times:

$$x_i^{old}(t) = (x_1(\tau_1^i(t)), \dots, x_n(\tau_n^i(t))), \forall t \in \mathcal{T}^i$$

Where $\tau_j^i(t)$ is the creation time of the information about coordinate j that agent i will use at a simultaneous or later time t. The difference $(t - \tau_j^i(t))$ between the current time and the time $\tau_j^i(t)$ when the j-th component is available at the agent updating $x_i(t)$ can be viewed as a form of communication delay. At all times $t \notin \mathcal{T}^i$, x_i is left unchanged, so the time evolution of the basic partially asynchronous gradient projection algorithm is controlled by:

$$x_i(t+1) = x_i(t), \forall t \notin \mathcal{T}^i \tag{8.12a}$$

$$x_i(t+1) = \left[x_i(t) - \gamma \frac{\partial f}{\partial x_i}(x_i^{old}(t)) \right]_{l_i}^{u_i}, \forall t \in \mathcal{T}^i \tag{8.12b}$$

Any particular choice of $\mathcal{T}^i$ and $\tau_j^i(t)$ values is called an *scenario*. For any fixed scenario, the values of $x(t), t > 0$ are uniquely determined by the initial conditions. We make the *partial asynchronous assumptions*, which means that there exists a positive integer B such that:

1. For every i and for every $t \geq 0$, at least one of the elements of the set $\{t, t+1, \dots, t+B-1\}$ belongs to $\mathcal{T}^i$.

---

[3] This can be generalized in the case when each agent handles a block of coordinates, with potentially more than one coordinate per block. In this case, the feasibility set should be separable per blocks.

2. There holds

$$t - B < \tau_j^i(t) \leq t$$

for all i and j, and all $t \geq 0$ belonging to $\mathcal{T}^i$

Note that according to previous assumptions, variable t does not have to be explicitly known to the agents. Actually, t does not necessarily correspond with real time and is just an artificial variable used to order the sequence of events.

Partially asynchronous assumptions mean that each agent performs an update at least once during any time interval of length B and the information used by any agent coming from any other agent is outdated by at most B time units. Then, the value of $x(t+1)$ depends only on $x(t), x(t-1), \ldots, x(t-B+1)$, and not on any earlier information. This such general assumptions can be used to model situations like the loss or out-of-sequence delivery of signaling messages to the agents, as long as old information is purged from the system after at most B time units.

The following proposition ([5], p. 529), establishes the convergence of the basic gradient projection algorithm in a distributed partially asynchronous implementation.

Proposition 8.4 In problem $\min_{x \in \mathcal{X}} f(x)$, we assume that $\mathcal{X}$ is a non-empty, box-like set (8.4), f is convex, continuously differentiable and bounded below in $\mathcal{X}$, and ∇f is Lipschitz continuous with constant K_1. Then, the distributed (partially asynchronous) basic gradient projection scheme (8.12) converges to the global optimum for $0 < \gamma < \frac{1}{K_1(1+B+nB)}$.

The next proposition ([5], p. 529) says that the diagonally scaled version of the algorithm for box-like constraints also enjoys convergence guarantees in the partially asynchronous case.

Proposition 8.5 Under the conditions of Prop. 8.4, the distributed (partially asynchronous) version of the diagonal scaled gradient projection scheme:

$$x(t+1) = P_{\mathcal{X}}(x(t) - \gamma B(t)^{-1} \nabla f(x(t)))$$

where $B(t)$ are diagonal matrices bounded above, and the minimum diagonal term of $B(t)$ along t is $K_2 > 0$, converges to the global optimum for $0 < \gamma < \frac{K_2}{K_1(1+B+nB)}$. In particular, if f is twice differentiable and strongly convex, the method converges for the scaling matrices (8.9).

The following example illustrates how the signaling delay B can make a distributed implementation of a gradient algorithm not converge, if γ values are not correctly dimensioned.

Example 8.3 Three traffic sources share a common link. The traffic volume generated by each source, denoted as $x_i, i = 1, 2, 3$, is optimized by the problem $\max_{x \geq 0} \sum_i x_i - 10(\sum_i x_i - 1)^2$, which intends to maximize the throughput, but adding a quadratic penalization term which tends to zero if the total traversing traffic equals to one. The optimum solution can be obtained by solving KKT conditions: $x_1 = x_2 = x_3 = 0.35$. A basic gradient projection algorithm for the problem is given by:

$$x_i(t+1) = \left[x_i(t) - \gamma \left(1 - 2 \left(\sum_i x_i^{old}(t) - 1 \right) \right) \right]_0$$

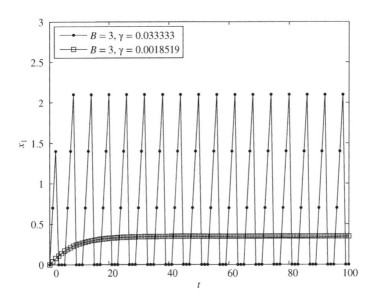

Figure 8.3   Convergence using delayed information, Example 8.3

We see that each source needs to know the sum of the traffic in the link $\left(\sum_i x_i\right)$ to perform the gradient iteration, but not the full individual x_i values. We consider an asynchronous implementation, where the $\sum_i x_i$ values are communicated to the links once every B' time slots, at times $t = B't'$, $t' = 0, 1, \ldots,$ and $x_i^{old}(t)$ vectors are constructed accordingly. In the synchronous case ($B' = 1$), convergence is guaranteed for $\gamma < \frac{2}{60} = 0.0333$, since the largest eigenvalue of the hessian matrix of the objective function is 60. Figure 8.3 plots the time evolution of source x_1 when $B' = 3$. We see that if the step γ computed for synchronous gradient is used, the algorithm does not converge. In its turn, convergence is obtained if $\gamma = \frac{1}{60(1+(B'-1)+n(B'-1))} = 0.00185$. Actually, it can be shown that convergence occurs also for some higher γ values.

Note that the γ values that provide sufficient convergence guarantees are quite restrictive, and approximately decrease proportionally with $n + 1$ and B. In [5], the reader is warned that these step values are fairly conservative and should not be taken at face value. Then, some analysis are provided that can yield to less restrictive sufficient conditions. Also, specific analysis tuned to particular problem instances can provide tighter convergence conditions.

8.5   Non-Smooth Functions

The application of gradient methods to non-smooth convex functions means using subgradients as search directions for those points when a gradient does not exist. This will be applied in this book in the maximization of the dual function of, for example, linear problems, for which the dual function is non-everywhere differentiable.

Let f be a convex non-everywhere differentiable function, and $\mathcal{X}$ a non-empty closed and convex set. The basic projected subgradient method for problem $\min_{x \in \mathcal{X}} f(x)$ is given by:

$$x(t+1) = P_{\mathcal{X}}(x(t) - \gamma(t)s(x(t))) \tag{8.13}$$

where $s(x(t))$ is a subgradient of the objective function f at point $x(t)$. Note that if f is differentiable, method (8.13) is the basic gradient projection. The convergence of the subgradient projection cannot be established by guaranteeing that in every step the function decreases, since in some cases a subgradient is an ascent direction (function f increases for any $\gamma(t)$ in (8.13)). Instead, convergence is established by showing that the distance $\|x(t) - x^*\|$ between $x(t)$ and an optimum solution x^* decreases with t.

Proposition 8.6 In problem (8.1), we assume that $\mathcal{X}$ is a non-empty, closed convex set, f is convex but may be non-everywhere differentable, and bounded below in $\mathcal{X}$. Subgradients $s(x) \in \partial f(x)$ are bounded:

$$\|s(x)\|_2 \leq S, \quad \forall x \in \mathcal{X}$$

Then, the following inequality holds for the basic subgradient projection iteration (8.13):

$$f_{best}(t) - f(x^*) \leq \frac{R^2}{2\sum_t \gamma(t)} + \frac{S^2 \sum_t \gamma(t)^2}{2\sum_t \gamma(t)} \tag{8.14}$$

where $f_{best}(t)$ is the best function cost found so far, until iteration t, and R is a bound to the distance from algorithm initial point to the optimum:

$$R \geq \|x(0) - x^*\|$$

Inequality (8.14) helps us to show that subgradient algorithms converge to the optimum if $\gamma(t)$ values satisfy:

$$\gamma(t) > 0, \quad \lim_{t \to \infty} \gamma(t) = 0, \quad \sum_{t=0}^{\infty} \gamma(t) = \infty$$

For instance, the step rules $\gamma(t) = 1/t$ and $\gamma(t) = 1/\sqrt{t}$ satisfy previous assumptions. However, for constant step sizes $\gamma(t) = \gamma$, the only guarantee is that iterations reach and stay in the *proximity* of the optimum, closer the smaller γ is. Indeed, the first sum in (8.14) vanishes when $t \to \infty$, but not the second, and we have:

$$\lim_{t \to \infty} f_{best}(t) - f(x^*) \leq \frac{S^2 \gamma}{2} \tag{8.15}$$

The basic subgradient method inherits the simplicity of the basic gradient iteration, extending its applicability. It works well for many type of problems, however, it can also work very badly in ill-conditioned instances, as happens in the gradient version. There are several variations and scaling methods to accelerate its convergence, although its applicability in a distributed scenario is a somewhat less investigated topic than the gradient counterpart, and we will not apply them in this book. The interested reader is referred to, for example, [2] or [7] for a more detailed view.

8.6 Stochastic Gradient Methods

In this section we describe the stochastic gradient method, consisting of applying the standard basic gradient projection to a problem where the gradients have measurement errors. This situation is of interest, since in many network algorithms gradient estimations come from observations and measures like, for example, average delays or loss probabilities that can be affected by inaccuracies.

The *stochastic (sub)gradient descent* for the problem $\min_{x \in \mathcal{X}} f(x)$ is an iteration of the form:

$$x(t+1) = P_{\mathcal{X}}(x_t - \gamma(t)\tilde{s}(x(t))) \tag{8.16}$$

where $P_{\mathcal{X}}$ stands for the orthogonal projection and $\tilde{s}(x(t))$ is an unbiased estimation of the subgradient of function f in point $x(t)$, sometimes referred to as a *noisy unbiased subgradient*. This means that $\tilde{s}(x(t))$ can be written as $s(x(t)) + w(t)$, where $s(x(t))$ is a subgradient of f in $x(t)$ and $w(t)$ is a random variable with zero mean, representing a sort of measuring error in the subgradient.

The following proposition from [8] is useful to provide sufficient convergence guarantees in some cases of interest.

Proposition 8.7 In problem (8.1), we assume that $\mathcal{X}$ is a non-empty, closed convex set, f is convex but may be non-everywhere differentable, and bounded below in $\mathcal{X}$. Noisy unbiased subgradients $\tilde{s}(x)$ can be obtained for any point $x \in \mathcal{X}$, meaning that $\mathbb{E}(\tilde{s}(x)) \in \partial f(x)$. Noisy subgradients have a finite variance:

$$\mathbb{E}(\|\tilde{s}(x)\|_2^2) \leq S^2, \quad \forall x \in \mathcal{X}$$

Then, the following inequality holds ("almost surely" should be understood here) for the basic stochastic subgradient projection iteration (8.16):

$$\mathbb{E}(f_{best}(t) - f(x^*)) \leq \frac{R^2}{2\sum_t \gamma(t)} + \frac{S^2 \sum_t \gamma(t)^2}{2\sum_t \gamma(t)} \tag{8.17}$$

where $f_{best}(t)$ is the best function cost found so far, until iteration t, and R is a bound to the distance from algorithm initial point to the optimum:

$$R \geq \|x(0) - x^*\|$$

Previous proposition establishes a convergence in expectation version of Prop. 8.6, under reasonable assumptions like unbiased and finite variance measurement errors. Analogous to the deterministic non-smooth case:

- Convergence (in expectation) occurs for non-summable and diminishing $\gamma(t)$.

- For constant step $\gamma(t) = \gamma$, convergence is guaranteed only to the proximity of the optimum:

$$\lim_{t \to \infty} \mathbb{E}(f_{best}(t) - f(x^*)) \leq \frac{S^2 \gamma}{2} \tag{8.18}$$

Next proposition states that for the smooth stochastic constant step gradient case, convergence (in expectation) can only be guaranteed to the *proximity* of the optimum. This is a difference with the deterministic iteration, which could in the smooth case converge to the optimum for a constant, but sufficiently small γ.

Proposition 8.8 ([9], Prop. 5) In problem (8.1), we assume that $\mathcal{X}$ is a non-empty, closed convex set, f is convex and differentiable with Lipschitz gradient with constant $K > 0$, strongly convex with constant $m > 0^4$. Let $\gamma \in \left(0, \frac{2}{K}\right)$, in the iteration (8.16). Then, we have

$$\mathbb{E}(\|x(t) - x^*\|^2) \leq q(\gamma)^t R^2 + \frac{1 - q(\gamma)^t}{1 - q(\gamma)} \gamma^2 S^2 \tag{8.19}$$

where R is an upper bound to the initial distance to the optimum $(R \geq \|x(0) - x^*\|)$, $q(\gamma) = 1 - m\gamma(2 - \gamma K) < 1$ and

$$\mathbb{E}(\|\tilde{s}(x)\|_2^2) \leq S^2, \forall x \in \mathcal{X}$$

The first sum in (8.19) is a transient error that vanishes as $t \to \infty$. However, the second sum is a *persistent error*, invariant to increasing the number of iterations, and thus would necessitate of a diminishing γ for eliminating it.

From a practical point of view, several observations and recommendations can be considered when dealing with stochastic gradients:

- Stochastic gradient methods are slower, and sometimes much slower in practice, than their deterministic counterparts. As an example, error drops like $\mathbb{E}(f(x(t) - f(x^*)) = \mathcal{O}(1/t)$ have been reported for differentiable and strongly convex functions. This is a logical consequence of the measurement error. The smaller the measurement noise (S), the better the convergence can be. Averaging techniques that take multiple samples of the gradient before iterating using its average as a gradient, can have a positive impact in convergence, helping to reduce both the zigzagging effect and the error variance[5].
- Scaling techniques has also been investigated to improve the convergence rate of stochastic gradient algorithms (e.g., see [10]). However, as pointed out in [11], the scaling does not reduce stochastic noise and therefore does not significantly reduce the noise variance. Then, in those cases when noise variance is high, scaling methods may be ineffective for improving convergence in practice.
- Stopping criteria should be adapted to the stochastic nature of the gradient. For instance, even when the optimum is reached, the algorithm still iterates randomly around the optimum and never gets stuck in it, challenging the detection of algorithm termination.

---

[4] This holds if f is twice differentiable, and all the eigenvalues of any hessian matrix are greater or equal than m.
[5] For instance, note that the average of n IID random variables of variance σ^2, has a variance of σ^2/n.

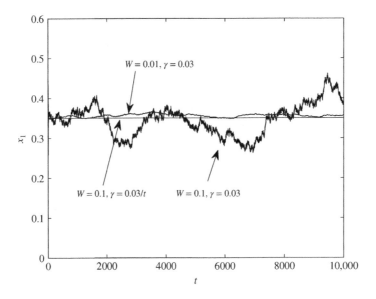

Figure 8.4   Example 8.4: Stochastic gradient realization

Example 8.4 In the gradient algorithm implementation of Example 8.3, let us assume that the link traversing traffic information $\sum_i x_i$ is signaled to each source without any delay, but each with a zero mean random error with uniform distribution $[-W, W]$. Figure 8.4 illustrates the evolution of the traffic of source one (x_1) in a realization of the stochastic algorithm for $W = \{0.01, 0.1\}$ and constant step size $\gamma = 0.03$, and $W = 0.1$ but diminishing step size $\gamma(t) = 0.03/t$. We see how in the constant step size, higher errors yield to a less tight convergence around the optimum, while diminishing γ can always converge reasonably fast to the optimum, without leaving it.

8.7   Stopping Criteria

In many situations in network optimization, gradient algorithm variants are used as guidelines for the design of, for example congestion control protocols, routing protocols, dynamic transmission power, or scheduling allocations. These protocols are running permanently, adapting the network to varying conditions like traffic or channel condition fluctuations. In such cases, there is no algorithm termination and thus no need to determine when the algorithm should be stopped.

In other situations, gradient algorithms are used for offline network design and are typically executed in a centralized equipment. For instance, as we will see in Chapter 11, gradient algorithms are used to maximize the dual function in some network design problem decompositions. In these cases, algorithms are fed with input data, run, and need to be stopped when the optimum solution is reached. Several stopping conditions can be used:

- *Gradient vanishing* (unconstrained optimization): In unconstrained convex problems, $\nabla f(x) = 0$ is an optimality condition, and $\|\nabla f(x)\| < \epsilon$ can be a reasonable termination.

If f is twice differentiable and K is the maximum eigenvalue of the hessian matrix in $\mathbb{R}^n$, we have that:

$$\|\nabla f(x)\| < \epsilon \Rightarrow \|x - x^*\| \leq \frac{\epsilon}{K}, f(x) - f(x^*) \leq \frac{\epsilon^2}{K}$$

where x^* is an optimum solution. Note that this stop criteria cannot be used in constrained optimization, since gradients can be arbitrarily large when the optimum is in the boundary of the constraint set (e.g., see Appendix B).

- *Comparison with cost lower bounds*: Stop when the best solution found so far f_{best} is close enough to a known lower bound: $f_{best} - L \leq \epsilon$. This lower bound can be precomputed or generated during the algorithm as a by-product. This is the case of primal-dual algorithms that generate in each iteration a primal and dual solution, so that evaluating the dual function produces a lower bound to the optimal cost. In these cases, a natural stopping criteria is that $f_{best} - L_{best} \leq \epsilon$, where L_{best} is the highest lower bound found so far. When the original problem enjoys strong duality, the algorithm can converge to a situation when $L_{best} = f_{best}$, certifying that the optimum has been reached. If strong duality does not hold, the condition $f_{best} - L_{best} \leq \epsilon$ is never met, even if the optimum solution is reached, when ϵ is greater than the duality gap.

- *Other*: Popular stop conditions in general optimization algorithms can also be applied, like stopping when the best function cost found so far f_{best} is not improved during sufficiently large amount of iterations, or when a maximum number of iterations or execution time is reached.

8.8 Algorithm Design Hints

As we will see in many examples throughout this book, gradient projection algorithms provide theoretical insight and guidelines to devise network algorithms, for example distributed protocols that dynamically adapt the routing, control the congestion, or the assignment of capacity to the network links. Theoretical convergence guarantees established for gradient algorithms, even in the presence of asynchronous distributed updates subject to delays, losses, or noisy gradient estimations, support this application.

Still, multiple difficulties can appear if we want to directly apply a gradient iteration into a real network protocol. We elaborate on some of them in the following subsections. Application examples of these techniques are scattered throughout in different case studies in the next chapters.

8.8.1 *Dimensioning the Step Size*

As we have seen in this chapter, convergence conditions in basic and scaled gradient projection algorithms depend on dimensioning the γ step and scaling matrices (if any) according to quantities like the largest and smallest eigenvalues of the hessian matrix, or upper bounds to the euclidean norm of the gradients/subgradients. In many network design problems, these quantities are unknown for the nodes implementing the gradient iterations. For instance, when applied to a congestion control problem they may require a demand source node to know the number of flows sharing the traversed links, the number of links in the longest path of other demands, or the utility functions employed. Using worst-case values for these quantities can yield to ridiculously small γ factors that make the algorithms too slow to be useful.

An alternative to this is fixing the γ step such that convergence is reached in normal or reasonable network conditions, according to some empirical tests. Then, extra techniques should be added to alleviate or attenuate the non-convergent behaviors if they occur. One of these techniques is limiting the maximum variation in a coordinate to a quantity Δx, a technique we call *bounded step length*.

Let $\min_{x \in \mathcal{X}} f(x)$ be the problem to solve, where f is convex and $\mathcal{X}$ is box-like (8.4). The standard basic gradient iteration with bounded step length is given by:

$$
x_i(t+1) = \begin{cases} \left[x_i(t) - \gamma \frac{\partial f}{\partial x_i}(t) \right]_{l_i}^{u_i} & \text{if } \gamma \frac{\partial f}{\partial x_i}(t) \leq \Delta x \\ [x_i(t) - \Delta x]_{l_i}^{u_i} & \text{otherwise} \end{cases}
$$

Then, the idea is that, if quantity $\gamma \frac{\partial f}{\partial x_i}(t)$ is above the maximum step length Δx, we replace it by Δx. As a result, each iteration can change the value of a coordinate in at most Δx units. Since Δx is measured in the same units as the problem variables, it may be easier to find a reasonable value to it within the problem context. For instance, in a capacity allocation algorithm, we may not want to change a link capacity in more than Δx Kbps in each iteration.

Applying Prop. 8.5 it is possible to show that bounded step length technique converges to the optimum solution for a sufficiently small γ (see Exercise 8.5). Exercise 8.6 also shows a counter-example of how the algorithm may not converge if constraints are not box-like.

8.8.2   Discrete Step Length

In multiple practical contexts, the update in the x_i coordinates should be performed in discrete steps, instead than on a continuum. For instance, if the optimization variable x has a coordinate per network link representing the link capacity, a typical technological constraint restricts the capacity adjustment to multiples of a fixed capacity slot.

Discretization of the step length can be modeled using the iteration:

$$
x_i(t+1) = \left[x_i(t) - f_\delta \left(\gamma_i(t) \frac{\partial f}{\partial x_i}(t) \right) \right]_{l_i}^{u_i} \tag{8.20}
$$

where $f_\delta(y)$ is a discretization function and we call δ the discretization step. The $\gamma_i(t)$ value is used to include a possible diagonal scaling. A suitable choice of f_δ is using the closest multiple of δ, with an absolute value lower or equal than y:

$$
f_\delta(y) = \text{sign}(y)\delta \left\lfloor \frac{|y|}{\delta} \right\rfloor \tag{8.21}
$$

By doing so, the convergence properties are:

- The convergence can be guaranteed to the *proximity of the optimum* for a sufficiently small γ factor. This behavior comes intuitively seeing that the discretized version always makes *smaller or equal steps* than the continuous version, which converges for a sufficiently small

γ step. If γ is not appropriately dimensioned, the algorithm can suffer from oscillations around the optimum, or simply diverge, as in the continuous version.

- If the discretization function is not based on the floor function $\lfloor y \rfloor$, but the ceiling function $\lceil y \rceil$ that rounds to the upper closest multiple of δ, the algorithm can still converge to the proximity of the optimum, but then *oscillations can occur* for any γ step we pick. The reason is that discretized jumps may be *longer* now than in the continuous version and the ratio $\lceil y \rceil / y$ can be arbitrarily large. Then, in the proximity of the optimum, these jumps may be non-improving, whatever the γ value is, and be followed by later jumps improving the solution, and repeat again. In Exercise 8.7 we show an example of this situation. If the function f_δ rounds to the closest multiple of δ (a sort of midway between floor and ceil functions), convergence to the proximity without oscillations can again we obtained, but requiring half the γ step than the floor function.

Example 8.5 We use a discretized gradient algorithm with step $\delta = 1$, for solving the problem $\min 10x^2$. The γ step is small enough to converge to the optimum in the continuous case (for example, $\gamma = \frac{1}{20}$). Then, if the initial solution is a fractional number, for example, $x(0) = 10.5$, the algorithm would never reach the optimum solution $x^* = 0$, but get stuck in the solution $x = 0.5$.

8.8.3 Heavy-Ball Methods

Heavy-ball methods are the most simple among the so-called multistep optimization techniques, which are those where the determination of next point $x(t + 1)$ depends not only on information regarding $x(t)$ (gradient, hessian at this point), but on some preceding steps $x(t - 1), x(t - 2), \ldots$.

The basic gradient projection two-step heavy-ball method, for a problem $\min_{x \in \mathcal{X}} f(x)$, takes the form:

$$x(t + 1) = P_{\mathcal{X}}(x(t) - \gamma \nabla f(x(t)) + \gamma_h(x(t) - x(t - 1))) \tag{8.22}$$

where $\gamma > 0$, and $\gamma_h \geq 0$ are parameters. When $\gamma_h = 0$, method (8.22) turns into the basic gradient projection.

The heavy-ball method owes its name to a physical analogy with the inertia suffered by a heavy ball that tends to continue in the same direction when moving. This is the effect of term $\gamma_h(x(t) - x(t - 1))$, which sums to the current gradient, the past search direction. Inertia term may increase convergence and reduce the zigzagging. This is because (i) the components of the gradient that make zigzagging in consecutive iterations tend to cancel, (ii) while the components of the search direction that do not zigzag, which may lead to the optimum, sum constructively. Proposition 8.9 characterizes the convergence benefits of heavy-ball methods in the unconstrained case.

Proposition 8.9 Let f be a convex function bounded below in $\mathbb{R}^n$, twice differentiable and strongly convex, with minimum x^*, and m and M the minimum and maximum eigenvalues of $\nabla^2 f(x^*)$. Then, if $0 \leq \gamma_h < 1$, and $0 < \gamma < \frac{2(1+\gamma_h)}{M}$, the two-step heavy-ball scheme (8.22) converges in the unconstrained case to x^*, for any initial point $x(0)$. The convergence rate is geometric. The best geometric progressing rate $\beta = \frac{\sqrt{M} - \sqrt{m}}{\sqrt{M} + \sqrt{m}}$, is obtained when γ and γ_h is

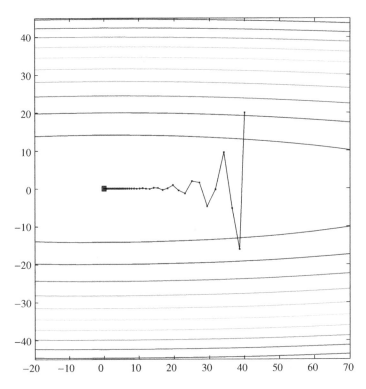

Figure 8.5 Application of the heavy-ball method with $\beta = 0.5$, to same example as Fig, 8.2 (same γ step also). The method needs 130 iterations (instead of 290) to arrive at a distance of 10^{-3} units of the optimum $x^* = (0,0)$

chosen to be:

$$\gamma = \frac{4}{(\sqrt{M} + \sqrt{m})^2}, \gamma_h = \left(\frac{\sqrt{M} - \sqrt{m}}{\sqrt{M} + \sqrt{m}}\right)^2$$

The convergence rate factor in the heavy-ball method $\beta = (\sqrt{M} - \sqrt{m})/(\sqrt{M} + \sqrt{m})$, shown in Prop. 8.9, is lower (better) than the factor $\beta = (M - m)/(M + m)$ in the basic gradient case (Prop. 8.1). Also, convergence is guaranteed for higher γ values ($\gamma < 2(1 + \gamma_h)/M$). This suggests that heavy-ball variations are effective techniques to apply to gradient schemes. Figure 8.5 illustrates this with an an example. However, it has been shown in [6] that heavy-ball methods are relatively less effective under noisy gradients than standard gradient methods, since (intuitively) the use of old information tends to increase the uncertainty region. Then, its use when gradient noise is significant is discouraged.

Aside of the potential convergence benefits, our interest in heavy-ball methods is in the simplicity. In particular, if β parameters are constants fixed in advance, applying the heavy-ball principle does not hinder a distributed implementation of a gradient method: the same node or agent computing a coordinate of the solution vector, just needs to store it to be used in future $s - 1$ iterations, but does not need to signal any new information to other nodes.

8.9 Notes and Sources

The basic steepest descent gradient algorithm dates to Louis Augustin Cauchy in his *Compte Rendu à l'Académie des Sciences* of October 18, 1847, who attributes to Newton the origins of what we call today the Newton's method. The modern theory of gradient and subgradient methods evolved from the 1960s.

Many good sources exist that cover gradient and subgradient algorithms in a more or less exhaustive form, the material used is mostly from [1–6]. The reader is referred to those sources for a comprehensive view of the topic, outside the scope of this chapter.

The description of the partially asynchronous operation that can accommodate gradient algorithms where different agents interact asynchronously using potentially delayed or out-of-order signaling messages and the convergence properties in this context, are extracted from [5]. Results on subgradient algorithms has been collected from [6, 7, 12]. The effect of using noisy gradient/subgradient observations in the method convergence is presented using material from [8–11, 13]. The effect of discretization steps in convergence is derived from some properties in [5]. Results on heavy-ball methods are from [6].

Finally note that this and the next three chapters are based on the application of gradient projection algorithms to network problems. Gradient projection schemes have been chosen for various reasons. First, they are the customary approach in the network optimization literature, since their simplicity has a didactic value and ease distributed implementations. Also, strong convergence results exist under asynchronous executions and with random gradient observations. As a recent alternative, *proximal algorithms* are a generalization of gradient projection algorithms, based on solving a so-called proximal operator of a function. These methods have received a significant attention in recent years, due to their potential for solving large-scale problems in a distributed form, and their convergence properties in noisy and asynchronous environments, suitable for practical network implementations, have been explored. The reader is referred to excellent surveying sources in the topic like [14, 15], the lecture notes [16], and the monograph [17].

8.10 Exercises

8.1 Compute the Lipschitz constant of the gradient of the objective function of the NUM modeling of the congestion control problem (6.1) in Chapter 9, considering α-fair utility functions for the demands. Repeat the exercise for the case when each demand is constrained to receive a minimum of two units of traffic.

8.2 Repeat the previous exercise, assuming the utility functions come from TCP Reno and TCP Vegas sources.

8.3 Compute the Lipschitz constant of the gradient of the objective function of problem (5.16) in Chapter 5, given by:

$$f(u_e, e \in \mathcal{E}) = \sum_e \frac{(e^{u_e})^{1-\alpha}}{1-\alpha}$$

for $\alpha > 1$.

8.4   Find the optimum solution of the problem:

$$\min_{x_1, x_2} \quad 2(x_1 - 1)^2 + (x_2 - 1)^2, \text{ subject to: } x_1 + x_2 \leq 1$$

Show that a diagonally scaled gradient projection algorithm, using the hessian matrix of the objective function as the scaling matrix, does not converge for any constant step $\gamma > 0$. In particular, show that if the algorithm hits the optimum in an iteration, it jumps out of it in the next one. Indicate a set of γ steps that guarantee convergence in the basic projection case, without diagonally scaling.

8.5   Apply Proposition 8.5 to show that bounded step length technique in the basic projection algorithm converges to the optimum solution for sufficiently small γ, in the case when constraints are box-like.

8.6   Find the optimum solution of the problem:

$$\max_{x \geq 0} 10x_1 + x_2, \text{ subject to } x_1 + x_2 \leq 1$$

Consider the application of a basic gradient projection scheme with the bounded step length technique. Show the ranges of $\gamma > 0$ for which the method gets stuck in the suboptimum point $(0.5, 0.5)$ if reached.

8.7   We use a gradient algorithm (8.20) with discretized step $\delta = 1$, for solving a problem $\min x^2$. The discretization function used is:

$$f_\delta(y) = \text{sign}(y)\delta \lceil \frac{|y|}{\delta} \rceil$$

Show that, if algorithm initial solution is $x(0) = 2.5$, the algorithm enters in a persistent cycle $\{0.5, -0.5, 0.5, -0.5, \ldots\}$ for any small γ step.

References

[1] M. Minoux, *Mathematical programming: theory and algorithms*, ser. Wiley-Interscience series in discrete mathematics and optimization. Wiley, 1986.

[2] D. P. Bertsekas, *Nonlinear Programming*. Bertsekas: Athena Scientific, 1999.

[3] L. Lasdon, *Optimization theory for large systems*, ser. Dover books on Mathematics. Dover Publications, 2002.

[4] S. Boyd and L. Vandenberghe, *Convex Optimization*. New York, NY, USA: Cambridge University Press, 2004.

[5] D. P. Bertsekas and J. N. Tsitsiklis, *Parallel and Distributed Computation: Numerical Methods*. Athena Scientific, 1997.

[6] B. T. Polyak, *Introduction to Optimization*. Optimization Software New York, 1987.

[7] A. Nedić, "Subgradient methods for convex minimization," Ph.D. dissertation, Massachusetts Institute of Technology, Department of Electrical Engineering and Computer Science, 2002.

[8] A. Nemirovski, A. Juditsky, G. Lan, and A. Shapiro, "Robust stochastic approximation approach to stochastic programming," *SIAM Journal on Optimization*, vol. 19, no. 4, pp. 1574–1609, 2009.

[9] F. Yousefian, A. Nedić, and U. V. Shanbhag, "On stochastic gradient and subgradient methods with adaptive steplength sequences," *Automatica*, vol. 48, no. 1, pp. 56–67, 2012.

[10] D. P. Bertsekas and J. N. Tsitsiklis, "Gradient convergence in gradient methods with errors," *SIAM Journal on Optimization*, vol. 10, no. 3, pp. 627–642, 2000.

[11] L. Bottou, "Stochastic gradient descent tricks," in *Neural Networks: Tricks of the Trade*. Springer, 2012, pp. 421–436.

[12] S. Boyd and A. Mutapcic, "Subgradient methods," *Lecture notes of EE364b, Stanford University, Winter Quarter*, vol. 2007, 2006.

[13] S. Boyd and A. Mutapcic, "Stochastic subgradient methods. lecture notes," 2007.

[14] B. Lemaire, "The proximal algorithm," *International series of numerical mathematics*, vol. 87, pp. 73–87, 1989.

[15] A. Iusem, "Augmented lagrangian methods and proximal point methods for convex optimization," *Investigación Operativa*, vol. 8, no. 11–49, p. 7, 1999.

[16] L. Vandenberghe, "Optimization methods for large-scale systems," *Lecture Notes*, 2010.

[17] N. Parikh and S. Boyd, "Proximal algorithms," *Foundations and Trends in optimization*, vol. 1, no. 3, pp. 123–231, 2013.

9

Primal Gradient Algorithms

9.1 Introduction

In this chapter we describe a comprehensive set of case studies where gradient projection algorithms are applied to the *primal* of network design problems. The outcomes are solution methods that provide theoretical support to the design of distributed protocols that optimize resource allocations.

We focus on problems of the form $\min_{x \in \mathcal{X}} f(x)$, for which the basic gradient projection iterations are:

$$x(t + 1) = P_{\mathcal{X}}(x(t) - \gamma(t)\nabla f(x(t))) \tag{9.1}$$

We remark that, in order for the gradient iteration (9.1) to inherit the convergence, robustness and stability properties described in Chapter 8, $\mathcal{X}$ should be a convex set and f a convex function, such that the overall problem is convex.

The major difficulty in the application of the gradient scheme (9.1) to the primal problem, is dealing with non-separable feasible sets $\mathcal{X}$, for which the projection operation $P_{\mathcal{X}}(x)$ is not easy to compute, or cannot be implemented in a distributed form. Section 9.2 describes the two main alternatives to address this issue: the utilization of interior or exterior *penalty methods* that permit removing those constraints in $\mathcal{X}$ that complicate the projection step, and sum them (with some modifications) to the objective function. Afterwards, remaining sections in the chapter describe the case studies analyzed. Table 9.1 enumerates them, indicating the optimization techniques applied.

Scattered along the case studies, we provide empirical tests illustrating the application of some of the convergence improving techniques described in Chapter 8. Also, we observe the effect of delayed and noisy information in the gradient iteration and the oscillation or divergence situations that can bring.

Optimization of Computer Networks – Modeling and Algorithms: A Hands-On Approach,
First Edition. Pablo Pavón Mariño.
© 2016 John Wiley & Sons, Ltd. Published 2016 by John Wiley & Sons, Ltd.
Companion Website: www.wiley.com/go/PavonMarinoSol16

Table 9.1   Case studies in Chapter 9

Problem type	Algorithm	Section
Adaptive bifurcated routing	Primal gradient and interior penalty	Section 9.3
Congestion control	Primal gradient and interior and exterior penalty	Section 9.4
Persistence probability adjustment	Primal gradient	Section 9.5
Power control in wireless networks	Primal gradient	Section 9.6

9.2   Penalty Methods

9.2.1   Interior Penalty Methods

Interior penalty methods, or *barrier methods*, apply to problems of the form:

$$\min_{x} f(x) \quad \text{subject to:} \tag{9.2a}$$

$$g_i(x) \leq 0, \quad i = 1, \ldots, m \tag{9.2b}$$

$$x \in \mathcal{X} \tag{9.2c}$$

where f and g_i are convex, $\mathcal{X}$ is a closed set, and the interior of the feasibility set relative to $\mathcal{X}$ is not empty:

$$S = \{x \in \mathcal{X} : g_i(x) < 0, \forall i = 1, \ldots, m\} \neq \emptyset$$

Barrier methods consist in moving the constraints $g_i(x) \leq 0$ in problem (9.2) to the objective function, inside the argument of a so-called *barrier function B(x)*, which should be continuous, bounded below, and go to ∞ as x approaches zero from negative values. The resulting optimization problem becomes:

$$\min_{x \in \mathcal{X}} f(x) + \epsilon \sum_{i=1}^{p} B(g_i(x)) \tag{9.3}$$

The role of barrier functions is penalizing with increasing cost those solutions that tend to violate the constraints. Since the barrier cost goes to infinity when any $g_i(x) \to 0$, any feasible solution of (9.3) is feasible for the original problem. That is, barrier methods guarantee feasibility. Constant $\epsilon > 0$ plays the role of weighting the importance of the barrier in the objective function. The lower the value of ϵ, the lower the effect that barrier functions have in distorting the original objective function f. This is formally stated in the following proposition.

Proposition 9.1 ([1], p. 372, [2] p. 199) Under the conditions described previously, lower $\epsilon > 0$ values produce solutions in (9.3) with lower or equal optimum costs. When ϵ tends to zero, the optimum solution of (9.3) tends to the global optimum of the original problem (9.2). In addition, the value $\epsilon \sum_{i=1}^{p} B(g_i(x))$ also tends to zero.

The most common examples of barrier functions are:

$$B(x) = -\sum_{i=1}^{p} \log(-g_i(x)), \quad \text{logarithmic barrier}$$

$$B(x) = -\sum_{i=1}^{p} \frac{1}{-g_i(x)}, \quad \text{inverse barrier}$$

The main difficulty of barrier methods is that using small ϵ values tends to produce ill-conditioned problems, since the objective function in (9.3) will grow steeply in the directions of the tight constraints and be almost constant in the other directions. Then, applying gradient methods to such modified problems can suffer from slow convergence and zigzagging. In centralized implementations, it is possible to alleviate this difficulty by solving a sequence of problems of the form (9.3) (i) with decreasing values of ϵ, such that the starting point of a minimization problem with $\epsilon(k+1)$ is the optimum solution of the previous problem with factor $\epsilon(k) > \epsilon(k+1)$ and (ii) using powerful Newton-like methods in each iteration. When the logarithmic barrier is used with these schemes, they are commonly called *interior point methods*. Interior point methods are extensively and successfully applied today in linear and convex programming.

Unfortunately, the application of barrier methods in network optimization problems where a distributed implementation is pursued, is hindered by the difficulty of varying the ϵ value during the algorithm, or using sophisticated Newton steps. Then, if a constant ϵ value is to be used, a trade-off appears between accuracy in the optimum solution (low ϵ) and convergence speed (high ϵ).

Example 9.1 The problem $\min_{x_1 \geq 1}(x_1 + x_2)^2$ has the optimum solution $x_1 = 1, x_2 = 0$. Applying a logarithmic barrier method we have that

$$\min(x_1 + x_2)^2 - \epsilon \log(x_1 - 1)$$

has an optimum solution $x_2 = 0$, and

$$x_1 = \arg\min_{x_1 > 1}(x_1 + x_2)^2 - \epsilon \log(x_1 - 1) = \frac{1 + \sqrt{1 + 2\epsilon}}{2}$$

which tends to $x_1 = 1$ as $\epsilon \to 0$.

9.2.2 Exterior Penalty Methods

Exterior penalty methods are a family of algorithms applicable to problems of the form:

$$\min_x f(x) \quad \text{subject to:} \tag{9.4a}$$

$$g_i(x) \leq 0, \quad i = 1, \dots, m \tag{9.4b}$$

where f is a continuous function and the feasibility set is closed. Note that linear equality constraints $h(x) = 0$ can be expressed in (9.4) as two inequality constraints $h(x) \leq 0, -h(x) \leq 0$.

Exterior penalty methods consist in moving the constraints $g_j(x) \leq 0$ in problem (9.4) to the objective function, inside the argument of a so-called penalty function $P(x)$, converting (9.4) into an unconstrained optimization problem:

$$\min f(x) + \mu \sum_{i=1}^{p} P(g_i(x)) \tag{9.5}$$

The role of exterior penalty functions is (i) adding a zero cost to the feasible solutions and (ii) adding a positive and large cost to those solutions which violate the constraints. When the optimum solution of the original problem is a boundary point, the solution of the penalized problem (9.5) is often unfeasible. The role of the μ parameter is adding a sufficiently high cost to the penalty, to make the optimum solutions of (9.5) only *slightly* unfeasible. For increasing values of μ, we can obtain solutions that approximate the optimum of the original problem coming from the *exterior* of the feasibility set[1]. This is formally stated under mild assumptions in the following proposition.

Proposition 9.2 ([2] p. 196) Let $P(x)$ be an exterior penalty function satisfying

- $P(x) \geq 0, \forall x \in \mathbb{R}^n$.
- $P(x) = 0 \Leftrightarrow x$ feasible.
- $P(x)$ is continous.

We also assume that f in (9.4) is continuous, the feasibility set is closed, and at least one of these conditions hold: (i) $f(x) \to \infty$, when $\|x\| \to \infty$ or (ii) the feasibility set is bounded and $P(x) \to \infty$ when $\|x\| \to \infty$. Then, when the penalty coefficient μ tends to infinity: (i) the optimum solution of (9.5) tends to be an optimum solution of the original problem (9.4), (ii) $\mu \sum_{i=1}^{m} P(g_i(x)) \to 0$ and (iii) if for any $\mu > 0$ the optimum of (9.5) is feasible, it is a global optimum.

Naturally, exterior penalty methods are useful in those cases in which some unfeasibility is allowed before the algorithm ends. We are interested in smooth penalty functions that are differentiable everywhere, so that gradient methods can be applied to the penalized problem. Typically the quadratic penalty function is used:

$$P(x) = \begin{cases} 0, & \text{if } x \leq 0 \\ x^2, & \text{if } x > 0 \end{cases} \tag{9.6}$$

The main drawback of exterior penalty methods is that the problem becomes ill-conditioned as the penalty factor μ increases. In particular, it can be shown (e.g., see [2], p. 237) that the hessian matrix of the objective function to minimize has m eigenvalues that tend to infinity as μ does. As was observed in Chapter 8, basic gradient algorithms suffer of zigzagging and slow convergence in these circumstances. In centralized implementations of penalty methods, it is possible to apply Newton-like scalings that alleviate this problem. However, this is not possible in general when a distributed implementation is pursued. Then, if a constant μ value is to be used, a trade-off appears between accuracy in the optimum solution/unfeasibility permitted (high μ) and convergence speed (low μ).

---

[1] This approach is the basis of the so-called *Sequential Unconstrained Minimization Techniques* (SUMT) methods [3].

9.3   Adaptive Bifurcated Routing

Let us consider a basic routing problem in the flow-path formulation, for a network $\mathcal{G}(\mathcal{N}, \mathcal{E})$, where $\mathcal{N}$ is the set of nodes and $\mathcal{E}$ the set of links. Offered traffic is given by a demand set $\mathcal{D}$. For each demand $d \in \mathcal{D}$, h_d denotes the offered traffic and P_d the set of candidate paths. For each path $p \in P = \bigcup_{d \in \mathcal{D}} P_d$, x_p is the amount of traffic to carry through p.

The optimal routing is given by:

$$\min_x \sum_e F_e \left(\sum_{p \in P_e} x_p \right) \quad \text{subject to:} \tag{9.7a}$$

$$\sum_{p \in P_d} x_p = h_d, \quad \forall d \in \mathcal{D} \tag{9.7b}$$

$$x_p \geq 0, \quad \forall p \in P \tag{9.7c}$$

where function F_e associates a cost to a link e, depending on the amount of traffic $y_e = \sum_{e \in P_e} x_p$ carried in it. F_e functions are convex with respect to y_e and thus also convex with respect to x_p (see Section A.2.4 in Appendix A). As an example, F_e functions can be given by a M/M/1 delay estimation ($F_e = \frac{1}{u_e - y_e}$) that assigns an increasing convex cost with y_e, which tends to infinity as the traffic in the links approaches the link capacity u_e and is infinite when $y_e > u_e$. By doing so, F_e acts as a barrier function and link capacity constraints are enforced penalizing those solutions that violate it.

The gradient of f, the objective function (9.7a), is given by:

$$\frac{\partial f}{\partial x_p}(x) = \sum_{e \in p} \frac{\partial F_e}{\partial y_e}(x), \quad \forall p \in P$$

where we apply that:

$$\frac{\partial F_e}{\partial x_p}(x) = \begin{cases} \frac{\partial F_e}{\partial y_e}, & \text{if } e \in p \\ 0, & \text{otherwise} \end{cases}, \quad \forall e \in \mathcal{E}$$

The value $\frac{\partial F_e}{\partial y_e}$ can be seen as a weight assigned to each link. Then, by applying a basic gradient projection algorithm to (9.7) directly, we have the iteration:

$$x_p(t + 1) = P_{\mathcal{X}_d}\left(x_p(t) - \gamma \left(\sum_{e \in p} \frac{\partial F_e}{\partial y_e}(x(t)) \right) \right), \quad \forall d \in \mathcal{D}, p \in P_d \tag{9.8}$$

where the set $\mathcal{X}_d$ is composed of the vectors $\{x_p, p \in P_d\}$ that satisfy (9.7bc). That is, if we denote as $x'_p(t)$ to the potentially unfeasible solution coming from the gradient update before projection in (9.8), the next solution is unique and given by:

$$x_p(t + 1) = \underset{\sum_{p \in P_d} x_p = h_d, x_p \geq 0}{\arg \min} \| x_p - x'_p \|^2, \forall p \in P_d \tag{9.9}$$

The previous projection operation is not immediate, since constraints are not box-like. Good news is that each demand source node can perform this projection efficiently using only local information (see Exercise 9.13, and its Net2Plan implementation), jointly deciding on all the paths of the demand.

Overall, a distributed implementation of the adaptive routing method (9.8) is possible, as shown in Algorithm 4. Each link e is in charge of monitoring the value $\frac{\partial F_e}{\partial y_e}$, as a function of the traffic traversing the link. Then, this information is periodically signaled to every network node. In turn, source nodes of the demand modify the routing according to (9.8).

Algorithm 4 Adaptive routing for (9.7)

1: *Link's algorithm*: At times $t = t_1(e), t_2(e), \ldots$, link e:
2:    Computes $\frac{\partial F_e}{\partial y_e}$ using link monitored information.
3:    Signals this information to all network source nodes.
4: *Source node's algorithm*: At times $t' = t_1'(d), t_2'(d), \ldots$, source node of demand d:
5:    Collects the most updated $\frac{\partial F_e}{\partial y_e}$ weights received from network links.
6:    Updates x_p for $p \in P_d$, applying (9.8), using the information signaled.

9.3.1 Removing Equality Constraints

As an alternative, in this section we use problem (9.7) as an example of how *equality* constraints can sometimes be eliminated to simplify the projections. We concentrate on an iterative algorithm that, in each iteration t, solves a variation of problem (9.7) where constraint (9.7a) is removed. Let $x_p(t)$ denote the current routing and $w_e(t)$ be the current values of the weights of the links:

$$w_e(t) = \frac{\partial F_e}{\partial y_e}(x(t)), \quad \forall e \in \mathcal{E}$$

For each demand d, we choose a path $\bar{p}_d \in P_d$ with a weight that is minimum among P_d. We call these paths (one per demand) the shortest paths according to weights w. Then, problem (9.7) is modified eliminating the decision variables associated to these shortest paths, applying equality constraint (9.7b):

$$x_{\bar{p}_d} = h_d - \sum_{p \in P_d - \bar{p}_d} x_p, \quad \forall d \in D$$

The new problem at iteration t is now given by

$$\min_x f(\bar{x}) \quad \text{subject to:} \tag{9.10a}$$

$$\bar{x}_p \geq 0, \quad \forall d \in D, p \in P_d - \bar{p}_d \tag{9.10b}$$

where $\bar{f}$, denotes the original objective function (9.7a), now expressed with respect to the reduced vector of variables $\bar{x}$. To apply the gradient projection algorithm to (9.10), we need first to obtain the gradient of $\bar{f}$:

$$\frac{\partial \bar{f}}{\partial \bar{x}_p} = \frac{\partial f}{\partial x_p} - \frac{\partial f}{\partial x_{\bar{p}_d}}, \quad \forall d \in D, p \in P_d - \bar{p}_d$$

Then, iteration of the algorithm becomes:

$$\bar{x}_p(t+1) = [\bar{x}_p(t) - \gamma(d_p - d_{\bar{p}_d})]_0, \quad \forall d \in D, p \in P_d - \bar{p}_d \tag{9.11}$$

where d_p and $d_{\bar{p}_d}$ are the path lengths according to:

$$d_p = \sum_{e \in p} w_e(t), \quad d_{\bar{p}_d} = \sum_{e \in \bar{p}_d} w_e(t)$$

Note that for every iteration, all the nonshortest paths that carry traffic will reduce traffic by $\gamma(d_p - d_{\bar{p}_d})$ units at most, which thus will be shifted to $\bar{p}_d$. The nonshortest paths that do not carry traffic, remain in the same situation.

The algorithm can be implemented in a similar distributed form as Algorithm 4. The only modification would be how each demand source node computes the next routing, now using (9.11) instead of (9.8). The convergence can be accelerated in practice using a diagonal scaling, if the information $\frac{\partial^2 F_e}{\partial y_e^2}(x(t))$ is also periodically signaled to the network nodes, together with the first derivatives. See Exercise 9.2 for details.

9.3.2   Optimality and Stability

Optimality and stability for any initial conditions are guaranteed in previous algorithms, for sufficiently small γ steps, inherited from the convergence properties of the gradient iteration. In this respect, it is important to remark that:

- Objective function (9.7) is convex but *not* strictly convex (and thus not strongly convex). Then, there is no guarantee of a convergence rate better than sublinear.
- If $F_e(y_e)$ functions are such that $F_e(y_e) \to \infty$ as $y_e \to u_e$, the gradient of the objective function will grow steeply as $y_e \to u_e$ in some links. Then, if the problem is feasible but in the optimum the traffic in some links is close to its capacity ($y_e \approx u_e$), the second derivative eigenvalues will be large, and convergence will require small γ steps.

9.3.2.1   Instability of Adaptive Non-Bifurcated Routing

A relevant remark is that *convexity of the feasibility set is a required condition for having both optimality and stability*. In our case, this means that it should be possible to bifurcate the routing among different paths in arbitrary fractions. For instance, convexity is lost if we force the routing to be non-bifurcated, since, for example, this means adding the discrete constraints $x_p \in \{0, h_{d(p)}\}$ to the problem formulation (9.7).

The routing in the Internet is a prominent example of the difficulties brought by non-convex routing problems. In particular, link-state protocols like OSPF assign a weight to each network link. Then, the demands are routed through the shortest path according to those weights and bifurcation can only occur in *equal fractions* (not arbitrary), when more than one shortest paths exist.

In this context, let us analyze an adaptive routing scheme that periodically adjusts the link weights as described in Algorithm 4, but that then carries *all* the traffic through the shortest path according to those weights[2]. Intuitively, our intention is adapt to traffic variations, shifting

---

[2] Mathematically, this means that in Algorithm 4 the projection is now performed in a discrete set of valid routings, which is not convex.

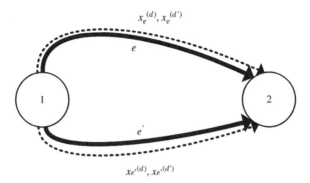

Figure 9.1   Example: Unstable routing. Demands d, d', two potential routings each through links, e or e', of the same capacity. If the routing is non-bifurcated, the minimum delay routing is unstable for some initial conditions if $h_d = d_{d'}$ and always unstable if $h_d \neq h_{d'}$

the traffic from those links that become congested by increasing their weights. However, it is easy to see how, because of the non-convex nature of the problem, such an approach can create dangerous interactions that produce an unstable network, performing constant routing changes even under a constant traffic demand.

A simple example where equilibrium may not be reached consists of a network of two nodes connected by two links of the same capacity, with two offered traffic demands (see Fig. 9.1).

- If both demands have the same volume, then equilibrium and optimality can be reached or not, depending on the initial conditions. If both demands are initially routed through different links, the solution remains unchanged and is both optimal and stable. However, if both demands are initially routed through the same link, in the next routing update the link carrying traffic will be assigned a higher weight than the idle link, and both demands change their route. Then, neither equilibrium nor optimality is ever reached, since the routing eternally alternates between both links.
- If both demands have different volumes the problem has now two optimum solutions (the ones that route each demand in different links). Then, if the initial solution consists of the two demands being carried in the same link, the algorithm oscillates between two non-optimal solutions. Still, if the two demands are initially routed in two different links (one of the optimum solutions), the algorithm would still oscillate, now between both optimum solutions, permitting the routings of both in each iteration, and thus would have optimality but not equilibrium.

The dangerous and undesired interactions described in the previous example are brought by the non-convex nature of the Internet shortest path routing policy. They have been part of a long disputed debate over the Internet since their very beginning [4], discussing and even benchmarking in trial tests the practical utilization of adaptive routing techniques in IP networks. Eventually, the difficulty of enforcing stability and optimality at a practical level has prevented the application of adaptive routing policies that react in short time scales to traffic variations. In turn, the link weights are usually static or updated a reduced number of times per day, such that the immediate effects of a routing change do not trigger a new change. In this

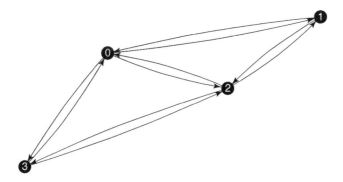

Figure 9.2   Test network. Link capacities $u_e = 25, \forall e$

latter case, so-called multihour routing, updates are computed in a centralized form based on a long-term network-wide monitoring or estimations of the offered traffic.

9.3.3   Implementation Example

In this section we present some empirical convergence tests of Algorithm 4 for solving routing problem (9.7). The data is taken from simulations conducted in the Net2Plan tool for a network shown in Fig. 9.2 and non-uniform traffic demands with the traffic matrix:

$$
h_{st} = \begin{pmatrix} 0 & 26.75 & 11.26 & 7.9 \\ 26.75 & 0 & 2.98 & 0.81 \\ 11.26 & 2.98 & 0 & 0.3 \\ 7.9 & 0.81 & 0.3 & 0 \end{pmatrix}
\tag{9.12}
$$

To avoid the difficulties brought by functions that are not Lipschitz continuous, we use link cost functions F_e of the form:

$$
F_e(y_e) = \begin{cases} \frac{1}{u_e - y_e}, & \text{if } y_e/u_e \leq 0.99 \\ \frac{y_e - 0.99 u_e}{(0.01 u_e)^2} + \frac{1}{0.01 u_e}, & \text{if } y_e/u_e > 0.99 \end{cases}
$$

As shown in Fig 9.3, the F_e function equals $1/(u_e - y_e)$ when link utilization is below 0.99 and continues as its linear interpolation when utilization is higher than 0.99. As a result, the first and second derivatives with respect to y_e are:

$$
\frac{\partial F_e}{\partial y_e} = \begin{cases} \frac{1}{(u_e - y_e)^2}, & \text{if } y_e/u_e \leq 0.99 \\ \frac{1}{(0.01 u_e)^2}, & \text{if } y_e/u_e > 0.99 \end{cases}, \qquad
\frac{\partial^2 F_e}{\partial y_e^2} = \begin{cases} \frac{2}{(u_e - y_e)^3}, & \text{if } y_e/u_e \leq 0.99 \\ 0, & \text{if } y_e/u_e > 0.99 \end{cases},
$$

And $F_e(y_e)$ has a Lipschitz continuous gradient, with a Lipschitz constant given by the maximum value of its second derivative: $\frac{2}{(0.01 u_e)^3}$. The gradient and second derivatives of the

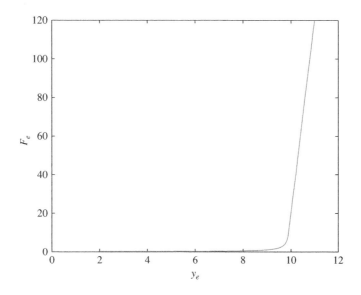

Figure 9.3 F_e function for a link with $u_e = 10$

objective function f with respect to decision variables x_p are:

$$\frac{\partial f}{\partial x_p} = \sum_{e \in p} \frac{\partial F_e}{\partial y_e}, \quad \forall p \in \mathcal{P}$$

$$\frac{\partial^2 f}{\partial x_{p_1} x_{p_2}} = \sum_{e \in p_1 \cap p_2} \frac{\partial^2 F_e}{\partial y_e^2}, \quad \forall p_1, p_2 \in \mathcal{P}$$

The Lipschitz constant of the objective function f can be computed as the modulus of the largest eigenvalue of its hessian matrix $\nabla^2 f$, which is also the norm two (spectral radius) of matrix $\nabla^2 f$: $||\nabla^2 f||_2$. Since the second derivative is not continuous when a link has an utilization of exactly 0.99, the hessian matrix at these points may not be symmetric. Leaving aside these points, the spectral radius of the hessian satisfies the following inequalities:

$$||\nabla^2 f||_2 \le ||\nabla^2 f||_1 = \max_i \sum_j |\nabla^2 f_{ij}| \le \sum_p K \frac{2}{(0.01 u_e)^3} \le |\mathcal{P}| K \frac{2}{(0.01 u_e)^3} \qquad (9.13)$$

where K is the maximum number of links traversed by a path. The last inequality in (9.13) is satisfied with an equality, in the (quite useless) case where all the paths in $\mathcal{P}$ have the same sequence of K links (and thus overlap), and all these links have a 0.99 utilization. A better approximation requires full knowledge of all the paths and the number of links that share each couple of paths.

Expression (9.13) could be used to dimension the γ step in a gradient algorithm, such that full theoretical convergence guarantees are provided, even in worst-case scenarios. However, the resulting γ steps would be quite small, making the method too slow in practice. As an example, a four node network like the one in our example, assuming, for example $|\mathcal{P}| = 16$

paths, $u_e = 25$, and $K = 4$ links, would yield to a γ step:

$$\gamma < \frac{2}{|\mathcal{P}|K\frac{2}{(0.01u_e)^3}} = \frac{(0.01u_e)^3}{|\mathcal{P}|K} = 0.000244$$

In the following subsections, we see that convergence is obtained in practice for much higher γ steps.

9.3.3.1 Convergence in the Asynchronous Case

We assume that each router computes all the loopless paths to any destination and uses Algorithm 4 to update the traffic carried in each route. The routing update is performed asynchronously: each router independently computes the routing using the most updated information link weight it has. Then, it waits a random time between 0.5 and 1.5 units before the next routing update. Also, independent from the routing update, each router computes the weight of its outgoing links, signals them to every other network node, and waits a random time between 0.5 and 1.5 time units until the next link weight for computation and signaling. We consider that signaling information arrives instantly at each other node, but a fraction of 5% signaling messages are randomly lost.

These assumptions reflect a realistic asynchronous and distributed application of the algorithm in a case where signaling delays are negligible. Figure 9.4 illustrates the algorithm

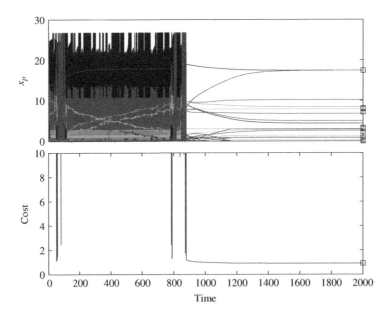

Figure 9.4 Evolution of Algorithm 4 in asynchronous case with signaling losses, $\gamma = 2$. The upper graph plots the x_p evolution for each path. The lower graph plots the cost evolution (9.7a). Optimum solutions are marked with squares on the right hand-side

evolution in the case $\gamma = 2$. As we see, the algorithm starts with a relatively long period of chaotic routing fluctuations up to $t = 850$, followed by a period where convergence swiftly occurs. This behavior usually witnesses a γ step that is very close to the convergence limit, such that the algorithm fluctuates until it is captured in a trajectory that results in convergence. Actually, in our tests fluctuations remained at $t = 5000$ when $\gamma = 3$ and disappeared when $\gamma = 1$. Finally, note that when the convergent phase starts, a solution with a close to optimal cost is found very fast (after a couple of iterations). However, after this happens, the algorithm smoothly changes the routing with small cost improvements, requiring around 800 iterations to reach equilibrium.

9.3.3.2 Limiting the Routing Change

In Section 8.8.1 it was shown that by limiting the change in a coordinate $(\gamma \frac{\partial f}{\partial x_p})$ to a maximum value Δx convergence guarantees were kept, as long as projection was performed onto a box-like set. This is not the case in our problem, since projection (9.9) includes the constraint $\sum_{p \in P_d} x_p = h_d$. Still, we can keep convergence guarantees if we change the algorithm iteration as follows:

- The routing $\hat{x}_p(t + 1), p \in P_d$ for all the paths of the demand are computed using (9.8), as in the original algorithm.
- If the maximum variation $\hat{x}_p(t + 1) - x_p(t)$ is below the maximum allowed change Δx, $\bar{x}(t + 1)$ becomes the new routing. If not, the new routing is:

$$x_p(t + 1) = x_p(t) + (\bar{x}_p(t + 1) - x_p(t)) \frac{\Delta x}{\max_{p \in P_d} |\bar{x}_p(t + 1) - x_p(t)|}$$

The result is that in the new algorithm no route changes its carried traffic by more than Δx units. This technique has the effect of permitting use of a higher γ producing a faster convergence. As an example, Fig. 9.5 illustrates the convergence in the same asynchornous case as before for $\gamma = 10$ and a maximum route variation $\Delta x = 0.1$. As we can see, convergence and equilibrium without fluctuations occurs at around one order of magnitude faster than in the previous example.

9.3.3.3 Effect of Signaling Delays

As predicted by theory, higher signaling delays mean using outdated information in the gradient iteration, requiring smaller γ steps to converge. Figure 9.6 illustrates this in a modification of the previous tests ($\gamma = 10, \Delta x = 0.1$), where each signaling message now takes a random time uniformly distributed in the interval $B = 50 \pm 0.5$. As shown, oscillations occur and neither equilibrium nor optimality are reached. In Exercise 9.1, the reader is asked to analyze the coupled effects that B, γ, and Δx have in algorithm convergence. For instance, the reader can see how lower Δx thresholds provide robustness for higher delays B and how the increase in the network traffic worsens convergence, since gradients become larger at higher link utilizations.

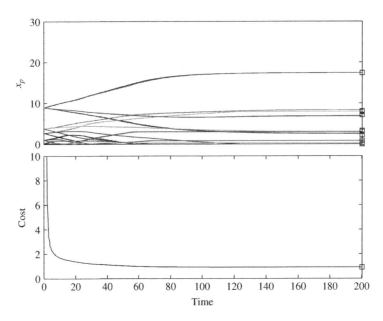

Figure 9.5   Evolution of Algorithm 4 in the asynchronous case with signaling losses, $\gamma = 10$, but limiting the routing changes to $\Delta x = 0.1$. The upper graph plots the x_p evolution for each path. The lower graph plots the cost evolution (9.7a). Optimum solutions are marked with squares on the right hand-side

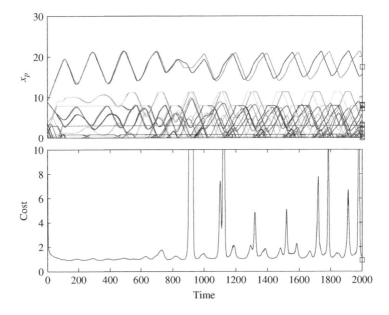

Figure 9.6   Evolution of Algorithm 4 in the asynchronous case with signaling losses, $\gamma = 10$, $\Delta x = 0.1$, random signaling delay in the interval 50 ± 0.5. The upper graph plots the x_p evolution for each path. The lower graph plots the cost evolution (9.7a). Optimum solutions are marked with squares on the right hand-side

9.4 Congestion Control using Barrier Functions

In this section we show a primal approach for solving the NUM congestion control problem (9.14) in a network $\mathcal{G}(\mathcal{N}, \mathcal{E})$ with a set D of traffic demands. The decision variables h_d reflect the rate of demand $d \in D$, p_d is the (known) sequence of links traversed by traffic of d, and u_e is the known capacity in link e.

$$\max_h \ \sum_d U_d(h_d) \quad \text{subject to:} \tag{9.14a}$$

$$\sum_{d:e \in p_d} h_d \le u_e, \quad \forall e \in \mathcal{E} \tag{9.14b}$$

$$m_d \le h_d \le M_d, \quad \forall d \in D \tag{9.14c}$$

The primal approach followed consists of moving constraints (9.14b) to the objective function through a barrier function $B(y)$ based on the M/M/1 estimation of the average queuing delay in link e:

$$B(y_e) = \begin{cases} \frac{y_e}{u_e - y_e}, & \text{if } u_e > y_e \\ \infty, & \text{if } y_e \ge u_e \end{cases}$$

where $y_e = \sum_{d:e \in p_d} h_d$ is the traffic traversing e. The resulting problem is:

$$\max_{m \le h \le M} \ \sum_d U_d(h_d) - \epsilon \sum_e \frac{\sum_{d:e \in p_d} h_d}{u_e - \sum_{d:e \in p_d} h_d} \tag{9.15}$$

The barrier function satisfies the requisites in Prop. 9.1 and thus the optimum of (9.15) approximates the optimum of the original problem (9.14) as $\epsilon \to 0$. Applying a basic gradient projection algorithm to (9.15) we have the iteration:

$$h_d(t+1) = \left[h_d(t) + \gamma \left(\frac{\partial U_d}{\partial h_d}(h_d(t)) - \epsilon \sum_{e \in p_d} \frac{u_e}{(u_e - \sum_{d:e \in p_d} h_d(t))^2} \right) \right]_{m_d}^{M_d} \tag{9.16}$$

Algorithm 5 Congestion control for (9.15)

1: *Link's algorithm*: At times $t = t_1(e), t_2(e), \dots$, link e:
2:   Computes $\frac{u_e}{(u_e - y_e)^2}$ using link monitored information.
3:   Signals this information to all network source nodes.
4: *Source node's algorithm*: At times $t' = t_1'(d), t_2'(d), \dots$, source node of demand d:
5:   Collects the most updated signaled values from network links.
6:   Updates h_d for ingress demands, applying (9.16), using the information signaled.

A distributed implementation of (9.16) is described in Algorithm 5. Link algorithm computes the weights $\frac{u_e}{(u_e - y_e)^2}$ using only local information. Then, these weights have to be *explicitly* signaled to the traversing demands source nodes. For instance, in ATM networks under the Available Bit Rate (ABR) control mode, this can be implemented by storing the weights in the Resource Management (RM) cells.

9.4.1   Implementation Example

We present some empirical tests of Algorithm 5. The network topology is the one shown in Fig. 9.2, and a demand exists between each node pair, carried through the shortest path route in km between the end nodes. We are interested in the algorithm convergence in an asynchronous case, a more realistic situation. Each demand independently computes its injected traffic h_d using the most updated link weight information it has. Then, it waits a random time between 0.5 and 1.5 units before the next update. Also, independent from the demand update, each link computes its weight, signals it to every network node, and waits a random time between 0.5 and 1.5 time units until the next link weight for computation and signaling. We consider that signaling information arrives instantly to each other node but a fraction of 5% signaling messages are randomly lost.

The utility functions of the demands are the α-utility functions (3.19) with $\alpha = 2$. Minimum and maximum demand rates are set to $m_d = 0.1, M_d = \infty \forall d$.

An implementation aspect of practical importance is that the gradient of the barrier function is not Lipschitz continuous. Then, the gradient of the objective function (in absolute value) can largely grow when the link utilization approaches 100%. High link utilizations are a normal state for the network, especially for low ϵ values ($\epsilon = 10^{-5}$ in our case). The reason is that, unless the utilization is very close to 100%, the algorithm pushes the demands to inject more and more traffic. Only when the link is close to saturate, the barrier cost *abruptly* appears. When this happens, the gradient can take a large (negative) value and drastically reduce the rate.

As a result, it is important to limit the maximum change in each demand rate or otherwise violent fluctuations of the traffic are produced. Figure 9.7 illustrates a reasonable algorithm convergence, when the maximum demand change is limited to $\Delta h = 1$ traffic units, according to the technique described in Section 8.8.1 (step $\gamma = 20$). As can be seen, convergence occurs

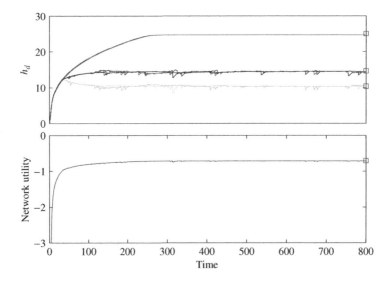

Figure 9.7   Evolution of Algorithm 5 in the asynchronous case with signaling losses, $\gamma = 20$, $\Delta h = 1$. The upper graph plots the h_d evolution for each path. The lower graph plots the network utility evolution (9.14a). Optimum solutions are marked with squares on the right-handside

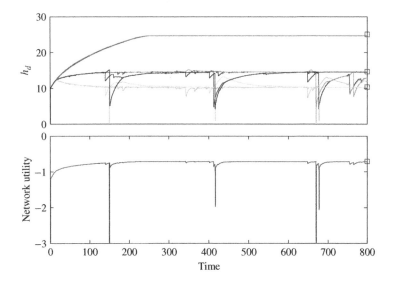

Figure 9.8   Same example as Fig. 9.7, $\Delta h = 10$

in practice to the optimum solution, but the fluctuations caused by the non-Lipschitz continuity of the gradient result in small oscillations. Lower Δh values reduce the fluctuations even more. In contrast, setting larger Δh values can produce drastic changes in the demand carried traffics. For instance, Fig. 9.8 shows such effects at a value $\Delta h = 10$. Higher Δh values result in permanent and large oscillations without equilibrium.

9.4.1.1   Diagonal Scaling

Since the problem constraints are box-like, it is possible to apply diagonal scaling techniques without harming the algorithm convergence (aside of the difficulty because of the non-Lipschitz continuity of the gradient). The second partial derivative of the objective function f of (9.14) is:

$$\frac{\partial^2 f}{\partial h_d^2}(h) = -\alpha h_d^{-\alpha-1} - \epsilon \sum_{e \in p_d} \frac{2u_e}{(u_e - \sum_{d:e \in p_d} h_d(t))^3} \tag{9.17}$$

Then, the gradient iteration is modified by dividing the γ step of each demand by the absolute value of the previous quantity (9.17). Note that to apply the diagonal scaling, links should compute the quantity $\frac{2u_e}{(u_e - y_e)^3}$, which does not require storing per-flow information, and signal it together with the link price to the initial nodes of the traversing demands.

In our case study, the diagonal scaling makes it tricky to find a good balance of the γ step to avoid drastic fluctuations when gradients and second derivatives become close to infinity. To illustrate this, Fig. 9.9 shows the algorithm evolution for $\gamma = 0.5$ and $\Delta h = 1$, which is remarkably fast. However, the reader can check using the provided implementation in Net2Plan how other factors, like $\gamma = 0.4$ or $\gamma = 0.6$, can yield to long erratic phases before convergence is reached.

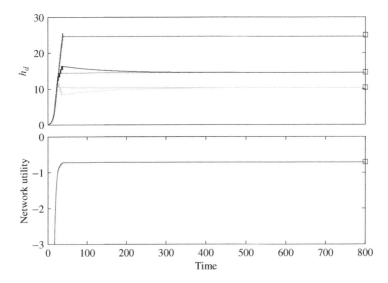

Figure 9.9 Evolution of Algorithm 5 in asynchronous case with signaling losses, diagonal scaling, and $\gamma = 0.5$, $\Delta h = 1$

The interested reader can observe the interplay of signaling delay, gradient noise, and other implementation aspects in algorithm convergence using the Net2Plan implementation provided.

9.4.2   Exterior Penalty

In this section we present a primal algorithm for (9.14), where link capacity constraints are moved to the objective function through a differentiable exterior penalty function $P(y)$ given by:

$$P(y_e) = \begin{cases} 0, & \text{if } y_e \leq u_e \\ (y_e - u_e)^2, & \text{if } y_e > u_e \end{cases}$$

where:

$$\frac{\partial P(y_e)}{\partial h_d} = \begin{cases} 0, & \text{if } y_e \leq u_e \\ 2(y_e - u_e), & \text{if } y_e > u_e \end{cases}, \quad \forall e \in \mathcal{E}$$

The resulting problem is:

$$\max_{m \leq h \leq M} \sum_d U_d(h_d) - \mu \sum_e P(\sum_{d:e \in p_d} h_d) \tag{9.18}$$

We see that $P(y_e)$ functions satisfy the requisites of Prop. 9.2 and thus the optimum of (9.18) approximates the optimum of the original problem (9.14) from the exterior as $\mu \to \infty$. Denoting $\pi_e(t)$ as the gradient of the penalty function at link e and time t, the basic gradient projection iteration is:

$$h_d(t+1) = \left[h_d(t) + \gamma \left(\frac{\partial U_d}{\partial h_d}(h_d(t)) - \mu \sum_{e \in p_d} \pi_e(t) \right) \right]_{m_d}^{M_d} \tag{9.19}$$

Algorithm 5 can be trivially modified to have a distributed implementation of (9.19). This is left as an exercise for the reader (see Exercise 9.5).

9.5 Persistence Probability Adjustment in MAC Protocols

In this section we devise a distributed algorithm based on a gradient projection primal approach, to optimize the persistence probabilities in a random-access (Aloha type) wireless network $\mathcal{G}(\mathcal{N}, \mathcal{E})$, adopting the model described in Section 5.4.1. A MAC protocol is assumed, such that time is slotted, and in every time slot each node randomly decides to use the channel with a probability of q_n. One link e out of its outgoing links is randomly chosen to transmit traffic at the link nominal rate $\bar{u}_e$. We call p_e to the link persistence probability, the probability of finding a particular link e active at any time.

Collisions can occur and we denote $\mathcal{N}_{to}^I(e)$ as the set of nodes whose transmission interferes to link e, such that if any node in $\mathcal{N}_{to}^I(e)$ is using the channel when e is transmitting, the traffic in e is lost. From this information, it is easy to construct the sets $\mathcal{E}_{from}^I(n)$, as the set of links that are interfered by a node n, excluding outgoing links of n:

$$n \in \mathcal{N}_{to}^I(e) \Leftrightarrow e \in \mathcal{E}_{from}^I(n)$$

The capacity of a link e depends on the average amount of collision-free traffic that is able to transmit, which depends on the persistence probabilities of e and its interfering nodes according to:

$$u_e = \bar{u}_e p_e \prod_{n \in \mathcal{N}_{to}^I(e)} q_n, \quad \forall e \in \mathcal{E}$$

The reader is referred to Section 5.4.1 for a full description of the problem. The persistence probabilities (p_e, q_n) that result in a capacity assignment that maximizes the network α-fairness are optimized by (9.20):

$$\max_p \frac{1}{1-\alpha} \sum_e \left[\bar{u}_e p_e \prod_{n \in \mathcal{N}_{to}^I(e)} \left(1 - \sum_{e'' \in \delta^+(n)} p_{e''} \right) \right]^{1-\alpha} \quad \text{subject to:} \tag{9.20a}$$

$$\sum_{e \in \delta^+(n)} p_e \leq 1, \quad \forall n \in \mathcal{N} \tag{9.20b}$$

$$0 \leq p_e \leq 1, \quad \forall e \in \mathcal{E} \tag{9.20c}$$

Note that formulation (9.20) is a simplified version of (5.15) in Section 5.4.1 where:

- Variables u_e are eliminated, together with constraints $u_e^{min} \leq u_e \leq u_e^{max}$.
- We replace variables q_n by its expression $\sum_{e'' \in \delta^+(n)} p_{e''}$.

It is possible to show (see Exercise 9.8) that for $\alpha \geq 1$ fairness factors, problem (9.20) involves the maximization of a concave function with linear constraints and thus we can apply a gradient projection algorithm for solving it, without any variable change. The average link capacity u_e, given by the collision-free traffic received by the end node of link e is:

$$u_e = \bar{u}_e p_e \prod_{n \in \mathcal{N}_{to}^I(e)} \left(1 - \sum_{e'' \in \delta^+(n)} p_{e''}\right), \quad \forall e \in \mathcal{E}$$

The gradient of the objective function f (9.20a) is:

$$\frac{\partial f}{\partial p_e} = \frac{u_e^{1-\alpha}}{p_e} - \sum_{e' \in \mathcal{E}_{from}^I(a(e))} \frac{u_{e'}^{1-\alpha}}{1 - \sum_{e'' \in \delta^+(a(e'))} p_{e''}}, \quad \forall e \in \mathcal{E}$$

And the algorithm iteration, applying a basic gradient projected approach with a constant step of sufficiently small γ is:

$$p_e(t+1) = P_{\mathcal{X}_n}\left(p_e(t) + \gamma \frac{\partial f}{\partial p_e}(p(t))\right), \quad \forall n \in \mathcal{N}, e \in \delta^+(n) \tag{9.21}$$

where $\mathcal{X}_n$ is the set of points $(p_e \forall e \in \delta^+(n))$ limited by constraints (9.20bc). This projection can be solved efficiently (see Exercise 9.13 and its Net2Plan implementation) and using only local node information.

A distributed implementation of iteration (9.21), is sketched in Algorithm 6. The key aspects to consider are:

- $u_e(t)$ values are the time average of collision-free traffic in link e in the last observation period. This quantity can be computed by the receiver end of e.
- To update the persistence probabilities of a link e, the link origin node (n) should know the link capacity (u_e) and the capacity of all the links e' that are interfered with by n ($e' \in \mathcal{E}_{from}^I(n)$). Also, node n should know the persistence probability $q_{n'}$ for all the nodes n' that have an outgoing link interfered by n.
- The previous point requires establishing a signaling mechanism that permits to each link e destination node ($b(e)$), delivering monitored u_e information to (i) link origin node (its neighbor) $a(e)$ and to (ii) all the nodes that interfere e ($n \in \mathcal{N}_{to}^I(e)$). Also, a node n should deliver its q_n probability to all the nodes that interfere with it.
- For those cases when $\alpha = 1$ (proportional fairness), there is no need to monitor or signal u_e estimations, since they apply in the gradient as $u_e^{1-\alpha} = u_e^0 = 1$.

Algorithm 6 Persistence probability adjustment for (9.20)

1: *Node's algorithm, signaling part*: At times $t = t_1(n), t_2(n), \ldots$, node n:
2:   Estimates capacity u_e of incoming links.
3:   Signals u_e of incoming links and own q_n to all nodes that interfere me.
4: *Node's algorithm, update part*: At times $t' = t_1'(n), t_2(n)', \ldots$, node n:
5:   Collects the most updated signaled values of u_e of outgoing links, and of links that interfere me.
6:   Collects the most updated $q_{n'}$ of nodes n' that n interferes to.
7:   Updates $p_e(t+1)$, for outgoing links e, using (9.21).

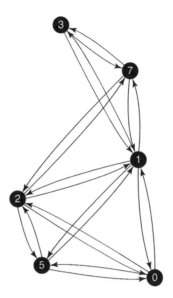

Figure 9.10 Example figure

9.5.1 Implementation Example

We test Algorithm 6 in a wireless network like the one shown in Fig. 9.10, with a nominal rate of $\bar{u}_e = 1$ in all links. As an interference model, we assume that reception in a node n is interfered by any simultaneous transmission of any other node n' that can send traffic to n and also interfered by itself (a node cannot receive and transmit simultaneously):

$$\mathcal{N}_{to}^{I}(e) = \{n \neq a(e), \text{ such that exists a link } n \to b(e)\} \bigcup a(e)$$

As an example, in the topology of Fig. 9.10:

- Link $e_{50} = (5 \to 0)$ is interfered by nodes $\mathcal{N}_{to}^{I}(e_{50}) = \{n_0, n_1, n_2\}$.
- Links that are interfered by node n_3 are: $\mathcal{E}_{from}^{I}(n_3) = \{e_{01}, e_{21}, e_{13}, e_{71}, e_{17}, e_{27}, e_{73}, e_{51}\}$

We also consider bidirectional topologies, so if a link $n \to n'$ exists, link $n' \to n$ also does. In these circumstances, a node n, in order to update the persistence probabilities of its outgoing links according to (9.21), needs to know:

- The u_e estimations of those links ending in any neighbor node of n. This can be accomplished if every node estimates the capacity u_e of its incoming links and broadcast this information to all its neighbors.
- The persistence probabilities $q_{n'}$ of those nodes n' that are at a two-hop distance from n in at least a path. This requires every node n' to (i) broadcast its $q_{n'}$ value, together with (ii) the $q_{n''}$ values received from its neighbors.

Signaling and persistence probabilities updates occur asynchronously:

- Each node independently updates its persistence probabilities using the gradient iteration and the most recent u_e and q_n information it has and waits a random time between 0.5 and 1.5 time units before the next update.
- Each node independently broadcasts the u_e values of its incoming links, its q_n value and the most update ones learned from its in neighbors, and waits a random time between 0.5 and 1.5 units before the next signaling procedure. Broadcast messages arrive instantly.

The duration of the time slot is assumed to be arbitrarily small. Then, the signaling and gradient updates happen in average at the same periodicity, which is much *slower* than the slotted and synchronous periodicity of the MAC protocol. In other words, many time slots are produced applying the persistence probabilities computed. Thus, we can consider u_e estimations as an unbiased and precise observation of the link capacities.

The utility functions of the capacities are the α-utility functions (3.19), with $\alpha = 2$. The gradient of the objective function is not Lipschitz continuous and grows to infinity when a link has $p_e \to 0$, or a node n has a persistence probability $q_n \to 1$. In practice, both situations do not happen in the proximity of optimal solutions: (i) when $p_e = 0$, link e is allocated no capacity and the solution has an infinite gradient coordinate and (ii) a node can have persistence probability one in the optimum only if it is isolated and thus does not have to share any medium. Still, in our tests we avoid previous issues by setting a minimum and maximum values of the persistence probabilities modifying constraints (9.20bc):

$$\sum_{e \in \delta^+(n)} p_e \le 0.99, \quad \forall n \in \mathcal{N}$$

$$0.03 \le p_e \le 0.99, \quad \forall e \in \mathcal{E}$$

The interested reader can observe the interplay of signaling delay, update frequencies, and other implementation aspects in algorithm convergence using the Net2Plan implementation provided.

We use a variation of the bounded step technique to avoid drastic changes in the persistence probability updates. In particular, we limit the p_e change in an iteration to $\Delta p_e = \pm 1\%$, so if the p_e variation in the output links of a node exceeds $\pm 1\%$, the variations of all the output links are scaled down by the same factor to satisfy the Δp_e limit. Note that this variation is compatible with the distributed implementation of the algorithm. Figure 9.11 shows the algorithm convergence using $\gamma = 0.00001$ as the algorithm step. As can be seen, convergence occurs reasonably fast in practice, to a close to optimal solution (calculated solving (9.20) with JOM). Note that although small differences exist between the two solutions found, both have a similar network utility. The reader can check how higher Δp_e values or γ steps can very easily produce algorithm oscillations, although still with close to optimal solutions.

9.5.1.1 Effect of Estimation Noise

Algorithm 6 relies on the estimation of link capacities u_e used in the gradient computations. In our previous tests, we assumed that these estimations were perfectly precise. This is a reasonable assumption when the number of time slots monitored is large, which means that the

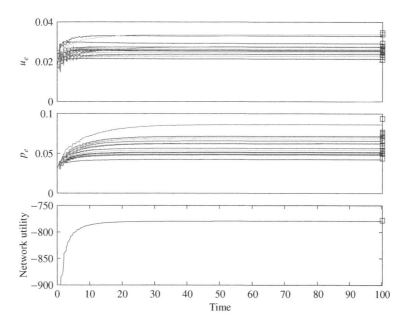

Figure 9.11 Evolution of Algorithm 6 in the asynchronous case with signaling losses, $\gamma = 0.00001$, $\Delta p_e = 0.01$. The upper graphs plot the u_e and p_e evolution for each link. The lower graph plots the network utility evolution (9.20a). Optimum solutions are marked with squares on the right-hand side

signaling intervals are much longer than the MAC time slots. When this is not the case, measurement errors occur in the u_e predictions and gradient updates use noisy information.

In Chapter 8, we presented the main convergence properties of the stochastic gradient projection algorithm. We saw that under mild assumptions (e.g., unbiased noisy gradient with finite variance), convergence in expectation to the *proximity* of the optimum occurred. In this section, we show how these convergence properties apply to Algorithm 6. In particular, we consider the same network and algorithm setup as in Fig. 9.11 in the case when capacity observations are subject to an unbiased error uniformly chosen in the range of $\pm 20\%$ of the true link capacity. Results are shown in Fig. 9.12. We see that (i) convergence in expectation seems to occur reasonably fast, (ii) although naturally the persistence probabilities do not become stable since gradient observations are noisy and (iii) still, the algorithm fluctuates among solutions with close to optimal network utilities.

9.6   Transmission Power Assignment in Wireless Networks

In Section 5.5, we modeled a wireless network in a soft-interference scenario where all network links can transmit simultaneously thanks to an appropriate multiplexing scheme. However, the receiver side of a link e, sees the incoming power from other links as a noisy interfering signal that limits the maximum rate that e can attain. In particular, we saw that the theoretical limit to the capacity u_e achievable by a link, is (up to a scaling factor) given by:

$$u_e = \log(1 + \text{SNR}_e) \approx \log(\text{SNR}_e) \tag{9.22}$$

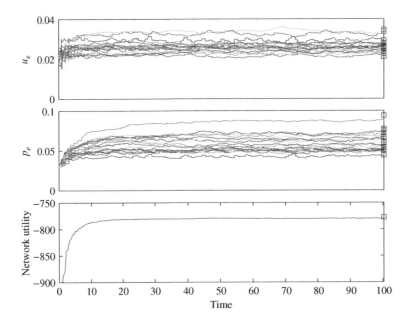

Figure 9.12 Same example as Fig. 9.11, but capacity observations are subject to a noise uniformly chosen in the range ±20% of the true link capacity

where SNR$_e$ is the Signal-to-Noise Ratio, including the interference power as noise, at the receiver end of e, and the approximation is valid when SNR$_e$ values are large enough. The SNR of a link e is given by:

$$\text{SNR}_e = \frac{p_e G_{ee}}{\sigma_e^2 + \sum_{e' \neq e} p_{e'} G_{e'e}} \tag{9.23}$$

In (9.23), p_e is the transmission power at link e in *linear units*, σ_e^2 is the thermal noise power at the receiver end, and $G_{ee'}, e, e' \in \mathcal{E}$ is the interference map between links: the fraction of the power transmitted in p_e that reaches the receiver of e'. In general, $G_{ee'} \ll G_{ee}$, when $e \neq e'$, since the multiplexing scheme is able to further attenuate the effects of the interfering signals.

In this section, we attempt a primal approach to solve the power allocation problem that enforces a fair capacity assignment to the network links:

$$\max_{\tilde{p}\text{min} \leq \tilde{P} \leq \tilde{p}\text{max}} \sum_e U_e \left(\log \left(\frac{e^{\tilde{P}_e} G_{ee}}{\sigma_e^2 + \sum_{e' \neq e} e^{\tilde{P}_{e'}} G_{e'e}} \right) \right) \tag{9.24}$$

We note that:

- $\tilde{P}_e$ is the transmission power in link e expressed in logarithmic units: $\tilde{P}_e = \log p_e$. Recall that this transformation is required to guarantee concavity in the objective function of (9.24).
- Problem (9.24) is a modification of problem (5.27) in Chapter 5, where variables u_e are replaced by expression $u_e = \log(\text{SNR}_e)$.

- Applying Prop. 5.5 in Chapter 5, the objective function is concave with respect to $\tilde{P}_e$ variables.
- The constraints $\tilde{P}^{min}$ and $\tilde{P}^{max}$ parameters are the minimum and maximum transmission powers in logarithmic units.

Problem (9.24) involves the maximization of a concave function under linear box-like constraints and can be solved using a gradient projection algorithm. The gradient of the objective function f of (9.24) is:

$$\frac{\partial f}{\partial \tilde{P}_e} = \frac{\partial U_e}{\partial \tilde{P}_e} + \sum_{e'' \ne e} \frac{\partial U_{e''}}{\partial \tilde{P}_e} \tag{9.25}$$

For α-fair utilities of the form (9.26):

$$U_e(u_e) = \begin{cases} w_e \log\ u_e & \text{if } \alpha = 1 \\ w_e \dfrac{y_e^{1-\alpha}}{1-\alpha} & \text{if } \alpha \ge 0, \alpha \ne 1 \end{cases} \tag{9.26}$$

we have that (taking $u_e = \log\ SNR_e$):

$$\frac{\partial U_e}{\partial \tilde{P}_e} = \frac{\partial U_e}{\partial u_e} \frac{\partial u_e}{\partial \tilde{P}_e} = w_e u_e^{-\alpha}$$

$$\frac{\partial U_{e''}}{\partial \tilde{P}_e} = \frac{\partial U_{e''}}{\partial u_{e''}} \frac{\partial u_{e''}}{\partial \tilde{P}_e} = -w_{e''} u_{e''}^{-\alpha} \frac{SNR_{e''}}{e^{\tilde{P}_{e''}} G_{e''e''}} e^{\tilde{P}_e} G_{ee''}$$

using $m_{e''} = w_{e''} u_{e''}^{-\alpha} \frac{SNR_{e''}}{e^{\tilde{P}_{e''}} G_{e''e''}}$, we can rewrite the gradient of the objective function as:

$$\frac{\partial f}{\partial \tilde{P}_e} = w_e u_e^{-\alpha} - e^{\tilde{P}_e} \sum_{e'' \ne e} m_{e''} G_{ee''}$$

Then, the gradient projection iteration is given by:

$$\tilde{P}_e(t+1) = \left[\tilde{P}_e(t) + \gamma \frac{\partial f}{\partial \tilde{P}_e} \right]_{\tilde{P}^{min}}^{\tilde{P}^{max}} \tag{9.27}$$

Note that m_e quantity of a link can be computed from information that wireless network links typically monitor: the current link capacity u_e, its SNR, and the gain of the direct channel G_{ee}. In addition, a link e should also estimate by external procedures the fractions $G_{ee''}$ of the signal transmitted that reaches other links receiver ends, to compute the iteration (9.27). Algorithm 7 sketches a distributed implementation of the iteration under these assumptions.

9.6.1 Implementation Example

We test Algorithm 7 to control the uplink of a cellular network shown in Fig. 9.13. This means the communications from the mobile phone to the base station central node. The base station is permanently monitoring the link capacity, its SNR and its received power, and thus can compute the m_e factor for all the links. Since links are permanently informing the base station about its transmission power, it can also estimate the full interference map $G_{ee'}$.

Algorithm 7 Transmission power allocation for (9.24)

1: *Link's algorithm, signaling part*: At times $t = t_1(e), t_2(e), \ldots$, link e:
2:    Estimates its capacity u_e, m_e and interference map $G_{ee''}, \forall e''$.
3:    Signals m_e to all network nodes.
4: *Link's algorithm, update part*: At times $t' = t_1'(n), t_2'(n), \ldots$, node n:
5:    Collects the most updated signaled/monitored values of u_e, $m_{e''}, G_{ee''}, \forall e'' \neq e$.
6:    Updates $\tilde{P}_e(t+1)$, using (9.27).

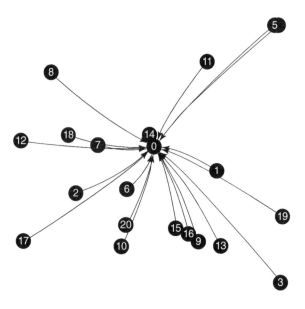

Figure 9.13 Example network: uplink channels in a cell

We assume that mobile phones update their transmission power asynchronously and independently. After a power update applying (9.27), the phone waits a random time between 0.5 and 1.5 time units before the next update. In addition, the base station signals each phone the m_e and u_e values asynchronously. The time between two signaling messages sent to a phone is random, and uniformly distributed between 0.5 and 1.5 time units. In addition, a 5% of the messages is lost.

The wireless channel has a path-loss exponent equal to 3. The multiplexing system is able to attenuate the power of an interfering signal by 60 dB (10^6). Finally, the $G_{ee'}$ values are normalized so that its maximum coordinate is 1, and satisfying that:

$$G_{ee'} \propto \begin{cases} d_e^{-3}, & \text{if } e = e' \\ d_e^{-3} 10^{-6}, & \text{if } e \neq e' F \end{cases}$$

where d_e is the length of link e. The maximum and minimum power on each phone in logarithmic units are $\tilde{P}^{max} = 3, \tilde{P}^{min} = 0$ and the thermal noise σ^2 at the base station for all the channels is the same. We compute it such that it is 10 times smaller than the worst-case interfering power

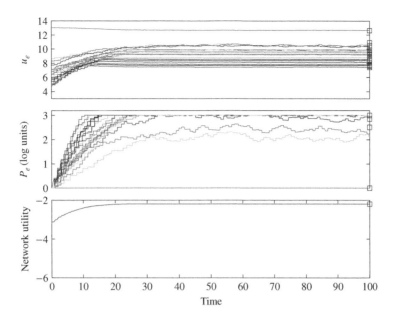

Figure 9.14 Evolution of Algorithm 7 in the asynchronous case with signaling losses, $\gamma = 10$. The upper graphs plot the u_e and p_e evolution for each link. The lower graph plots the network utility evolution (9.24). Optimum solutions are marked with squares on the right-hand side

in any link:

$$\sigma^2 = \frac{\tilde{P}^{max}\max_e \sum_{e' \neq e} G_{e'e}}{10}$$

According to this, the worst-case so-called Rise Over Thermal (ROT) parameter among the links is 10, a reasonable assumption in cellular networks.

We assume a $\gamma = 10$ step in the gradient iteration. Also, because of the monitoring processes involved, we consider that gradient values are actually noisy gradients. To model this, every phone in every iteration adds a random noise uniformly sampled in the interval $[-0.01, 0.01]$ to the gradient. Note that since $\gamma = 10$, this can result in a power variation of up to ± 0.1 in one iteration.

Figure 9.14 illustrates the algorithm convergence in the described case and a fairness factor $\alpha = 2$. As we see, the objective function converges to the optimum in about 20 iterations. Fluctuations in transmission power after that do not reflect in large link capacity deviations, and actually do not impact the overall network utility.

As predicted by theory in Section 8.8.3, convergence can be improved using a heavy-ball technique like:

$$\tilde{P}_e(t+1) = \left[\tilde{P}_e(t) + \gamma \frac{\partial f}{\partial \tilde{P}_e} + \gamma_h(x(t) - x(t-1)) \right]_{\tilde{p}min}^{\tilde{p}max} \tag{9.28}$$

where $\gamma_h(x(t) - x(t-1))$ is an inertia term that tends to make this iteration variation, similar to the previous one. Figure 9.15 shows the algorithm convergence in the same setting as in

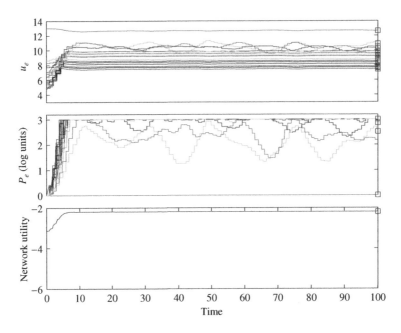

Figure 9.15   Same example as Fig. 9.14, with a heavy-ball variation (9.28) with $\gamma_h = 0.9$

Fig. 9.14, but applying the heavy-ball variation (9.28) to with $\gamma_h = 0.9$. Convergence is significantly faster (e.g., 6–7 iterations). However, as also predicted by theory, the heavy-ball iteration amplifies the effect of noise, resulting in larger deviations around the optimum, and higher fluctuation in transmission powers. Still, these oscillations produced no effect in the total network utility.

9.7   Notes and Sources

The mathematical background of primal gradient methods is accessible from multiple sources, like [1, 2, 5–8]. The material for interior and exterior penalty methods exposed in this chapter is quite basic and is mainly extracted from [2] and [1]. Barrier penalties have become crucial in the development of interior point methods for convex programs, the state-of-the-art for many modern solvers. Exterior penalty methods are the base of SUMT (Sequential Unconstrained Minimization Techniques) algorithms first proposed in [3]. [1] and [8] are good starting points for the reader interested in these topics.

An evolution of exterior penalty techniques not covered in this book are the *augmented Lagrangian* and *multiplier methods*. In this case, a constrained problem is replaced by a series of unconstrained problems adding a per constraint penalty term to the objective, plus another extra term designed to mimic a Lagrange multiplier (in that sense, it can be seen as an hybrid of a primal and dual approach). The interest of the method is that constraint multipliers are obtained as a by-product, and that convergence does not need the μ term to reach infinity.

However, the convergence of these methods has been somewhat less studied under asynchronous executions and noisy observations. Interested readers are referred to [9].

The primal algorithm for adaptive routing case study in Section 9.3 removing equality constraints has been extracted from [10] [11].

A continuous version of the primal algorithm for congestion control in Section 9.4 based on a barrier function was initially presented in [12] and appears in other sources like [13].

A dual algorithm for adjusting the persistence probabilities in random-access wireless networks is presented in [14] and [15]. In Section 9.5 we present a primal approach to a simplified version of the problem presented in [14] and in Chapter 5.

The model for optimizing the transmission power in soft-interference networks in Section 9.6 is analogous to that in other works like [15–17] or [18]. In [18] a dual approach for the cross-layer optimization of congestion control and transmission power yields a distributed algorithm with similarities, in the transmission power side, to the primal approach presented in Section 9.6, which is not published elsewhere.

9.8 Exercises

9.1 Use the Net2Plan implementation available in the Net2Plan repository for the adaptive routing algorithm in Section 9.3.2 to observe the coupled effects that signaling average delay B, γ step and maximum route change Δx, have in algorithm convergence. Find the γ and Δx parameters that provide empirical convergence in the case of signaling delays $B = 20 \pm 5$ time units, in two cases: (i) when the optimal routing has a bottleneck utilization of 0.5 and (ii) when the bottleneck utilization is 0.95.

9.2 Modify the Net2Plan implementation available in the Net2Plan repository of the primal algorithm for adaptive routing, to implement the algorithm version when equality constraints are removed (Section 9.3) with or without diagonal scaling. Repeat the tests described in previous exercise and comment on the empirical convergence improvements of diagonal scaling in these tests. *Hint*: the diagonal values of the hessian of the objective function $\bar{f}$ are given by:

$$\frac{\partial^2 \bar{f}}{\partial \bar{x}_p^2}(x(t)) = \sum_{e \in L_p} \frac{\partial^2 F_e}{\partial y_e^2}(x(t))$$

where L_p is the path composed of the links belonging to either p or $\bar{p}_d$, but not both.

9.3 Apply an exterior penalty method using a quadratic penalty to devise a primal algorithm for the adaptive routing problem in Section 9.3.2. Implement the algorithm in Net2Plan using the provided implementation for the interior penalty as a template. Repeat the results for the case of a logarithmic barrier.

9.4 Use the Net2Plan implementation available in the Net2Plan repository for the congestion control algorithm in Section 9.4, and empirically study the γ and Δh trade-offs in a network like the one in Fig. 9.10.

9.5   Devise a primal algorithm for solving the congestion control problem in Section 9.4, using an exterior penalty function for the link capacity constraints. As a penalty function of link e, use:

$$P(y_e) = \begin{cases} 0, & \text{if } y_e \leq u_e \\ (y_e - u_e)^2, & \text{if } y_e > u_e \end{cases}$$

Show that the convergence conditions are satisfied with this penalty function. Use the Net2Plan implementation available in the Net2Plan repository and repeat the empirical studies in Section 9.4.1, tailoring the γ, Δh and μ parameters. Compare the convergence results of the interior and exterior penalty methods.

9.6   The multi-path congestion control problem consists of optimizing the demand rates, in the case when a demand traffic d can be bifurcated in several paths P_d. We use P to denote the set of all paths in the network, P_e the paths traversing link e, and x_p for the decision variable representing the traffic carried in path p. The multi-path NUM problem formulation is given by:

$$\max_{x \geq 0} \sum_d U_d \left(\sum_{p \in P_d} x_p \right) \quad \text{subject to:} \tag{9.29a}$$

$$\sum_{p \in P_e} x_p \leq u_e, \quad \forall e \in \mathcal{E} \tag{9.29b}$$

Devise a primal algorithm for solving the multi-path congestion control problem using a logarithmic barrier for constraints (9.29b), with a maximum gradient step Δp. Implement it in Net2Plan and empirically tailor the γ, Δp, and ϵ parameters for a set of selected topologies.

9.7   Repeat Exercise 9.6 using exterior penalty (9.6) for constraints (9.29b).

9.8   Show that the objective function of problem (9.20) is concave with respect to p_e variables for α utilities with $\alpha \geq 1$. *Hint*: Apply the relation $x = e^{\log x}$.

9.9   Use the Net2Plan implementation available in the Net2Plan repository for the persistence probability adjustment algorithm in Section 9.5 and repeat the empirical studies in Section 9.4.1 for other selected topologies. Find a setting γ, Δp_e that is robust for all the topologies and report on the convergence speed in them with this setting.

9.10   Repeat the previous exercise for α-fairness values $\{0.5, 1, 2, 3\}$. Comment on the effect of α in empirical convergence and provide a theoretical explanation for it.

9.11   Modify the Net2Plan implementation available in the Net2Plan repository of the primal algorithm for transmission power adjustment, discretizing the power change in units of 0.2 dB, and using the floor function based discretization function (8.21) in Chapter 8. Tailor empirically the γ parameter in this case, assuming that $\gamma_h = 0$ (without heavy-ball inertia).

9.12 Use the Net2Plan implementation available in the Net2Plan repository for the transmission power adjustment algorithm in Section 9.6 and repeat the empirical studies in Section 9.5.1 for other selected topologies. For the case when $\gamma_h = 0$, find a γ setting robust for all the topologies and report on the convergence speed in them with this setting.

9.13 Apply KKT optimality conditions to find an efficient algorithm for solving the problem in $\mathbb{R}^n$:

$$\min_{x \geq 0} \sum_i (x_i - c_i)^2, \text{ subject to: } \sum_i x_i = C$$

References

[1] D. P. Bertsekas, *Nonlinear Programming*. Bertsekas: Athena Scientific, 1999.

[2] M. Minoux, *Mathematical Programming: Theory and Algorithms*, ser. Wiley series in discrete mathematics and optimization. New York, NY, USA: John Wiley & Sons, Inc., Wiley, 1986.

[3] A. Fiacco and G. McCormick, *Nonlinear Programming: Sequential Unconstrained Minimization Techniques*. New York, NY, USA: John Wiley & Sons, Inc., 1968.

[4] J. M. McQuillan and D. C. Walden, "The ARPA network design decisions," *Computer Networks (1976)*, vol. 1, no. 5, pp. 243–289, 1977.

[5] B. T. Polyak, *Introduction to Optimization*. Optimization Software New York, 1987.

[6] D. P. Bertsekas and J. N. Tsitsiklis, *Parallel and Distributed Computation: Numerical Methods*. Athena Scientific, 1997.

[7] L. Lasdon, *Optimization Theory for Large Systems*, ser. Dover books on Mathematics. Dover Publications, 2002.

[8] S. Boyd and L. Vandenberghe, *Convex Optimization*. New York, NY, USA: Cambridge University Press, 2004.

[9] D. P. Bertsekas, *Constrained optimization and Lagrange multiplier methods*. New York, NY, USA: Academic Press, 2014.

[10] J. N. Tsitsiklis and D. P. Bertsekas, "Distributed asynchronous optimal routing in data networks," *Automatic Control, IEEE Transactions on*, vol. 31, no. 4, pp. 325–332, 1986.

[11] D. Bertsekas and R. Gallager, *Data Networks*. Englewood Cliffs, NJ: Prentice Hall, 1992.

[12] F. P. Kelly, A. K. Maulloo, and D. K. Tan, "Rate control for communication networks: shadow prices, proportional fairness and stability," *Journal of the Operational Research Society*, pp. 237–252, 1998.

[13] R. Srikant and L. Ying, *Communication Networks: An Optimization, Control, and Stochastic Networks Perspective*. Cambridge, UK: Cambridge University Press, 2013.

[14] J.-W. Lee, M. Chiang, and A. R. Calderbank, "Utility-optimal random-access control," *Wireless Communications, IEEE Transactions on*, vol. 6, no. 7, pp. 2741–2751, 2007.

[15] M. Chiang, S. H. Low, J. C. Doyle *et al.*, "Layering as optimization decomposition: A mathematical theory of network architectures," *Proceedings of the IEEE*, vol. 95, no. 1, pp. 255–312, 2007.

[16] D. O'Neill, D. Julian, and S. Boyd, "Seeking Foschini's genie: optimal rates and powers in wireless networks," *IEEE Transactions on Vehicular Technology*, 2003.

[17] D. C. ONeill, D. Julian, and S. Boyd, "Adaptive management of network resources," in *Vehicular Technology Conference, 2003. VTC 2003-Fall. 2003 IEEE 58th*, vol. 3. IEEE, 2003, pp. 1929–1933.

[18] M. Chiang, "Balancing transport and physical layers in wireless multihop networks: Jointly optimal congestion control and power control," *Selected Areas in Communications, IEEE Journal on*, vol. 23, no. 1, pp. 104–116, 2005.

10

Dual Gradient Algorithms

10.1 Introduction

In this chapter we present a selected set of case studies to illustrate the so-called *dual approach* for devising network design algorithms. In brief, a dual approach means that a gradient projection algorithm is applied to find the optimum multipliers of a Lagrange relaxation of the problem, as an indirect form to also find its primal solution. Methods obtained in the case studies are amenable to distributed implementation, and inherit the convergence properties for asynchronous executions and noisy gradient observations from the standard gradient iteration.

Let (10.1) be the primal problem to solve.

$$\min_{x \in \mathcal{X}} f(x) \quad \text{subject to:} \tag{10.1a}$$

$$\pi_i : g_i(x) \leq 0, \quad i = 1, \ldots, M \tag{10.1b}$$

where set $\mathcal{X}$ represents an arbitrary set of constraints which are not relaxed, $g_i(x) \geq 0$ are the relaxed constraints and π_i their associated non-negative multipliers. Given a vector of feasible multipliers $\pi \geq 0$, the set of associated primal solutions is given by:

$$\mathcal{X}^*(\pi) = \arg \min_{x \in \mathcal{X}} \left\{ f(x) + \sum_i \pi_i g_i(x) \right\} \tag{10.2}$$

Under mild assumptions (e.g., $\mathcal{X}$ being a compact set), sets $\mathcal{X}^*(\pi)$ are never empty and at least a minimizer $x^*(\pi) \in \mathcal{X}(\pi)$ exists for every $\pi \geq 0$ vector. Then, the dual function:

$$w(\pi) = \min_{x \in \mathcal{X}} \left\{ f(x) + \sum_i \pi_i g_i(x) \right\}$$

is well defined and is concave for all its domain $\pi \geq 0$. A key idea of the dual approach, is that for any multiplier π, a subgradient $s(\pi)$ of the dual function in π is given by (see Prop. B.8 in Appendix B)[1]:

$$s_i = g_i(x^*(\pi)), \quad \forall i = 1, \ldots, M \tag{10.3}$$

[1] The word supergradient actually describes better $s(\pi)$, since it defines an overestimator of a concave function. As established in Appendix A and is customary in the literature, we use the word subgradient to refer to both subgradients of convex functions and supergradients of concave functions.

Optimization of Computer Networks – Modeling and Algorithms: A Hands-On Approach,
First Edition. Pablo Pavón Mariño.
© 2016 John Wiley & Sons, Ltd. Published 2016 by John Wiley & Sons, Ltd.
Companion Website: www.wiley.com/go/PavonMarinoSol16

and this holds for *any* minimizer $x^*(\pi)$ of the multipliers. That is, from a vector of multipliers π we can obtain a minimizer $x^*(\pi)$, which can violate the relaxed constraints, and observing the slack of the constraints $g(x^*(\pi))$ we form a subgradient.

The rationale of the method is then that the optimum multipliers π^* that maximize the dual function can be computed using any flavor of subgradient projection algorithm. When a distributed implementation is sought, a basic iteration with a constant γ step is the usual choice:

$$\pi_i(t + 1) = \left[\pi_i(t) + \gamma g_i\left(x^*(\pi(t))\right)\right]_0, \quad \forall i = 1, \ldots, M \tag{10.4}$$

As shown in Appendix B, optimization theory tells us that if the original problem has the property of strong duality (e.g., a convex problem where Slater conditions hold), once the optimum π^* multipliers are found, *at least one* among their associate minimizers $x^*(\pi^*)$ is an optimal primal solution. In summary, the general scheme of the dual approach is described in Algorithm 8.

Algorithm 8 General dual algorithm for (10.1)

1: *Initialization*: $t = 0$, $\pi(0)$ is any non-zero vector.
2: *Primal iteration*: Compute a minimizer $x^*(\pi(t))$ solving (10.2).
3: *Dual iteration*: Compute $\pi(t + 1)$ multipliers with (10.4).
4:    If optimum π^* is not reached, $t \leftarrow t + 1$, go to the Primal iteration.

The following considerations should be made:

- *Equality constraints*: To simplify the writing, previous explanations assumed only inequality constraints. When some constraints are in the equality form, the whole scheme is the same. The only difference occurs in the gradient update (10.4). Since π_i multipliers for equality constraints do not need to be non-negative, the projection is skipped for them in the subgradient iteration. Then, if we denote $\mathcal{I}_=$ and $\mathcal{I}_\leq$ to be the sets of indexes of the equality and inequality constraints, respectively, the subgradient update (10.4) is replaced by:

$$\pi_i(t + 1) = \left[\pi_i(t) + \gamma g_i(x(\pi(t)))\right]_0, \quad \forall i \in \mathcal{I}_\leq \tag{10.5}$$

$$\pi_i(t + 1) = \pi_i(t) + \gamma g_i(x(\pi(t))), \quad \forall i \in \mathcal{I}_= \tag{10.6}$$

- *Differentiability of the dual function*: For those cases when the dual function is differentiable, we know that subgradients (10.3) are actually gradients. This is good news, since the iteration (10.4) is a gradient projection algorithm that is known to converge to the optimum multipliers for a constant γ step, as long as it is sufficiently small. Recall that when the dual function is not differentiable, a constant γ step guarantees convergence just to a *proximity* of the optimum. In problems with strong duality, the dual function is differentiable everywhere if the objective function f is strictly convex, g_i are convex functions, and $\mathcal{X}$ a convex set (see Prop. B.9). Instead, when strong duality does not hold, the dual function is never differentiable in any dual optimum π^* (see Prop. B.10).

- *Uniqueness of the minimizers*: For any feasible multipliers π, the set of associated minimizers $\mathcal{X}^*(\pi)$ contains the solutions to the relaxed problem (10.2). If the solution of (10.2) is unique, $\mathcal{X}^*(\pi)$ has a single element $(x^*(\pi))$. This is a desirable property and an important requirement in many algorithms. The reason is that when optimum multipliers

π^* are reached, at least one associated minimizer is guaranteed to be primal optimum. But if more than one minimizer exists, the rest can be non-optimum and even unfeasible. A sufficient condition to have uniqueness in the minimizer computation is that the problem objective function f is strictly convex, the constraints $g(x)$ are convex and $\mathcal{X}$ is a convex set. Then, the relaxed problem becomes the minimization of a strictly convex function in a convex set, which has a unique optimum.

- *Problem regularization.* From previous points, it becomes evident that strict convexity of the objective function is a valuable feature to improve convergence of dual algorithms. For those problems (10.1) where the objective function f is convex but not strictly convex, strict convexity can be achieved by adding to f a so-called *regularization term* like $\epsilon \sum_j x_j^2$, being x_j the j-th coordinate of vector x. For sufficiently small $\epsilon > 0$, the resulting objective function $f(x) + \epsilon \sum_j x_j^2$ is in practice equal to the original one, and strictly convex[2]. However, as will be shown in a later case study, numerical difficulties can prevent the use of too small ϵ factors. Other regularization techniques exist, for instance, the so-called proximal minimization algorithms, consisting of adding the term $\epsilon \|x(t)\|^2$ in the step t of the gradient iteration [1].

- *Feasiblity of intermediate solutions.* One of the disadvantages of the dual approach, is that it can produce unfeasible solutions that violate the dualized constraints in all the iterations but the last. In contrast, recall that in the primal gradient methods, every iteration produced a feasible solution.

- *Asynchronous operation, use of outdated information:* The dual approach described in Algorithm 8 enjoys the convergence properties of the standard gradient iteration in the presence of asynchronous updates and also the use of delayed information in the gradient computation. These aspects will be explored in empirical tests in the case studies.

- *Dual approach in $\mathcal{NP}$-hard problems.* In contrast to the primal methods described in Chapter 9, the dual approach can be applied to $\mathcal{NP}$-hard problems, or in general, for problems for which strong duality does not need to hold. In such cases, the convergence of the method has the following issues:

  – The dual function is not differentiable, at least in any optimum π^* multipliers. Then, (10.4) is a subgradient iteration and convergence requires a diminishing step rule, difficult to realize in distributed implementations.

  – Even if the optimum multipliers π^* are reached, there is no guarantee that any associated minimizer is optimum (which would mean that strong duality holds for that particular problem instance) or even feasible.

Still, dual approaches can be valuable methods for many $\mathcal{NP}$-hard problems. The rationale is that given some multipliers π, maybe in the proximity of π^*, the associated minimizers are optimum solutions to problems that can be small variations of the original problem (see Prop. B.19 in Appendix B for details). This aspect will be further explored in Chapter 11.

The creative part in the algorithm design, is finding for our target problem the set of constraints to relax such that the resulting primal (10.2) and dual iterations (10.4) can be implemented in a distributed form or require a significantly lower computation. In this respect, a major advantage of the dual approach is that the projection in the gradient iteration is always

---

[2] Recall that for any $\epsilon > 0$, $\epsilon \sum_j x_j^2$ is strictly convex and that a strictly convex function summed to a convex function is strictly convex.

Table 10.1 Case studies in Chapter 9.

Problem type	Algorithm	Section
Adaptive bifurcated routing	Dual gradient and regularization	Section 10.2
Center-free backpressure routing	Dual subgradient without regularization	Section 10.3
Distributed congestion control	Dual gradient	Section 10.4
Backoff adjustment in CSMA protocols	Dual gradient	Section 10.5

performed in a very simple constraint set $\pi \geq 0$ and thus each multiplier can be projected without knowing the value of the rest. In turn, one of the disadvantages is that a problem regularization may be needed if the objective function is not strictly convex and this can further complicate the primal iteration.

What follows is a comprehensive set of case studies trying to cover the main techniques involved in dual-based network algorithms (see Table 10.1 for an index). Case studies include empirical tests to expose some trade-offs when tailoring the algorithm parameters.

10.2 Adaptive Routing in Data Networks

In this section we present a dual algorithm for the minimum average hop routing problem in a network $\mathcal{G}(\mathcal{N}, \mathcal{E})$, with a given traffic demand $\mathcal{D}$. We focus on a path-flow formulation (10.7), being $\mathcal{P}$ the set of paths and l_p the number of links traversed by a path $p \in \mathcal{P}$.

$$\min_x \sum_p l_p x_p \quad \text{subject to:} \tag{10.7a}$$

$$\sum_{p \in \mathcal{P}_e} x_p \leq u_e, \quad \forall e \in \mathcal{E} \tag{10.7b}$$

$$\sum_{p \in \mathcal{P}_d} x_p = h_d, \quad \forall d \in \mathcal{D} \tag{10.7c}$$

$$x_p \geq 0, \quad \forall p \in \mathcal{P} \tag{10.7d}$$

Dualizing the link capacity constraints (10.7b) using π_e as link multipliers, we have the dual problem $\max_{\pi \geq 0} w(\pi)$, where $w(\pi)$ is the dual function that assigns to each link weight vector π the minimum cost of its relaxed routing, given by:

$$w(\pi) = \min_{x \in (10.7c,d)} \left\{ \sum_p l_p x_p + \sum_e \pi_e \left(\sum_{p \in \mathcal{P}_e} x_p - u_e \right) \right\}$$

$$= \min_{x \in (10.7c,d)} \left\{ \sum_p x_p \sum_{e \in p} (\pi_e + 1) - \sum_e \pi_e u_e \right\}$$

The dual function may be non-differentiable, since the objective function is not strictly convex. A subgradient s of the the dual function is given by the slack of relaxed constraints, and this yields to the dual iteration:

$$\pi_e(t+1) = \left[\pi_e(t) + \gamma \left(\sum_{p \in P_e} x_p - u_e \right) \right]_0, \quad \forall e \in \mathcal{E} \tag{10.8}$$

We notice that each link e is able to compute its subgradient coordinate using local information: its capacity u_e and the traffic traversing the link $\sum_{p \in P_e} x_p$. For this, no per-flow information should be kept. Note also that when γ steps are constant, π_e values evolving as in (10.8) become *proportional to the length of the queue* of packets in link e.

Periodically, the $\pi_e(t+1)$ values computed should be signaled to all the network nodes[3]. Then, each source d adapts the routing to the current weights making:

$$\{x_p, p \in P_d\} = \underset{x_p \geq 0, \sum_{p \in P_d} x_p = h_d}{\arg \min} \left\{ \sum_{p \in P_d} x_p \sum_{e \in p} (\pi_e + 1) \right\} \tag{10.9}$$

which means that the traffic of a demand d should be *carried only through paths which are shortest paths* using $\pi_e + 1$ as link weights. This is consistent with the optimality conditions in the routing problem seen in Chapter 4, which states that the optimal routing is a shortest path routing when the link weights are the optimal multipliers.

Unfortunately, the described scheme yields two main difficulties:

- Since the dual function may be non-differentiable, the subgradient iteration (10.8) with a constant step length is guaranteed to converge just to the proximity of the optimum link weights.
- Even if the optimum weights are attained, there may be infinite associated routings solving (10.9). This happens when more than one shortest path route exists. Theory guarantees that at least one splitting of the traffic among the shortest path routes is optimal, but others may be non-optimal and even unfeasible. The algorithm provides no information on which one to choose. Example 10.1 illustrates this case.

In this case study, we address the previous issues by adding a regularization term $\epsilon \sum_p x_p^2$ to the objective function, which now becomes strictly convex for any $\epsilon > 0$. By choosing an ϵ value sufficiently small, the regularized problem is in practice equal to the original one, but now the dual function is differentiable and only one minimizer exists for each multipler π.

Regularization changes nothing in the dual update (10.8), but complicates the routing update performed by each demand, which now requires solving a somewhat more complex problem:

$$\{x_p, p \in P_d\} = \underset{x_p \geq 0, \sum_{p \in P_d} x_p = h_d}{\arg \min} \left\{ \sum_{p \in P_d} x_p \sum_{e \in p} (\pi_e + 1)) + \sum_{p \in P_d} x_p^2 \right\} \tag{10.10}$$

The good side is that (10.10) can still be efficiently solved (see Exercise 10.2, and its Net2Plan implementation) and only requires using demand local information and signaled π_e values. Algorithm 9 illustrates a distributed implementation of the complete scheme.

---

[3] This can be implemented in multiple forms in current networks. For instance, using protocols like BGP-LS (BGP Link-State) or OSPF-TE (OSPF Traffic Engineering) in IP/MPLS networks. Software Defined Networking (SDN) controllers can also collect and disseminate this information.

Algorithm 9 Adaptive routing for regularized (10.7)

1: *Link's algorithm*: At times $t = t_1(e), t_2(e), \ldots$, link e:
2:      Estimates link carried traffic and weight π_e.
3:      Signals π_e to all network nodes.
4: *Demand's algorithm*: At times $t' = t_1'(d), t_2'(d), \ldots$, demand d:
5:      Collects the most updated signaled π_e values.
6:      Updates $(x_p, p \in P_d)$, using (10.10).

Example 10.1 Consider the network in Fig. 10.1, with one single traffic demand of offered volume 1.5 units, which should be carried through two paths, p, and p'. All the links have a capacity $u_e = 1$ and thus both paths should carry traffic to satisfy the demand. According to the optimality conditions of the problem, this means that both paths are shortest paths when the π_e weights are optimal. Let us assume that the dual iteration reached such optimal link weights. Applying (10.9), we have no information on which splitting of the traffic between two paths could be chosen. One of these solutions ($x_p = 1, x_{p'} = 0.5$) is optimal. However, other solutions are suboptimal (e.g., $x_p = 0.75, x_{p'} = 0.75$), or can be even unfeasible ($x_p = 1.25, x_{p'} = 0.25$).

10.2.1   Optimality and Stability

Optimality and stability for any initial conditions can be guaranteed in the regularized version of the dual algorithm described, for sufficiently small γ steps that make the algorithm converge to the optimal link multipliers. This may be no longer true if the original problem (10.7) is not convex for example, if non-bifurcated discrete constraints like $x_p \in \{0, h_{d(p)}\}$ are added. In this case, oscillations and instabilities can occur, analogous to those described in Section 9.3.2.

10.2.2   Implementation Example

In this section we present some experimental tests of Algorithm 9. The network topology (Fig. 9.2), link capacities ($u_e = 25, \forall e$) and traffic matrix (9.12) are the same as the ones in Section 9.3.3. The candidate paths for each demand are all the loopless paths between the demand end nodes. The link weight and routing updates are performed asynchronously. Each link updates its weight using the gradient iteration, signals it to the rest of network nodes, and waits a random time between 0.5 and 1.5 time units to the next update. Signaling messages arrive immediately to all network nodes, except for a 5% of the messages that are randomly lost.

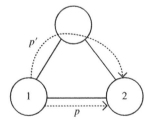

Figure 10.1   Example network

Each router updates the routing of all its outgoing demands using the most recent link weight information it has, and then waits a random time between 0.5 and 1.5 time units before the next routing update.

10.2.2.1 Dimensioning of the ϵ Regularization Parameter

Figure 10.2 illustrates the algorithm convergence for the case of $\epsilon = 10^{-3}$ and a step $\gamma = 0.001$. We observe that, during the first iterations, some links are oversubscribed and their link weights grow, while others are not and their weights decrease. This first phase is slow given the small γ step. However, when a critical point is reached, the routing is successfully adapted in a few iterations and remains stable in the optimum. In our experiments, higher γ steps could fail to do that. We argue that the long first phase should not be seen as a big issue in an adaptive routing algorithm, which is supposed to be permanently reacting to relatively slow variable network conditions.

Parameter ϵ plays a very important role to keep algorithm stability. Using lower ϵ values means that small variations in the link weights can produce drastic changes in the routing, since the reaction is closer to "non-bifurcation, shortest path takes all". In our example, this happens, for example for $\epsilon = 10^{-5}$, as shown in Fig. 10.3. See that links can be oversubscribed.

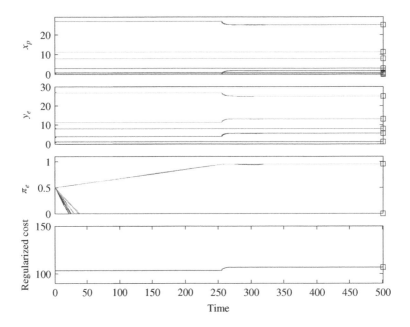

Figure 10.2 Evolution of Algorithm 9 in the asynchronous case with signaling losses, $\gamma = 0.001$, $\epsilon = 10^{-3}$. The upper graph plots the x_p evolution for each path, the y_e graph shows the traffic in each link, π_e graph is the link weight evolution, and the lower graph plots the regularized cost $\sum_p l_p x_p + \epsilon \sum_p x_p^2$. Optimum solutions are marked with squares on the right-hand side. Regularized costs are lower than the optimal occuring since in the first phase, some links are oversubscribed, and the solution is unfeasible

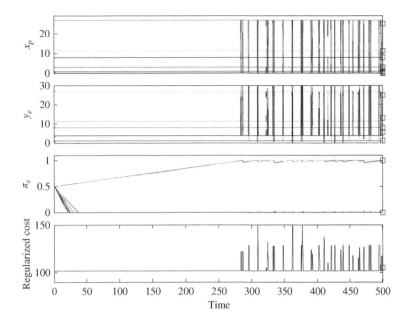

Figure 10.3   Same example as Fig. 10.2, effect of reducing ϵ to $\epsilon = 10^{-5}$

10.2.2.2   Effects of Signaling Delays

Higher signaling delays means using outdated information in the link iteration that can hinder or even prevent convergence. As an example, Fig. 10.4 shows the algorithm evolution in the same case as Fig. 10.2, but assuming that each signaling message takes a random delay uniformly distributed in the interval $B = 10 \pm 0.5$. As is shown, oscillations occur and neither equilibrium nor optimality are reached (actually, solutions violate link capacity constraints). The reader can easily check how smaller γ values can counteract this effect, as predicted by theory.

10.2.2.3   Effects of Gradient Noise

Inaccuracies in the computation of the link traffic and/or the link capacity (when it is not fixed) mean that link weights can have measurement errors. As predicted by theory, convergence is then guaranteed to the proximity of the optimum, as long as the noise variance remains finite. As an example, Fig. 10.5 shows the algorithm evolution when link available capacities $(u_e - y_e)$ are measured with an unbiased error of ± 2.5 traffic units, which is $\pm 10\%$ of link capacity. As can be seen, the link measurement errors are translated into small fluctuations in some of the routes, which can cause link oversubscriptions. The routes of non-bifurcated demands are insensitive as long as the link weight noise is not large enough to change the routing decision.

10.3   Backpressure (Center-Free) Routing

We refer to a center-free routing algorithm as an scheme where routing decisions in a node are determined by signaling information from their neighbor nodes and do not rely on network

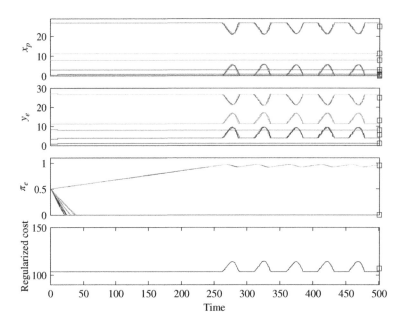

Figure 10.4   Same example as Fig. 10.2, effect of signaling delay $B = 10 \pm 0.5$

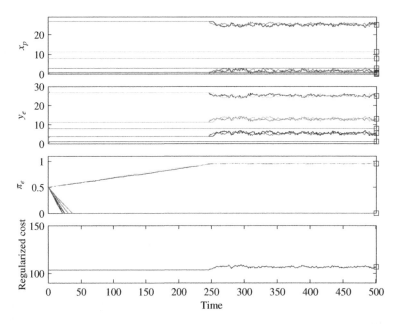

Figure 10.5   Same example as Fig. 10.2, effect of measurement error in idle link capacity of $\pm 10\%$ of link capacity

wide information (e.g., link weights of all network links traversed). Moreover, routing decisions are taken hop-by-hop, in contrast to the previous section where nodes bifurcate the traffic among a precomputed set of end-to-end paths.

Center-free algorithms can be obtained from flow-link or destination-link formulations, by relaxing *flow conservation constraints* and moving them to the objective function. In this section we illustrate this, presenting a flavor of a well-known center-free scheme called the *backpressure algorithm*.

We focus on a flow-link formulation of the minimum average hop routing problem in a network $\mathcal{G}(\mathcal{N}, \mathcal{E})$, with offered traffic given by demand set $\mathcal{D}$:

$$\min_{x} \Delta P \sum_{de} x_{de} \quad \text{subject to:} \tag{10.11a}$$

$$\sum_{e \in \delta^+(n)} x_{de} - \sum_{e \in \delta^-(n)} x_{de} \begin{cases} \geq h_d, \text{ if } n = a(d) \\ \geq 0, \text{ otherwise} \end{cases}, \quad \forall d \in \mathcal{D}, n \in \mathcal{N} - b(d) \tag{10.11b}$$

$$\sum_{d} x_{de} \leq u_e, \quad \forall e \in \mathcal{E} \tag{10.11c}$$

$$x_{de} \geq 0, \quad \forall d \in \mathcal{D}, e \in \mathcal{E} \tag{10.11d}$$

Factor $\Delta P > 0$ in the objective function is the *pressure difference threshold*. Note that the optimum routing of (10.11) is the same whatever value ΔP takes. The importance of ΔP factor to explain the algorithm behavior will be clarified later.

Compared to the traditional flow-link formulation, we have eliminated the flow conservation constraints (10.11b) for demand end nodes ($n = b(d)$) since they are redundant. Also, we have replaced equalities with inequalities. Both problems are equivalent in the sense that, in the optimum flow, conservation constraints are satisfied with equalities, since injecting more traffic in the links than strictly needed is not optimal.

We dualize constraints (10.11b), using q_{nd} as multipliers associated to node n and demand d. Note that these multipliers are now non-negative. The relaxed problem is thus:

$$\min_{x \in (10.11c,d)} \Delta P \sum_{de} x_{de} + \sum_{d,n \neq b(d)} q_{nd} \left(h_{nd} - \sum_{e \in \delta^+(n)} x_{te} - + \sum_{e \in \delta^-(n)} x_{te} \right) \tag{10.12a}$$

$$= \min_{x \in (10.11c,d)} \sum_{de} x_{de} \left(\Delta P + q_{b(e)t} - q_{a(e)t} \right) + \sum_{d,n \neq b(d)} q_{nd} h_{nd} \tag{10.12b}$$

To simplify the notation, we used h_{nd} to denote the traffic of demand d that is generated by node n:

$$h_{nd} = \begin{cases} h_d, \text{ if } n = a(d), \\ 0, \text{ otherwise} \end{cases}, \quad \forall d, n \neq b(d)$$

In the dual approach, the evolution of the q_{nd} multipliers is controlled by the gradient projection iteration:

$$q_{nd}(t + 1) = \left[q_{nd}(t) + \gamma \left(h_{nd} - \sum_{e \in \delta^+(n)} x_{de}(t) + \sum_{e \in \delta^-(n)} x_{de}(t) \right) \right]_0 \tag{10.13}$$

The following considerations are made:

- The variation of q_{nd} in the dual iteration (10.13) is proportional to the difference between the traffic of d that the node should forward, minus the traffic it actually forwards, and becomes zero if the accumulation of these quantities along time vanishes. Then, for fixed step γ, q_{nd} values are *proportional* to the *queue length* Q_{nd} associated to the traffic of demand d, that is at node n, waiting to be forwarded:

$$q_{nd} = \gamma Q_{nd}$$

- The algorithm requires that each node n is signaled the $q_{n'd}$ values for all demands d, and from all neighbor nodes n' to whom it can transmit traffic to ($n' = b(e), \forall e \in \delta^+(n)$). There-fore, each node n' should just collect the sizes of its internal queues (one per demand) and signal this information to its incoming neighbors (the nodes from which it receives traffic).
- For a given set of q_{nd} multipliers signaled, each node n determines the traffic to forward solving the relaxed problem (10.12). Interestingly, this problem can be solved independently for each link: a link e must forward the traffic of demand d for which the quantity $(\Delta P + q_{b(e)d} - q_{a(e)d})$ is minimum, assuming that the quantity is negative. If $\Delta P + q_{b(e)d} - q_{a(e)d}) > 0$ for all demands, the link does not forward traffic.
- According to the previous operation, the multiplier q_{nd} can be interpreted as a *pressure* that node n feels to forward traffic of demand d. The forwarding condition $q_{a(e)d} - q_{b(e)d} > \Delta P$ means that the pressure felt by the origin node of e to transmit traffic of d minus the pressure felt by the destination node should be higher than the pressure difference threshold ΔP.

Algorithm 10 summarizes the described backpressure scheme. A similar dual approach can be applied to destination-link formulation. This is left as an exercise (Exercise 10.6).

Algorithm 10 Center-free backpressure routing for (10.11)

1: *Node's algorithm*: At times $t = t_1(n), t_2(n), \ldots$, node n:
2:    Collects queue length information for its queues $\{q_{nd}, \forall d\}$.
3:    Signals it to incoming neighbor nodes.
4: *Link's algorithm*: At times $t' = t_1'(e), t_2'(e), \ldots$, link e:
5:    Collects the most updated q_{nd} values signaled from outgoing neighbors.
6:    Transmits traffic of demand d with highest pressure difference, as long as it is $> \Delta P$.

10.3.1 Relation between γ, ΔP, and Average Queue Sizes, Q_{nd}

The following relations hold:

- The optimum routing does not depend on the ΔP parameter, but the optimum multipliers q_{nd} do depend on it. Actually, if we denote $q_{nd}(\Delta P)$ to the optimum q_{nd} multipliers of problem (10.11) using the ΔP threshold, we have:

$$q_{nd}(\Delta P) = \Delta P q_{nd}(1)$$

In other words, multipliers are proportional to the ΔP parameter. The proof is left for Exercise 10.7. Intuitively, the dual function of (10.11) is proportional to ΔP and so are

q_{nd} multipliers as subgradients of the dual function. Interestingly, this also means that if a γ step is sufficiently small to guarantee convergence for a problem with ΔP factor, an $\alpha\gamma$ step will also be sufficiently small to guarantee convergence in a problem with $\alpha\Delta P$.

- If a constant γ step is used, and *if the algorithm converges* to the optimum multipliers q_{nd}, then, according to (10.13):

$$q_{nd} = \gamma Q_{nd}$$

Putting together previous points, we have that *if convergence is achieved*, the resulting queue sizes $Q_{nd}(\Delta P, \gamma)$, for a problem with pressure difference ΔP and a constant step γ, satisfies the relation:

$$Q_{nd}(\Delta P_1, \gamma_1) = \frac{\Delta P_1 \gamma_2}{\Delta P_2 \gamma_1} Q_{nd}(\Delta P_2, \gamma_2) \tag{10.14}$$

And the convergence guarantees depend on the ratios $\Delta P_1/\gamma_1$ and $\Delta p_2/\gamma_2$. Then, γ and ΔP parameters are coupled and, for analyzing convergence guarantees and queue evolution of Algorithm 10 it is enough to focus just on the ratio $\Delta P/\gamma$.

10.3.2 Implementation Example

In this section we present experimental results of Algorithm 10 to illustrate some of the trade-offs appearing in it. We focus on a network like the one in previous section (topology in Fig. 9.2), traffic matrix (9.12). All links have a capacity of $u_e = 25$ units.

We assume that time is slotted and assume without loss of generality a slot duration of 1 time unit. New traffic is generated every time slot, according to the traffic matrix (9.12). Also, for every time slot, each link repeats for u_e times the forwarding procedure:

- Choose the demand d with highest pressure difference $q_{a(e)d} - q_{b(e)d}$. If more than one exists, pick one arbitrarily.
- If the pressure difference is above $\Delta P = 1$, forward one packet of the queue for demand d and reduce appropriately the queue size in $a(e)$ and thus the $q_{a(e)d}$ pressure.

The signaling processes occur asynchronously for every node and asynchronously to the slotted operation. A node collects its queue information, signals it in a message to each incoming neighbor node, and waits a random time between 0.5 and 1.5 time units until the next signaling event. Some 5% of the signaling messages are lost. Then, a node uses its local queues $q_{a(e)d}$ information for the forwarding decision, but possibly outdated $q_{b(e)d}$ sizes are signaled from the neighbor nodes.

10.3.2.1 Non-Differentiability of the Dual Function

Since the objective function (10.11a) is not strictly convex, the dual function is not differentiable. Thus, (10.13) is a subgradient algorithm, with guarantees of converging just to the proximity of the optimum for constant γ steps. In addition, even if the optimum q_{nd} multipliers are found, many minimizers can exist. This reflects in the possibility of having several demands with the same maximum pressure difference in a node. Algorithm 10 picks one arbitrarily.

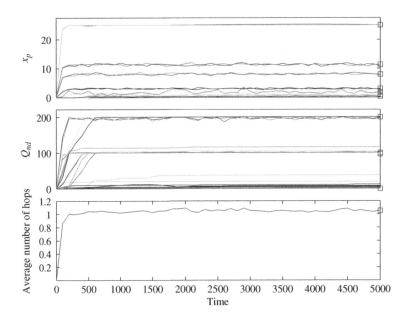

Figure 10.6   Evolution of Algorithm 10 in asynchronous case with signaling losses, $\gamma = 0.01$, $\Delta P = 1$. The upper graph plots the resulting route evolution, the medium graph the queue sizes in the average number of packets, and the lower graph the average number of hops

In this case, we do not address the lack of differentiability adding a regularization term $\epsilon \sum_{de} x_{de}^2$ to the objective function. The reason is that we cannot send fractions of packets through a link (something that could come up if a regularization term is used). Besides, we are targeting a scheduling algorithm where the forwarding decisions are as simple as possible to permit fast implementations to be used in time slots in the order of microseconds/miliseconds. As we will see, the resulting fluctuations that can appear at the packet level become relatively unimportant observing the aggregated traffic at higher time scales.

10.3.2.2   Trade-Off between Queue Sizes and Convergence

As predicted by theory, sufficiently small γ steps are needed to have algorithm convergence. According to relation (10.14), this means that resulting *queue sizes should be large enough*. Longer queues mean longer end-to-end delays and thus there is a trade-off between algorithm convergence and queueing delays: there is a minimum and unavoidable queueing delay that should be accepted to have algorithm convergence.

Figures 10.6–10.8 help us to illustrate these aspects. The three figures show the evolution of the routes, queue sizes (in number of packets), and average number of hops of the routing for different γ steps ($\Delta P = 1$). Each sample of these quantities is the result of averaging 100 contiguous slots. We see that:

- Convergence to the optimum routing is achieved in Fig. 10.6 ($\gamma = 0.01$) and Fig. 10.7 ($\gamma = 0.1$). The resulting average number of hops is close to 1, which is possible since the offered

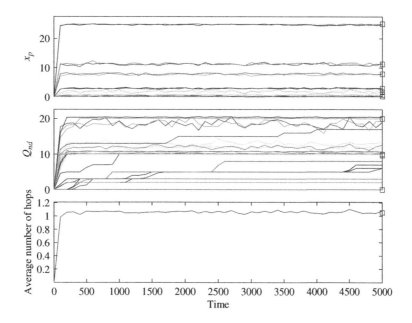

Figure 10.7   Same example as Fig. 10.6, $\gamma = 0.1$

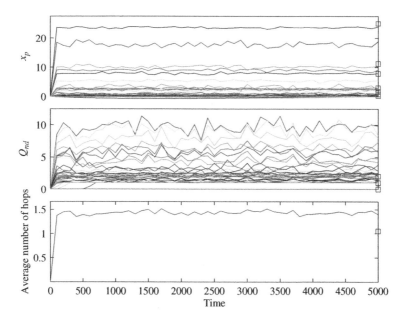

Figure 10.8   Same example as Fig. 10.6, $\gamma = 1$

traffic between the nodes separated two hops is small. The initial phase with a number of hops lower than 1 reflects the fact that the network is initially empty. In this transitory part, traffic enters the network but does not leave it, since queues are growing and the traffic is being stored instead of forwarded. This phase results in a significant delay to the initial traffic that may be queued a relatively long time and/or follow very long routes. It ends when queues stabilize and the resulting routes are optimum.

- As predicted by theory, average queue sizes when $\gamma = 0.1$ are 10 times smaller than when $\gamma = 0.01$. Then, they are also relatively more affected by the random fluctuations of the traffic. The optimum queue values plotted correspond to q_{nd}/γ.
- When $\gamma = 1$, the equilibrium reached is far from an optimum solution: the average number of hops is now ≈ 1.4 and queue sizes are much larger than q_{nd}/γ.

Exercise 10.8 observes the effects of increasing signaling delays into the algorithm. Interestingly, it can be seen how a backpressure algorithm is significantly robust to such delays.

10.4 Congestion Control

In this section, we show a dual approach to solving the NUM congestion control problem:

$$\max_h \sum_d U_d(h_d) \quad \text{subject to:} \tag{10.15a}$$

$$\sum_{d:e \in p_d} h_d \le u_e, \quad \forall e \in \mathcal{E} \tag{10.15b}$$

$$m_d \le h_d \le M_d, \quad \forall d \in D \tag{10.15c}$$

where decision variable h_d is the rate allocated to demand $d \in D$, restricted to the interval $[m_d, M_d]$. p_d is the known path associated to demand d and $U_d(h_d)$ a strictly concave utility function that determines the fairness properties of the optimum solution of (10.15).

In the dual approach, we relax link capacity constraints (10.15b) associating π_e multipliers to them. The dual function $w(\pi)$ returns the optimum relaxed cost for a set of multipliers π:

$$w(\pi) = \max_{h \ge 0} \sum_d U_d(h_d) + \sum_e \pi_e \left(u_e - \sum_{d:e \in p_d} h_d \right) \tag{10.16}$$

$$= \max_{h \ge 0} \sum_d U_d(h_d) + h_d \sum_{e \in p_d} \pi_e + \sum_e \pi_e u_e \tag{10.17}$$

The dual function is differentiable, since the objective function is strictly concave and problem (10.15) enjoys strong duality. Then, the only minimizer associated to the optimum multipliers is the primal optimum. Optimum multipliers are pursued using a basic gradient projection algorithm in the iteration:

$$\pi_e(t+1) = \left[\pi_e(t) - \gamma \left(u_e - \sum_{d:e \in p_d} h_d \right) \right]_0, \quad \forall e \in \mathcal{E} \tag{10.18}$$

The coordinates of the gradient of the dual function are the differences between the link capacities and the offered traffic to the links. Then, for constant γ steps, $\pi_e(t)$ evolution is proportional to the amount of traffic queued, pending to be transmitted through link e. Given a set of multipliers $\pi(t)$, the associated demand rates are computed finding the minimizer in (10.17) and thus the primal iteration is the h_d value for which:

$$
h_d(t) = \left[U'^{-1}_d \left(\sum_{e \in p_d} \pi_e(t) \right) \right]^{M_d}_{m_d}
$$

For α-fair utility functions (3.19), equation (10.19) becomes:

$$
h_d(t) = \left[\left(\sum_{e \in p_d} \pi_e(t) \right)^{-\alpha} \right]^{M_d}_{m_d}
\tag{10.19}
$$

The dual scheme described is amenable to a distributed implementation where links and sources cooperate as in Algorithm 11.

Algorithm 11 Congestion control for (10.15)

1: *Link's algorithm*: At times $t = t_1(e), t_2(e), \ldots$, link e:
2:     Observes the traffic carried in the link $\sum_{d:e \in p_d} h_d$.
3:     Computes its new price according to (10.18).
4:     Signals the new price $\pi_e(t+1)$ to the sources traversing e.
5: *Demand's algorithm*: At times $t' = t'_1(d), t'_2(d), \ldots$, demand d:
6:     Receives from the network the sum of link prices $\sum_{e \in p_d} \pi_e$.
7:     Chooses a new transmission rate according to (10.19).

The link iteration requires monitoring the aggregated link occupation, without storing any per-flow information. The signaling of the link weights to the demand source nodes is assumed to be implemented in a explicit form, using an external signaling mechanism (e.g., the one provided by the Resource Management (RM) cells ATM Available Bit Rate (ABR) control mode).

10.4.1   Optimality and Stability Conditions

An asynchronous operation of Algorithm 11 means that the traffic observed in the link e at a moment t depends on the sources' congestion control decisions performed asynchronously in the past. In addition, a variation in a source can take time to be observable in the links, for example because of non-negligible propagation times. Then, the gradients computed for adjusting the link weights may be based on outdated information. Still, the theory predicts that, for a sufficiently small γ step, the algorithm convergence is guaranteed. The following result in this line is extracted from [2]:

Proposition 10.1 In problem (10.15), we assume that utility functions are increasing, twice continuously differentiable and strongly convex, where:

$$|U_d''(h_d)| \geq \frac{1}{\eta}, \quad \forall h_d \in [m_d, M_d]$$

we denote L as the maximum number of hops a demand traverses and S the maximum number of demands sharing a link. Then, if gradient step γ satisfies:

$$0 < \gamma < \frac{2}{\eta LS}$$

the synchronous version of Algorithm 11 converges to the optimum (assuming the problem is feasible), for any initial network conditions.

As shown in [2], and confirmed by the experiments shown later in this section, convergence can be obtained in practice for significantly higher γ values than the ones in previous sufficient conditions.

10.4.2   Implementation Example

We present here some tests illustrating the convergence properties of Algorithm 11. We use the same topology as in the primal congestion algorithm: topology of Fig. 9.2, with a demand between each node pair, carried through the shortest path route in km between the end nodes. Each link e independently computes its weight π_e monitoring its traffic, signals it to the traversing demands, and waits a random time between 0.5 and 1.5 units before the next update. Signaling messages arrive instantly to demand source nodes, but a fraction of 5% messages are randomly lost. Independent from the demand update, each demand recomputes its injected traffic using the most updated link information it has, and waits a random time between 0.5 and 1.5 time units until the next computation.

The utility functions of the demands are the α-utility functions (3.19), with $\alpha = 2$. Minimum demand injected traffic is set to $m_d = 0.1, \forall d$, and the maximum $M_d = \infty$. Observing network topology (Fig. 9.2), the maximum number of hops of a demand is $L = 2$, and the maximum number of demands sharing a link is $S = 2$. Since $\alpha = 2$, we have that:

$$U_d''(h_d) = -2h_d^{-3}$$

The minimum value that U_d'' can take occurs when h_d gets its maximum value. Since $M_d = \infty$ the h_d limit is set by the link capacity ($u_e = 25$). Thus:

$$\eta = \max_d \left\{ \frac{h_d^3}{2} \right\} = \frac{25^3}{2} = 7812.5$$

Then, applying Prop. 10.1, convergence is guaranteed in the synchronous case for:

$$0 < \gamma < \frac{2}{\eta LS} = 6.4 \times 10^{-5}$$

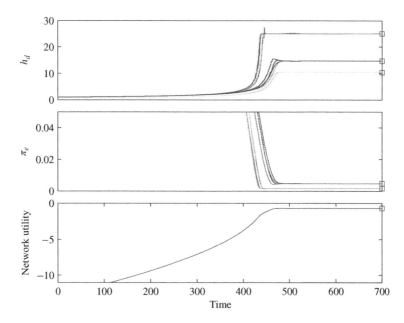

Figure 10.9 Evolution of Algorithm 11 in the asynchronous case with signaling losses, $\gamma = 0.0001$. The upper graph plots the resulting injected traffics h_d, medium graph the link multipliers and the lower graph the network utility to maximize

We have found that empirical convergence in the asynchronous executions occurs for $\gamma = 10^{-4}$, as shown in Fig. 10.9. We can see that traffic adaptation is relatively fast, once the π_e multipliers reach the proximity of the optimum (initial multipliers are set to one). In our tests, convergence did not occur for $\gamma = 5 \times 10^{-4}$. When signaling was subject to random delays in the interval 5 ± 0.5 time units, a γ step as small as $\gamma = 10^{-5}$ was needed to have convergence.

10.5 Decentralized Optimization of CSMA Window Sizes

In Section 5.4.2 of Chapter 5 we modeled MAC protocols based on carrier sense (CSMA) for wireless networks. In CSMA protocols, nodes sense the channel and refrain the transmission if they find it occupied. When the channel is sensed idle again, nodes wait a back-off time randomly chosen for each outgoing link. This timer is frozen when the channel is sensed as occupied. Eventually, the link transmits the message, but a collision can still occur if more than one node end its back-off timer at exactly the same time.

We modeled the case when link back-off times are randomly selected using an exponential distribution of average $1/R_e$ and called $r_e = \log R_e$ to the *transmission aggressiveness* (TA) of the link. The higher the TA r_e, the shorter are the average back-off times R_e, and the more aggressively that link e tries to transmit traffic.

Since the back-off times are chosen in a continuum, the model assumes that collisions never occur. Thus, the network permanently produces schedules (sets of simultaneously active links) that are valid, or collision-free. We denote $\mathcal{M}$ as the set of valid schedules, and $\mathcal{E}(m)$ the set of active links in schedule $m \in \mathcal{M}$. Given the TA in the links $r = \{r_e, e \in \mathcal{E}\}$, the fraction of time π_m that the network uses schedule $\pi_m(r)$ is given by:

$$\pi_m(r) = \frac{e^{\sum_{e \in \mathcal{E}(m)} r_e}}{\sum_m e^{\sum_{e \in \mathcal{E}(m)} r_e}}, \qquad \forall m \in \mathcal{M}$$

The resulting capacity in a link e $(u_e(r))$ is computed by aggregating the time fractions of those schedules where e is active $(\mathcal{M}(e))$, multiplied by the link nominal bit rate $\bar{u}_e$:

$$u_e(r) = \bar{u}_e \sum_{m \in \mathcal{M}(e)} \pi_m(r)$$

A relevant property introduced in the proof of Prop. 5.3 in Section 5.4.2, is that given a TA allocation r and strictly feasible capacities u, the resulting schedules $\pi(r)$ are those which maximize the negative entropy of the distribution among those schedules that attain u. This means that:

$$\pi(r) = \underset{\sum_m \pi_m = 1, \pi_m \geq 0}{\arg\max} \left\{ -\sum_m \pi_m \log(\pi_m) + \sum_e r_e \left(u_e - \bar{u}_e \sum_{m \in \mathcal{M}(e)} \pi_m \right) \right\} \qquad (10.20)$$

In this section, we exploit this property to devise an algorithm that *approximates* the optimum TA allocation which maximizes the network utility of the resulting capacity allocation. For this, we focus on the optimization problem:

$$\underset{\pi, u}{\max} -\sum_m \pi_m \log(\pi_m) + \beta \sum_e U_e(u_e) \quad \text{subject to:} \qquad (10.21a)$$

$$r_e : u_e \leq \bar{u}_e \sum_{m \in \mathcal{M}(e)} \pi_m \quad \forall e \in \mathcal{E} \qquad (10.21b)$$

$$\sum_m \pi_m = 1, \pi_m \geq 0, \quad \forall m \in \mathcal{M} \qquad (10.21c)$$

where U_e are increasing and strictly concave utility functions, and both u and π are problem variables. β is an input parameter, chosen to be as large as possible (limited by numerical inaccuracies), so that the importance of the utility maximization part makes the negative entropy effect negligible.

Problem (10.21) is convex, since it involves the maximization of a concave function under linear constraints. Moreover, the problem has a unique optimum solution, as the objective function is strictly concave. We attempt a dual approach to solve (10.21), using r_e as multipliers of dualized constraints (10.21b). As will be shown later, these multipliers are the TAs of the links in the CSMA protocol. The Lagrangian function is given by:

$$L(\pi, u, r) = -\sum_m \pi_m \log(\pi_m) + \beta \sum_e U_e(u_e) + \sum_e r_e \left(\bar{u}_e \sum_{m \in \mathcal{M}(e)} \pi_m - u_e \right)$$

Given a set of multipliers $r \geq 0$, the associated maximizers $\pi(r), u(r)$ can be computed independently:

$$\pi(r) = \underset{\sum_m \pi_m = 1, \pi_m \geq 0}{\arg \max} \left\{ -\sum_m \pi_m \log (\pi_m) + \sum_e r_e \bar{u}_e \sum_{m \in \mathcal{M}(e)} \pi_m \right\} \tag{10.22a}$$

$$u(r) = \arg \max \left\{ \beta \sum_e U_e(u_e) - r_e u_e \right\} \tag{10.22b}$$

Observing (10.20), we see that maximizer $\pi(r)$ is the schedule enforced by an ideal CSMA protocol that uses r as its TA. In turn, the link capacities $u(r)$ associated to such TA, may not satisfy the *relaxed* constraints (10.21b). This means that if r are *arbitrary* TAs in the network, the link e may not be able to carry a traffic given by $u_e(r)$. However, when TAs are the *optimum* TAs, associated $u_e(r)$ capacities are the unique solution of problem (10.21), they satisfy constraints (10.21a), and optimize the objective function. It can be shown (see [3]) that as β parameter increases ($\beta \to \infty$), the resulting capacities approximate the ones that optimize network utility.

Since the objective function is strictly concave, the dual function of problem (10.21) is differentiable and has a gradient given by the slack of relaxed constraints (10.21b). A basic project gradient iteration to find the optimum TAs r is given by:

$$r_e(t+1) = \left[r_e(t) - \gamma \left(\bar{u}_e \sum_{m \in \mathcal{M}(e)} \pi_m(r) - u_e(r) \right) \right]_0 \tag{10.23}$$

We see that:

- Expression $s_e(r) = \bar{u}_e \sum_{m \in \mathcal{M}(e)} \pi_m(r)$ is the amount of traffic served or carried by link e when TAs in the links are given by r. This can be monitored independently by each link, without the need of computing the π_m schedules.
- The associated capacities $u_e(r)$ are a sort of *intended capacities* computed solving (10.22b). This computation can be also made independently link by link, using link local information. Recall that before the optimum TAs are reached, it can happen that $s_e < u_e(r)$. Then, $u_e(r)$ has the meaning of *actual* link capacity, just when optimum TAs are reached.

Previous ideas are the guidelines for Algorithm 12, a distributed scheme that finds the optimum TAs that maximize the network utility of the resulting link capacity allocation. Since the dual function is differentiable, algorithm converges to the optimum for a sufficiently small γ step also in the asynchronous case. Note that no signaling information is exchanged between nodes. Coordination occurs thanks to the standard CSMA interactions, that produce the service rates s_e observed by the links. Finally, note that in real networks, service rates monitored $s_e(r)$ are actually random variables. A study of the convergence in expectation of the resulting stochastic algorithm can be consulted in [3].

Algorithm 12 CSMA window adjustment for (10.21)

1: *Link's algorithm:* At times $t = t_1(e), t_2(e), \ldots$, link e:
2:    Observes the traffic carried in the link s_e during previous interval.
3:    Computes its new TA r_e according to (10.23).
4:    Applies the TA during next interval.

10.5.1 Implementation Example

This section shows convergence tests validating the Algorithm 12, and illustrates a trade-off between convergence speed and algorithm accuracy related to the β factor. We perform our experiments in the wireless network of Fig. 9.10, with a nominal rate $\bar{u}_e = 1$ in all the links. We assume the standard wireless limitations in the CSMA model described: a node cannot receive simultaneous transmissions from different nodes, and cannot receive and transmit simultaneously. Nodes operate independently and asynchronously: the time between two node wake-ups for TA update is random and uniformly distributed between 0.5 and 1.5 time units. During an update event, the node recomputes its TA using the monitored information since its last update, and applies it until the next.

We assume that the ideal CSMA MAC protocol described in previous section is operating between any two consecutive r_e adjustments. This means that collisions are avoided, the network spends a fraction of time given by $\pi(r)$ in each scheduling state, and served traffic of each link s_e is determined by these $\pi(r)$ vectors. However, to reflect the random nature of the traffic in a real network, the monitored traffic $s_e(r)$ is the ideal one plus a noise uniformly picked in an interval of $\pm 10\%$ of $s_e(r)$.

Figure 10.10 shows the convergence tests for $\beta = 10, \gamma = 5$. The algorithm converges and obtains the optimal TAs r reasonably quickly, with close to optimal utilities achieved in ≈ 50 iterations. However, because of the random noise added to the r evolution, the small changes in the TAs produce a significant added noise in the link capacities that does not reflect into significant utility variations. It is easy to see that, as predicted by theory, lower γ steps can reduce these effects, at a cost of a slower convergence.

As a final remark, it is interesting to see how the link capacities obtained by the CSMA protocol are significantly larger (about one order of magnitude) than the ones obtained in Section 9.5 for the same network using a random access MAC protocol[4].

10.5.1.1 Effect of Increased β Factor

As a measure of the inaccuracy associated to the $\beta = 10$ factor, we observed that in the optimum cost, the weight of the negative entropy part $(-\sum_m \pi_m \log \pi_m)$ was approximately 100 times less than the weight of the network utility part $(\beta \sum_e U_e(u_e))$. Figure 10.11 shows the case

---

[4] Results in the random access case were obtained for a different fairness factor $\alpha = 2$. The reader can check that the differences in link capacities are similar if the same α factor is used.

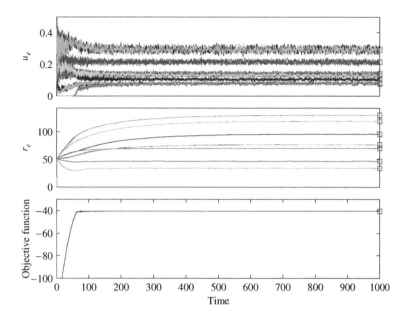

Figure 10.10 Evolution of Algorithm 12 in the asynchronous case with measurement noise, $\gamma = 5$. The upper graph plots the resulting link capacities u_e, the medium graph the TAs (link multipliers) and the lower graph network utility to maximize. Optimum solutions are marked with squares on the right-hand side

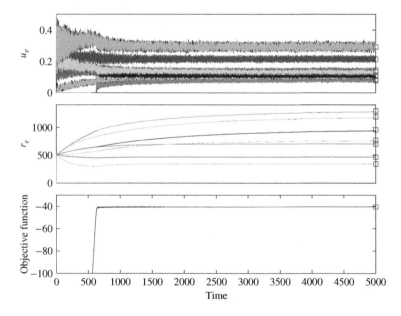

Figure 10.11 Same example as Fig. 10.10, $\beta = 100$

when $\beta = 100$. Now, the weight of the negative entropy falls to 0.1% in the objective function and network utility is actually the same in practice.

A trade-off exists between improved accuracy (higher β) and algorithm convergence. Higher β factors, mean higher TA values in the links. The explanation for this is that, as network utility dominates the objective function, multiplying it by, for example, 10 (by making $\beta = 100$ instead of $\beta = 10$), means scaling the multipliers of problem (10.21) by 10, which are the link TAs.

Intuitively, scaling up by 10, the TAs also require scaling γ steps by 10 in order to have similar convergence times. However, as the reader can easily check, this amplifies the effect of measurement noise making the link capacities drastically oscillate, even for small variations in the TAs. As a result, as shown in Fig. 10.11, using the same γ for higher β factors results in slower convergence times and even higher amplitudes of link capacity variation.

10.6 Notes and Sources

For a comprehensive mathematical background on dual algorithms, the reader can access many good sources like [1, 4–7] or [8].

The dual algorithm in Section 10.2 for minimum average hop adaptive routing is a straight-forward application of a dual scheme to the regularized problem, not published elsewhere to the best of the author's knowledge.

The original backpressure algorithm was presented by Tassiulas and Ephemerides in [9], and applied in multiple network problems later. The derivation of the backpressure routing as a dual algorithm relaxing flow conservation constraints is present in other works like [10, 11] or [12].

The dual algorithm for congestion control and the sufficient convergence conditions in Prop. 10.1 come from [2].

The dual algorithm for distributed adjustment of CSMA window sizes is adapted from the works [3, 13].

10.7 Exercises

10.1 In the adaptive routing case described in Section 10.2, assume that a link e changes its weight in $\Delta\pi$ units, keeping the rest of weights unchanged. Compute the maximum variation that a path x_p may suffer in this case.

10.2 Apply KKT optimality conditions to find an efficient algorithm for solving the problem in $\mathbb{R}^n$:

$$\min_{x \geq 0} \ \sum_i c_i x_i + \sum_i x_i^2, \text{ subject to: } \sum_i x_i = C$$

10.3 Modify the Net2Plan implementation available in the Net2Plan repository of the dual algorithm for adaptive routing so that the link weight change in an iteration is limited to a maximum value $\Delta\pi$, an input parameter to the algorithm. Find a setting γ, $\Delta\pi$, keeping $\epsilon = 10^{-3}$, which is robust enough for a set of different selected topologies and report on the convergence speed in them with this setting.

10.4   Modify the Net2Plan implementation available in the Net2Plan repository of the dual algorithm for adaptive routing, so that the primal iteration carries all the traffic through the shortest path, and if more than one exists, chooses the shortest path in km. Does this algorithm enjoy the convergence properties of the dual algorithm? Devise empirical tests to assess the algorithm convergence.

10.5   Devise empirical tests using the Net2Plan implementation provided to compare the primal version of the adaptive routing scheme provided in Chapter 9 with that in Section 10.2

10.6   Adapt the derivations in Section 10.3 to devise a dual based backpressure algorithm for destination-based routing by relaxing flow conservation constraints.

10.7   Show that the optimum routing in (10.11) does not depend on ΔP parameter, but that the optimum multipliers q_{nd} of the relaxed problem are proportional to ΔP.

10.8   Make empirical tests using the Net2Plan implementation provided to assess the robustness of backpressure algorithms under significant signaling delays.

10.9   Modify the Net2Plan implementation available in the Net2Plan repository of the dual backpressure algorithm in Section 10.3, applying a heavy-ball inertia term to the dual iteration. Find a setting for γ and γ_h terms that is robust enough for a set of different selected topologies and report on the convergence speed in them with this setting.

10.10   [2] Compute the hessian matrix of the dual function in problem (10.15) relaxing link capacity constraints. Use this expression to prove Prop. 10.1.

10.11   Modify the Net2Plan implementation available in the Net2Plan repository of the dual algorithm for congestion control including a diagonal scaling of the gradient update, estimating the second derivative values of the dual function w as [14]:

$$\frac{\partial^2 w}{\partial \pi_e^2} = -\sum_{d:e\in p_d} \frac{h_d(t) - h_d(t-1)}{\sum_{e\in p_d}\pi_e(t) - \sum_{e\in p_d}\pi_e(t-1)}$$

To avoid numerical instabilities limit the minimum scaling value to ϵ an input parameter of the algorithm.

10.12   In the dual congestion control case described in Section 10.2, assume that a link e changes its weight in $\Delta\pi$ units, keeping the rest of weights unchanged. Compute the maximum variation that a rate h_d may suffer in this case.

10.13   Modify the Net2Plan implementation available in the Net2Plan repository of the dual algorithm for congestion control, so that the link weight change in an iteration is limited to a maximum value $\Delta\pi$, computed according to Exercise 10.12 to enforce a maximum Δh variation, an input parameter to the algorithm. Find a setting γ, Δh that is robust enough for a set of different selected topologies and report on the convergence speed in them with this setting.

10.14   In Exercise 9.6 the multi-path congestion control problem was described, and a primal algorithm was provided for it. Comment on the convergence difficulties appearing when applying a dual algorithm to this problem, relaxing the link capacity constraints. Derive a distributed dual algorithm applying a regularization term to the objective function $\epsilon \sum_p x_p^2$. Implement it in Net2Plan and use empirical tests to tailor the ϵ parameter.

10.15   [15] In Exercise 10.14, use the strictly convex function $\epsilon \sum_p (x_p - y_p)^2$ for problem regularization, where ϵ is an algorithm constant and y_p values are constants that will be different in each algorithm iteration. In particular, in the primal iteration, computing the routing $x_p(t+1)$, the y_p values used are current routing values $x_p(t)$. Implement this scheme in a Net2Plan algorithm and empirically find a robust setting for γ and ϵ in a set of selected scenarios. *Note*: This algorithm is an application of the Proximal Minimization Algorithm ([6], p. 233). Its convergence in this multi-path congestion control variant was studied in [15].

10.16   Modify the Net2Plan implementation available in the Net2Plan repository of the dual algorithm for CSMA backoff window adjustment applying a heavy-ball inertial term. Find a setting for γ and γ_h terms that is robust enough for a set of different selected topologies and report on the convergence speed in them with this setting.

10.17   Use the Net2Plan implementation available for the persistence probability adjustment in Aloha-type networks, and the backoff window optimization in CSMA networks, to compare the network utilities and link capacities achieved with both MAC protocols in wireless networks of different selected topologies.

References

[1] D. P. Bertsekas, *Nonlinear Programming*. Bertsekas: Athena Scientific, 1999.

[2] S. H. Low and D. E. Lapsley, "Optimization flow controli: basic algorithm and convergence," *IEEE/ACM Transactions on Networking (TON)*, vol. 7, no. 6, pp. 861–874, 1999.

[3] L. Jiang and J. Walrand, "A distributed csma algorithm for throughput and utility maximization in wireless networks," *IEEE/ACM Transactions on Networking (TON)*, vol. 18, no. 3, pp. 960–972, 2010.

[4] M. Minoux, *Mathematical programming: theory and algorithms*, ser. Wiley series in discrete mathematics and optimization. New York, NY, USA: John Wiley & Sons, Inc., 1986.

[5] B. T. Polyak, *Introduction to Optimization*. Optimization Software New York, 1987.

[6] D. P. Bertsekas and J. N. Tsitsiklis, *Parallel and Distributed Computation: Numerical Methods*. Athena Scientific, 1997.

[7] L. Lasdon, *Optimization Theory for Large Systems*, ser. Dover books on Mathematics. Dover Publications, 2002.

[8] S. Boyd and L. Vandenberghe, *Convex Optimization*. New York, NY, USA: Cambridge University Press, 2004.

[9] L. Tassiulas and A. Ephremides, "Stability properties of constrained queueing systems and scheduling policies for maximum throughput in multihop radio networks," *Automatic Control, IEEE Transactions on*, vol. 37, no. 12, pp. 1936–1948, 1992.

[10] X. Lin and N. B. Shroff, "Joint rate control and scheduling in multihop wireless networks," in *Decision and Control, 2004. CDC. 43rd IEEE Conference on*, vol. 2. IEEE, 2004, pp. 1484–1489.

[11] X. Lin, N. B. Shroff, and R. Srikant, "A tutorial on cross-layer optimization in wireless networks," *Selected Areas in Communications, IEEE Journal on*, vol. 24, no. 8, pp. 1452–1463, 2006.

[12] M. Chiang, S. H. Low, J. C. Doyle *et al.*, "Layering as optimization decomposition: A mathematical theory of network architectures," *Proceedings of the IEEE*, vol. 95, no. 1, pp. 255–312, 2007.

[13] J. Liu, Y. Yi, A. Proutiere, M. Chiang, and H. V. Poor, "Towards utility-optimal random access without message passing," *Wireless Communications and Mobile Computing*, vol. 10, no. 1, pp. 115–128, 2010.

[14] S. Athuraliya and S. Low, "Optimization flow control with Newton-like algorithm," in *Global Telecommunications Conference, 1999. GLOBECOM'99*, vol. 2. IEEE, 1999, pp. 1264–1268.

[15] X. Lin and N. B. Shroff, "Utility maximization for communication networks with multipath routing," *Automatic Control, IEEE Transactions on*, vol. 51, no. 5, pp. 766–781, 2006.

11

Decomposition Techniques

11.1 Introduction

In this chapter, we focus on the application of decomposition techniques to network optimization problems. A problem decomposition is a transformation of a problem into a set of, potentially, many independent and simpler problems, coordinated by a so-called *master program*, such that the optimum solution of this overall scheme also solves the original problem.

A key idea is that *there can be many alternative decompositions to the same network problem* and that different decompositions yield to different schemes, network engineering decisions, and protocols. Section 11.2 is devoted to reviewing the most relevant baseline techniques called primal and dual decomposition, together with other mixed approaches that can exist. Then, Sections 11.3–11.7 include a set of case studies to illustrate the richness and strength of decomposition techniques in network design. Table 11.1 summarizes the case studies addressed.

First, in Sections 11.3–11.5 we put the emphasis on problem decomposition as a theoretical support to create so-called *cross-layer algorithms*. These are coordination schemes that make protocols at different network layers cooperate to achieve a common goal, typically to maximize the network utility perceived by the application-layer users. In this case, each network layer is a subproblem, which is solved by a different protocol (potentially itself a distributed protocol), and the master program defines the signaling to coordinate the layers. The three case studies chosen are good examples of a recent and successful research trend called *Network layering as optimization decomposition* (e.g., see [1] for a surveying contribution). Here, decomposition theory is the mathematical language to build an analytic and systematic study of the network architecture of protocols.

As a further application of problem decomposition, in Section 11.6 we show how it supports the coordination of different agents in the cooperative solving of a single-layer problem. In particular, we present a case study where multiple interconnected network carriers cooperate to globally optimize the routing, for example, in the Internet. The interest behind the decomposition is creating separated subproblems for each carrier, such that the information they need to exchange is minimum and not sensible.

Finally, Section 11.7 includes a case study for a $\mathcal{NP}$-hard joint capacity and routing design problem to be solved *offline*. In this case, the target of the decomposition technique is to reduce

Optimization of Computer Networks – Modeling and Algorithms: A Hands-On Approach,
First Edition. Pablo Pavón Mariño.
© 2016 John Wiley & Sons, Ltd. Published 2016 by John Wiley & Sons, Ltd.
Companion Website: www.wiley.com/go/PavonMarinoSol16

Table 11.1 Case studies in Chapter 11

Problem type	Decomposition technique	Section
Cross-layer congestion control and QoS capacity alloc.	Primal	Section 11.3
Cross-layer congestion control and backpressure routing	Dual	Section 11.4
Cross-layer congestion control and power allocation	Dual	Section 11.5
Multidomain routing design	Primal	Section 11.6
Joint capacity and routing offline planning	Dual	Section 11.7

the overall computational complexity and also permit parallel executions of the subproblems, at a cost of loosing some convergence properties.

As in the previous chapters, the code of the devised algorithms implemented in Net2Plan tool is accessible and algorithm convergence of the case studies is illustrated in empirical tests.

11.2   Theoretical Fundamentals

Decomposition techniques are commonly classified into primal and dual methods:

- *Primal decomposition* or *right-hand side allocation* methods are applied to the original (primal) problem. In each iteration, the master program decides a particular allocation of the shared resources among the subproblems. Then, each subproblem independently computes the optimal form of using them. After the subproblems complete their local optimization, the master obtains feedback from them to improve the shared resources allocation in the next iteration.
- *Dual decomposition* methods are based on dualizing or relaxing some constraints and using a (sub)gradient iteration to solve the dual problem, as an indirect form to also solve the original problem. This is just the approach already described in Chapter 10. In this chapter, we focus on how it can be used in a systematic form to create coordination mechanisms between layers or to create heuristic algorithms for $\mathcal{NP}$-hard problems.

We see both decomposition methods in further detail next. To finalize, we include hybrid decomposition strategies to produce different engineering solutions.

11.2.1   Primal Decomposition

Primal decomposition is appropriate when the target problem has a set of decision variables that, if fixed, permit separation of the rest of the variables into small and independent problems. We illustrate this in problem (11.1):

$$\min_{y,\{x_i\}} \sum_i f_i(x_i) \quad \text{subject to:} \tag{11.1a}$$

$$x_i \in \mathcal{X}_i \quad i = 1, \cdots, M \tag{11.1b}$$

$$y \in \mathcal{Y} \tag{11.1c}$$

$$h_i(x_i) \le y \quad i = 1, \cdots, M \tag{11.1d}$$

where x_i represents a set of decision variables that only appear in subproblem i, while y is the decision variables that couple all the candidate subproblems together, through constraints (11.1d). We can call x_i the *local* or *private* variables and y the *coupling* or *complicating* variables. The set $\mathcal{Y}$ is assumed to be such that if $y \in \mathcal{Y}$, then the problem (11.1) is always feasible.

Primal decomposition separates the problem (11.1) into two levels. On one hand, the lower level operates assuming some fixed values $y \in \mathcal{Y}$, and computes the optimum x_i values in such case. This can be done independently for each subproblem $i = 1, \cdots, M$ solving:

$$\min_{x_i} f_i(x_i) \quad \text{subject to:} \tag{11.2a}$$

$$x_i \in \mathcal{X}_i \tag{11.2b}$$

$$h_i(x_i) \le y \tag{11.2c}$$

We denote $x_i^*(y)$ as the optimum solution of the i-th subproblem (11.2), for coupling variables y, and we denote $f_i^*(y) = f_i(x_i^*(y))$ as its local cost. According to this, $f_i^*(y)$ is a *perturbation function* of (11.2) that returns how the ith subproblem optimum cost changes if we perturb the right-hand side of inequality (11.2c) (see Appendix B for details). Since subproblem (11.2) has an optimal solution for every $i = 1, \cdots, M$ and every y value ($\mathcal{Y}$ should be chosen in that way), perturbation functions $f_i^*(y)$ are always well defined.

The upper level or *master program* is in charge of iteratively finding the best y values that make the resulting solution $(y, x_i^*(y), i = 1, \cdots, M)$ optimize the original problem:

$$\min_{y \in \mathcal{Y}} \sum_i f_i^*(y) \tag{11.3}$$

If the original problem is convex, all the subproblems and master program are also convex. Then, according to Prop. B.20 in Appendix B, perturbation functions $f_i^*(y)$ are convex functions, and for any $y \in \mathcal{Y}$, it holds that:

$$s_i(y) = -\pi_i^*(y)$$

is a subgradient of f_i^* in point y, being $\pi_i^*(y)$ any optimal multipliers of constraints (11.2c). As a result, a subgradient of the objective function of the master problem $\sum_i f_i^*(y)$ can be obtained summing the subgradients of the perturbation functions:

$$f(y) = \sum_i f_i^*(y) \Rightarrow s(y) = \sum_i s_i(y) = -\sum_i \pi_i^*(y)$$

The subgradient $s(y)$ is a gradient when the perturbation function is differentiable. A sufficient condition for this to happen is that $f_i(x)$ functions are strictly convex and that the unique optimum solution in each subproblem is a regular point where the gradients of all the binding constraints are linearly independent. In general, there is no easy form of guaranteeing this.

Whether $s(y)$ are gradients or subgradients, if the master problem (11.3) is convex, it can be solved using, for example a (sub)gradient projection algorithm that iterates through y variables according to:

$$y(t+1) = P_y(y(t) - \gamma s(y(t))) = P_y \left(y(t) + \gamma \sum_i \pi_i^*(y(t)) \right) \qquad (11.4)$$

Note that using a constant step γ means that convergence to optimal y values is guaranteed for a sufficiently small γ only if $s(y(t))$ are gradients, and only to the proximity of the optimum if they are subgradients. The overall primal decomposition is summarized in Algorithm 13.

Algorithm 13 Primal decomposition algorithm for (11.1)

1: *Initialization*: Set $t = 0$, initialize $y(0)$ to any value in $\mathcal{Y}$
2: *Local iteration*: Solve (11.2) for each subproblem, to obtain $x_i^*(y(t))$ and $\pi_i^*(y(t))$.
3: *Master iteration*: Update resource allocation using (11.4), $t \leftarrow t+1$ and go to Step 2.

The stop condition in Algorithm 13 is any valid stop condition for the (sub)gradient iteration (11.4). In the cases in which the subproblems are solved using an inner iterative method, convergence and stability of the overall system is guaranteed if the subproblem iterations are faster than the master iteration (11.4), such that subproblems converge to the local optimum solution and communicate to the master the optimum π_i^* multipliers before $y(t)$ is updated. In such case, robustness of the master iteration under outdated, disordered or noisy π_i^* values is inherited from the properties of the (sub)gradient algorithm. In turn, if iterations of the master problem and the inner gradient iterations in the subproblems are at the same time scale, convergence is still possible under certain conditions (see [2]).

11.2.1.1 Primal Decomposition without Strong Duality

Primal decomposition principle requires each of the subproblems to be convex, such that $\pi_i^*(y)$ multipliers become subgradients of the perturbation functions. Also, this means that $f_i^*(y)$ are convex functions and so is the objective function of the original problem. Still, the primal may not be convex if set $\mathcal{Y}$ is not convex, for example, it has integer constraints. Then, the subgradient algorithm (11.4) would iterate in y variables, which could be unfeasible ($y(t) \notin \mathcal{Y}$).

Primal decomposition can still be useful in this case. The master iteration should be replaced by other, for example heuristic, approaches that smartly explore the set of valid allocations $y \in \mathcal{Y}$. In this case, we can take benefit of the knowledge that $s(y(t))$ vectors are still subgradients of the objective function and define a whole semispace of y' solutions that we can get rid of, since they are worse than the already explored solution $y(t)$. In particular, applying the subgradient definition, they are solutions y' for which:

$$s(t)^T(y' - y(t)) \geq 0 \Rightarrow \sum_i f_i^*(y') \geq \sum_i f_i^*(y(t)), \quad \forall y'$$

Previous inequality can be added to future (heuristic) iterations of the master problem, reducing the space of solutions to explore.

11.2.2 Dual Decomposition

Dual decomposition is the application of the gradient iteration to the dual function of a relaxed problem. The reader is referred to Chapter 10 and Appendix B for details on the mathematical background.

In this section, we are interested in showing the strength of the dual approach as a systematic form of creating successful problem decompositions. In this respect, dual decomposition can be naturally applied to problems that have a set of constraints such that, if dualized, produce a relaxed problem that can be separated into subproblems that can now be solved independently. Let us focus on the formulation:

$$\min_x \sum_i f_i(x_i) \quad \text{subject to:} \tag{11.5a}$$

$$x_i \in \mathcal{X}_i \quad \forall i = 1, \cdots, M \tag{11.5b}$$

$$\sum_i h_i(x_i) \leq 0 \tag{11.5c}$$

where $x_i, i = 1, \cdots, M$ are blocks of decision variables. Initially, we put no restrictions to the functions f_i, h_i, nor sets $\mathcal{X}_i$. In problem (11.5), we see that the objective function can be separated into the sum of one function per each variable block. Also, constraints (11.5b) are "easy" in the sense that they do not couple decision variables of different blocks. Actually, if constraints (11.5c) did not exist, problem (11.5) could be separated into M independent subproblems $\min_{x_i \in \mathcal{X}_i} f_i(x_i)$. So, we tag constraint (11.5c) as the "coupling" or "complicating" constraint.

The common dual strategy consists of dualizing the complicating constraints. Using π as the vector of multipliers for (11.5c), we have:

$$\min_{x_i \in \mathcal{X}_i, \forall i} \sum_i f_i(x_i) + \pi^T \sum_i h_i(x_i) \tag{11.6}$$

which decouples into M programs (11.7) that can be solved independently for fixed values of π:

$$\min_{x_i \in \mathcal{X}_i} f_i(x_i) + \pi^T h_i(x_i), \forall i = 1, \cdots, M \tag{11.7}$$

We denote $x_i^*(\pi)$ as a minimizer of the ith subproblem in (11.7) for particular prices π. If strong duality holds for (11.5), there exists a set of prices π^* such that at least one $x_i^*(\pi^*)$ solves the original problem. If the original problem is convex with strictly convex objective function, $x_i^*(\pi^*)$ is the unique primal optimum.

The original problem is now separated into two levels of optimization. At the higher level, the master program finds the optimum of the dual problem $\max_{\pi \geq 0} w(\pi)$, using a gradient projection iteration like:

$$\pi(t+1) = [\pi(t) + \gamma s(t)]_0 \tag{11.8}$$

where $s(t)$ is a subgradient of the dual function given by the slack of the constraint in any minimizer of $\pi(t)$: [1]:

$$s(t) = \sum_i h_i(x_i^*(\pi(t))) = \sum_i s_i(t)$$

The complete scheme is illustrated in Algorithm 14.

Algorithm 14 Dual decomposition algorithm for (11.5)

1: *Initialization*: Set $t = 0$, initialize $\pi(0)$ to any value in $\pi(0) \geq 0$
2: *Local iteration*: Solve the M subproblems (11.7) using $\pi(t)$.
3: *Master iteration*: Iterate the master using $s(t)$, e.g., (11.8) and go to Step 2.

On some occasions, subproblems are so simple that they can be solved analytically with a closed formula. This happened in the case studies in Chapter 10. In other cases found in this chapter, solving the subproblems (11.7) for given prices $\pi(t)$ involves an inner iterative method. In such situation, convergence of the decomposition scheme is guaranteed if the subproblem inner iterations go at a faster time scale than the price updates, such that they converge before the prices $\pi(t)$ are changed. If this does not hold, convergence of the overall system can still be guaranteed under certain technical conditions. The theoretical analysis of these cases is outside the scope of this book, the interested reader is referred to [2]. In this chapter, we will content with illustrating the convergence with empirical tests. Still, we note that the master iteration inherits the robustness and stability properties of gradient algorithms. This means that for a sufficiently small γ step in (11.8), the dual decomposition is robust against using outdated $s_i(t)$ information in some or all of its coordinates, and inherits the convergence in expectation properties of stochastic gradient algorithms, when $s(t)$ information is subject to, for example unbiased measurement errors.

Finally recall that in the dual approach there is no guarantee of having feasible solutions (satisfying also the relaxed constraints) until the optimal multipliers are found. In contrast, in the primal decomposition, every iteration produced a feasible solution.

11.2.2.1 Dual Decomposition without Strong Duality

If problem (11.5) does not satisfy sufficient conditions guaranteeing strong duality, we do not know beforehand if strong duality holds for a particular problem instance, and then if optimum multipliers have an associated minimizer that is primal optimal. Actually, it can be the case that all the minimizers of all the iterations are primal unfeasible. For this reason, in this case Algorithm 14 should be completed, adding an intermediate step that uses any heuristic technique that produces a feasible solution from the subproblem minimizers in the local iteration.

Proposition B.19 in Appendix B is the theoretical support to motivate the application of the dual approach when strong duality does not hold. It tells us that given some multipliers $\pi(t)$,

---

[1] As established in Appendix B, under mild assumptions (like a compact feasibility set of (11.5)), the dual function is concave in all its domain, and the vector $s(t)$ exists for any dual feasible $\pi(t)$. If $x_i^*(\pi^*)$ is unique, for example the original problem is a convex problem with strictly convex objective functions, then s is actually a gradient.

maybe in the proximity of π^*, the associated minimizers are optimum solutions to problems that can be small variations of the original problem. In addition, the dual approach has the advantage that every iteration t can produce a lower bound to the optimum cost by evaluating the dual function in $\pi(t)$.

11.2.3   Other Decompositions

According to previous sections, those problems composed of subproblems coupled by a set of common constraints are adequate for dual decompositions, while when the subproblems are coupled by some common decision variables, primal decomposition can be more convenient. However, this is not a strict rule. The same problem can be decomposed by primal or dual methods, producing different optimization algorithms. Moreover, optimization problems can be reformulated, for example, adding new auxiliary decision variables and constraints that make up an equivalent optimization problem, which do not change the optimum solution nor the feasibility set. Then, the reformulated problem can go through primal or dual decompositions that produce new different algorithms!

The basic techniques of problem reformulation are illustrated in two examples that follow. First, we apply a dual decomposition to solve the problem with coupling variables (11.1), for which we already described a primal decomposition. First, we reformulate the problem as follows:

$$\min_{y,\{y_i\},x} \sum_i f_i(x_i) \quad \text{subject to:} \tag{11.9a}$$

$$x_i \in \mathcal{X}_i \quad i = 1, \cdots, M \tag{11.9b}$$

$$h_i(x_i) \le y_i \quad i = 1, \cdots, M \tag{11.9c}$$

$$y_i = y, \forall i = 1, \cdots, M \tag{11.9d}$$

$$y \in \mathcal{Y} \tag{11.9e}$$

Reformulation consists of creating the auxiliary variables $y_i, i = 1, \cdots, M$, which should be seen as *local copies* for the i-th subproblem of the common variables y. This is enforced by adding constraints (11.9d) ($y_i = y, i = 1, \cdots, M$), the so called *consistency constraints*. Then, if we dualize the consistency constraints, the relaxed problem can now be decomposed into M subproblems, each one handling only local variables, and one common problem handling the consistency:

$$\min_{x_i,y_i} \sum_i f_i(x_i) + \sum_i \pi_i y_i \quad \text{subject to: (11.9b,c),} \quad i = 1, \cdots, M$$

$$\min_{y \in \mathcal{Y}} -y \left(\sum_i \pi_i \right)$$

These subproblems are coordinated by a master program that finds the optimum π_i multipliers using $\sum_i (y_i - y)$ as subgradients.

As a second example, we can apply a primal decomposition to problems with coupling constraints like (11.5), for which we applied dual decomposition in the previous section.

Introducing again auxiliary variables $\{y_i, i = 1, \cdots, M\}$, we reformulate the problem as follows:

$$\min_{x, \{y_i\}} \sum_i f_i(x_i) \quad \text{subject to:} \tag{11.10a}$$

$$x_i \in \mathcal{X}_i \quad \forall i = 1, \cdots, M \tag{11.10b}$$

$$\sum_i h_i(x_i) \leq y_i \tag{11.10c}$$

$$\sum_i y_i \leq 0 \tag{11.10d}$$

Then, applying primal decomposition to (11.10), using $\{y_i\}$ as master variables, we have subproblems of the form:

$$\min f_i(x_i) \quad \text{subject to: (11.10b,c),} \quad \forall i = 1, \cdots, M$$

And a master program:

$$\min \sum_i f_i^*(y_i) \quad \text{subject to:} \quad \sum_i y_i \leq 0$$

where $f_i^*(y_i)$ is the optimum cost in the i-th subproblem, for master variables y. According to the standard primal method, the master program is solved using a subgradient projection iteration, with a subgradient given by:

$$\pi(y) = -\sum_i \pi_i^*(y)$$

being $\pi_i^*(y)$ the optimum multipliers for constraints (11.10c) in the i-th subproblem.

As a conclusion, the reformulation-decomposition technique is a powerful strategy to create different network optimization algorithms, providing a theoretical support to a systematic approach on multiple cooperation schemes between and within network layers.

11.3   Cross-Layer Congestion Control and QoS Capacity Allocation

In this example, we consider a network with two *types* of traffic demands represented by sets D_1 and D_2. We denote $\mathcal{N}$ to the set of network nodes and $\mathcal{E}$ the set of network links. The capacity u_e of each link is known and fixed. However, in order to enforce a strict separation between the QoS of demands in D_1 and D_2, each link e capacity is split into two: (i) a bandwidth u_e^1 dedicated to demands of D_1 and (ii) a bandwidth $u_e^2 = u_e - u_e^1$ dedicated to those in D_2.

A congestion control scheme determines the rate h_d of each demand, such that the network utility $\sum_{d \in D_1 \cup D_2} U_d(h_d)$ is maximized. All utility functions are increasing, strictly concave, and differentiable. The path p_d followed by each demand d is known.

The joint optimization of the congestion control and the link capacity partitioning is accomplished solving the following network optimization problem:

$$\max_{h \geq 0, u^1 \geq 0, u^2 \geq 0} \sum_{d \in D_1} U_d(h_d) + \sum_{d \in D_2} U_d(h_d) \quad \text{subject to:} \tag{11.11a}$$

$$\sum_{d \in D_1 : e \in p_d} h_d \leq u_e^1 \quad \forall e \in E \tag{11.11b}$$

$$\sum_{d \in D_2 : e \in p_d} h_d \leq u_e^2 \quad \forall e \in E \tag{11.11c}$$

$$u_e^1 + u_e^2 = u_e \quad \forall e \in E \tag{11.11d}$$

We apply a primal decomposition approach, observing that if u^1 and $u^2 = u - u^1$ capacity vectors are fixed, problem (11.11) decouples into two separated congestion control problems (11.12), one for each traffic class:

$$\max_{h^1 \geq 0} \sum_{d \in D_1} U_d(h_d) \quad \text{subject to: (11.11b)} \tag{11.12a}$$

$$\max_{h^2 \geq 0} \sum_{d \in D_2} U_d(h_d) \quad \text{subject to: (11.11c)} \tag{11.12b}$$

We denote $\pi^1(u^1)$ and $\pi^2(u^2)$ as the optimal multipliers of the link capacity constraints in each subproblem. The master program adjusts the split of link capacities between the two classes, solving:

$$\max_{u^1 \geq 0, u^2 \geq 0} U_1^*(y^1) + U_2^*(y^2) \quad \text{subject to: (11.11d)} \tag{11.13a}$$

where functions $U_1^*(u^1)$ and $U_2^*(u^2)$ are the optimum utilities of each subproblem (11.12a,b), respectively, for a particular bandwidth split u^1, u^2. Since the original problem is convex, we have that:

$$s(u^1, u^2) = [\pi^1(u^1), \pi^2(u^2)]$$

is a subgradient of the objective function of the master, which thus can be solved using an iterative projection algorithm decoupled per each link, such as:

$$(u_e^1, u_e^2)(t + 1) = P_{y_e} \left((u_e^1, u_e^2)(t) + \gamma s(u^1(t), u^2(t)) \right), \quad \forall e \in \mathcal{E} \tag{11.14}$$

where the set $\mathcal{Y}_e$ represents the constraints $u_e^1 \geq 0, u_e^2 \geq 0, u_e^1 + u_e^2 = u_e$. The projection operation can be solved efficiently using local information (see Exercise 11.6). The pseudocode in Algorithm 15 illustrates the complete scheme.

Algorithm 15 describes a double iteration. First, the master algorithm updates the u^1, u^2 values at a slow time scale. For a given u^1, u^2 bandwidth split, the links apply a QoS policy that reserves the specified bandwidth to each class. Then, the standard congestion control algorithm is left enough time to converge to a maximum utility solution. Depending on how congestion control problem is solved, optimal $\pi_e^1(u^1)$ and $\pi_e^2(u^2)$ multipliers are generated. For instance, as we saw in Chapter 6, if congestion control is based on Reno-like sources, π_e^1 and π_e^2 multipliers can be approximated by the packet loss probability of connections of type 1 and 2 in link e. In TCP Vegas versions, the multipliers would be approximated by the average queue backlog. Also, a general congestion control algorithm based on a dual relaxation of link capacity constraints, like Algorithm 11 with explicit signaling, can be used. In either case, the link prices are *local* information to the links, that links can convey implicitly or explicitly to the demand sources, but that in any case each link uses to complete its master iteration.

Algorithm 15 Cross-layer congestion control and QoS capacity allocation for (11.11)

1: Set $t = 0$, initialize $u^1(0), u^2(0)$ to any non-negative vectors satisfying (11.11d).
2: *QoS adjustment (slow)*: Each link updates its bandwidth split asynchronously using (11.14).
3: *Congestion control (fast)*: Each class solves its congestion control subproblem, π_e^1 and π_e^2 multipliers are generated in each link.

11.3.1 Implementation Example

We present here some tests illustrating the convergence of Algorithm 15. We use the topology in Fig. 9.2, with two demands between each node pair, carried through the shortest path route in km between the end nodes. Demands implement the dual approach for congestion control algorithm described in Section 10.4. Demands of both types have the same fairness factor $\alpha = 1$, but utility of type 2 is weighted by a factor of two, and thus for the same injected traffic, contributes double to network utility than a type 1 demand. Each link e independently computes its weight π_e monitoring its traffic, signals it to the traversing demands, and waits a random time between 0.5 and 1.5 units before the next update. Signaling messages to demand sources are delayed randomly between 2.5 and 3.5 time units, but a fraction of 5% messages are randomly lost. Independently from the link update, each demand recomputes its injected traffic using the most updated link information it has and waits a random time between 0.5 and 1.5 time units until the next computation.

The QoS capacity split iteration is performed at a slower time scale than congestion control. Each link e updates its split u_e^1, u_e^2 asynchronously, using the most updated local link information, and the time between two updates is randomly chosen between 40 and 60 time units.

Figure 11.1 illustrates the algorithm convergence when the congestion control step $\gamma_c = 0.001$ and the QoS split step in (11.14) is $\gamma_Q = 5$. Interestingly, the reader can check how convergence of the joint optimization occurs for the same γ_c and γ_Q values, also when both congestion control and QoS updates are on similar time scales and when the QoS update is, for example, 10 times faster on average than the congestion control update.

11.4 Cross-Layer Congestion Control and Backpressure Routing

In this section, we present a cross-layer algorithm motivated by the application of a dual decomposition to the joint optimization of congestion control and backpressure center-free routing in a network. Let $\mathcal{N}$ and $\mathcal{E}$ be the set of network nodes and links, respectively, u_e the

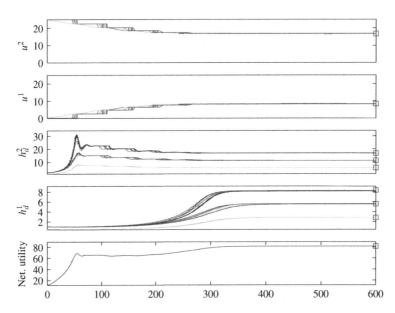

Figure 11.1   Evolution of Algorithm 15, $\gamma_c = 0.001$ (congestion control), $\gamma_Q = 5$ (QoS split). Plots from upper to lower are the capacity splits, demand rates, and network utility. Optimum values are plotted as squares on the right-hand side

link capacities and $\mathcal{D}$ the set of offered demands. Each demand d is associated to an utility function U_d, strictly increasing and strictly concave with respect to the flow rate h_d. The NUM modeling of the joint congestion and routing control problem optimizes both layers:

$$\max_{x \geq 0, h \geq 0} \sum_d U_d(h_d) \quad \text{subject to:} \tag{11.15a}$$

$$q_{nd} : \sum_{e \in \delta^+(n)} x_{de} - \sum_{e \in \delta^-(n)} x_{de} \begin{cases} \geq h_d, \text{ if } \quad n = a(d) \\ \geq 0, \text{ otherwise} \end{cases}, \quad \forall d \in D, n \in \mathcal{N} - b(d) \tag{11.15b}$$

$$\sum_d x_{de} \leq u_e, \quad \forall e \in \mathcal{E} \tag{11.15c}$$

The objective function targets maximizing the network utility, representing the performance observed at the application layer. (11.15b) are the flow conservation constraints in the flow-link formulation, (11.15c) the link capacity constraints. The reader is referred to Section 10.3 for a deeper explanation.

We see that congestion control (h) and routing (x) variables are coupled by flow conservation constraints. We apply a dual approach and relax them using q_{nd} as multipliers. For given q_{nd} values of the multipliers, the relaxed problem can be separated into two:

$$\min_{x \geq 0, \sum_d x_{de} \leq u_e} \sum_{de} x_{de}(q_{b(e)d} - q_{a(e)d}) \tag{11.16a}$$

$$\max_{h \geq 0} \sum_d U_d(h_d) - q_{a(d)d} h_d \tag{11.16b}$$

First subproblem (11.16a) determines the routing, and yields to a similar backpressure center-free scheme to that described in Section 10.3. In particular, note that this problem can be solved independently for each link: a link e must forward the traffic of demand d for which the quantity $(q_{b(e)d} - q_{a(e)d})$ is minimum, assuming that the quantity is negative. If $(q_{b(e)d} - q_{a(e)d}) > 0$ for all demands, the link does not forward traffic. Then, to make forwarding decisions, each node n needs to know the $q_{n'd}$ multipliers for all demands, but just in its neighbor nodes n'.

Second subproblem (11.16b) can be solved independently for each demand in one shot using local information $q_{a(d)d}$: the multiplier at the demand source node. For instance, if U_d are α-utility functions:

$$h_d = q_{a(d)d}^{-1/\alpha}, \quad \forall d \in \mathcal{D} \tag{11.17}$$

Note that the two layers are coordinated using very simple signaling. The routing side requires that each node n is signaled the $q_{n'd}$ values for all demands d and from all outgoing neighbor nodes. In turn, each demand d needs to know the multiplier q_{nd} for its origin node $n = a(d)$, which is local information.

The multipliers gradient update, using a constant γ step is:

$$q_{nd}(t + 1) = \left[q_{nd}(t) + \gamma \left(h_{nd} - \sum_{e \in \delta^+(n)} x_{de}(t) + \sum_{e \in \delta^-(n)} x_{de}(t) \right) \right]_0 \tag{11.18}$$

where h_{nd} is h_d when n is the origin node of d, and zero otherwise. Note that the subgradient coordinate for multiplier q_{nd} is the difference between the new traffic in n to transmit to d, and the traffic already transmitted. Then, for a constant γ step, q_{nd} becomes proportional to the queue size Q_{nd} of pending traffic at node n, of demand d, information easy to track:

$$q_{nd} = \gamma Q_{nd}, \quad \forall n \in \mathcal{N}, d \in \mathcal{D} \tag{11.19}$$

The complete scheme of the cross-layer algorithm, for an asynchronous and distributed operation, is shown in Algorithm 16.

Algorithm 16 Cross-layer congestion control and backpressure routing for (11.15)

1: *Node's algorithm*: At times $t = t_1(n), t_2(n), \ldots$, node n:
2:     Collects queue length information for its queues $\{q_{nd}, \forall d\}$.
3:     Signals it to incoming neighbor nodes.
4: *Link's algorithm*: At times $t' = t_1'(e), t_2'(e), \ldots$, link e:
5:     Collects the most updated q_{nd} values it has, signaled from outgoing neighbors.
6:     Transmits traffic of demand d with highest pressure difference, as long as it is > 0.
7: *Demand's algorithm*: At times $t'' = t_1''(d), t_2''(d), \ldots$, demand d:
8:     Adjusts its volume h_d using (11.17).

As a final remark, as happened in Section 10.3, there is an interesting connection between algorithm convergence and queue size. In particular, let us assume that a network instance has some particular optimum multipliers q_{nd}^* when solving (11.15). Then, if we apply Algorithm

16, the average Q_{nd}^* values observed in the queues when the algorithm converges is given by q_{nd}^*/γ. Higher γ values mean smaller queue sizes and thus smaller delays. However, we cannot have arbitrarily large γ steps, and thus arbitrarily small queue sizes, since the gradient iteration would not converge.

11.4.1 Implementation Example

We test Algorithm 16 on the same topology as the tests for the backpressure algorithm in Section 10.3.2, and a similar implementation of the routing process in the nodes in a time slotted fashion (slots of one time unit), together with an asynchronous signaling. A node collects its queue information, signals it in a message to each incoming neighbor node, and waits a random time between 0.5 and 1.5 time units until the next signaling event. Some 5% of the signaling messages are lost. Then, a node uses its local queues $q_{a(e)d}$ information for the forwarding decision, but using possibly outdated $q_{b(e)d}$ sizes signaled from the neighbor nodes.

We assume that a demand exists between each node pair, with an associated α-utility function (3.19), with $\alpha = 2$. We assume that the routing is performed by a fast scheduler (time slots are relatively small), while the congestion control has slower reactions. Each demand asynchronously updates its rate and waits a random time between 5 and 15 time units until its next update. As shown in Fig. 11.2, convergence to the optimum routes (x_p) and optimum demand rates (h_d) is achieved reasonably quickly for a step $\gamma = 5 \times 10^{-6}$.

Interested readers can use Net2Plan to observe the algorithm convergence in the presence of delays, losses, and so on, and check how the relation between queue sizes and the γ step is met while the algorithm converges.

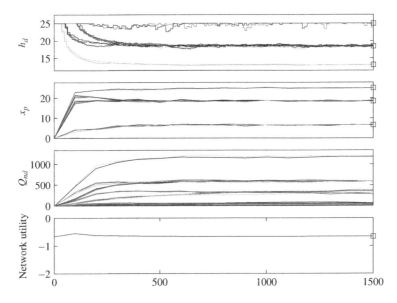

Figure 11.2 Evolution of Algorithm 16 in the asynchronous case with signaling losses, $\gamma = 5 \times 10^{-6}$. Plots from upper to lower are the demand rates, routes evolution, queue sizes in average number of packets, and network utility. Optimum values are plotted as squares on the right-hand side

11.5   Cross-Layer Congestion Control and Power Allocation

Herein we present a cross-layer distributed algorithm for joint optimization of the congestion control and transmission power allocation in a wireless network. Radio communications use a multiplexing scheme (e.g., CDM, OFDM), such that a link e receiver sees the power from its other end as a signal to detect that is interfered by the incoming power from other transmitters. The wireless constraints are the same as the ones described in Section 5.5, the reader is referred to that section for further details.

We denote $\mathcal{N}$ for the wireless network nodes and $\mathcal{E}$ to its links. The capacity u_e in each link e depends on its transmission power $\tilde{p}_e$ (in logarithmic units), and the transmission power of the rest of the network links interfering it. It is approximated by the expression:

$$u_e = \log \ (\text{SNR}_e) = \log \left(\frac{e^{\tilde{p}_e} G_{ee}}{\sigma_e^2 + \sum_{e' \neq e} e^{\tilde{p}_{e'}} G_{e'e}} \right) \tag{11.20}$$

where σ_e^2 is the thermal noise power at the receiver end (in linear units) and $G_{ee'}, e, e' \in \mathcal{E}$ is the interference map between links: the fraction of the power transmitted in e that reaches the receiver of e'. As shown in Prop. 5.5, u_e is a concave function of variables $\tilde{p}$.

The wireless network is fed with the traffic of a set D of demands. We are interested in the joint optimization of the transmission powers $\tilde{p}_e$ and rate allocations h_d, such that the network utility is maximized, solving the problem:

$$\max_{h \geq 0, \tilde{p}} \ \sum_d U_d(h_d) \quad \text{subject to:} \tag{11.21a}$$

$$\pi_e : \ \sum_{d:e \in p_d} h_d \leq \log \left(\frac{e^{\tilde{p}_e} G_{ee}}{\sigma_e^2 + \sum_{e' \neq e} e^{\tilde{p}_{e'}} G_{e'e}} \right), \quad \forall e \in \mathcal{E} \tag{11.21b}$$

$$\tilde{p}^{\min} \leq \tilde{p} \leq \tilde{p}^{\max}, \quad \forall e \in \mathcal{E} \tag{11.21c}$$

where U_d is the concave and increasing utility function of demand d, and $0 < \tilde{p}^{\min} < \tilde{p}^{\max} < \infty$ state the hardware constraints in the minimum and maximum transmission power. Problem (11.21) is convex with strong duality. We pursue a dual decomposition approach relaxing the link constraints. This decouples the relaxed problem into two:

$$\max_{h \geq 0} \ \sum_d U_d(h_d) - h_d \sum_{e \in p_d} \pi_e \tag{11.22a}$$

$$\max_{\tilde{p}^{\min} \leq \tilde{p} \leq \tilde{p}^{\max}} \ \sum_e \pi_e \log \left(\frac{e^{\tilde{p}_e} G_{ee}}{\sigma_e^2 + \sum_{e' \neq e} e^{\tilde{p}_{e'}} G_{e'e}} \right) \tag{11.22b}$$

For given π_e multipliers, congestion control problem (11.22a) can be solved in one shot. For instance, for α-fair utility functions of the form (3.19), the maximizer $h_d(\pi)$ is given by:

$$h_d(\pi) = \left[\left(\sum_{e \in p_d} \pi_e \right)^{-1/\alpha} \right]_0, \quad \forall d \in D \tag{11.23}$$

In turn, given π_e multipliers, transmission power optimization (11.22b) can be solved with a primal projection gradient algorithm. Repeating the steps in Section 9.6, for the particular case of $U_e(u_e) = \pi_e u_e$, we have that the gradient of the objective function f of (11.22b) is:

$$\frac{\partial f}{\partial \tilde{p}_e} = \frac{\partial U_e}{\partial \tilde{p}_e} + \sum_{e'' \neq e} \frac{\partial U_{e''}}{\partial \tilde{p}_e}$$

$$\frac{\partial U_e}{\partial \tilde{p}_e} = \frac{\partial U_e}{\partial u_e} \frac{\partial u_e}{\partial \tilde{p}_e} = \pi_e$$

$$\frac{\partial U_{e''}}{\partial \tilde{p}_e} = \frac{\partial U_{e''}}{\partial u_{e''}} \frac{\partial u_{e''}}{\partial \tilde{p}_e} = -\pi_{e''} \frac{SNR_{e''}}{e^{\tilde{p}_{e''}} G_{e'' e''}} e^{\tilde{p}_e} G_{e e''}$$

using $m_{e''} = \pi_{e''} \frac{SNR_{e''}}{e^{\tilde{p}_{e''}} G_{e'' e''}}$, we can rewrite the gradient of the objective function as:

$$\frac{\partial f}{\partial \tilde{p}_e} = \pi_e - e^{\tilde{p}_e} \sum_{e'' \neq e} m_{e''} G_{e e''}$$

The dual decomposition presented, creates two separated subproblems that use congestion prices π_e as the coordination information. The link prices evolution depends on both layers. The subgradient coordinate for link e is obtained monitoring the difference between the link capacity and the offered traffic.

$$\pi_e(t+1) = \left[\pi_e(t) - \gamma(t) \left(u_e(t) - \sum_{d: e \in p_d} h_d(t) \right) \right]_0 \tag{11.24}$$

The upper problem can be solved, for example, by a standard TCP algorithm, which receives the link prices π_e implicitly, or a dual congestion control algorithm like the one described in Section 10.4 that explicitly conveys the π_e multipliers to the demand source nodes. The lower problem can be solved using the primal gradient projection iteration described in Section 9.6 (Algorithm 7). Algorithm 17 illustrates the higher layer scheme.

Algorithm convergence requisites and robustness under asynchronous implementations are investigated in detail in [3]. Intuitively, it is shown that problem (11.21) can be made equivalent to a problem maximizing a strictly concave function under convex constraints and that both subproblems (11.22) maximize a strictly concave function. Then, iteration (11.24) is actually a gradient iteration, and given some link weights π, the associated primal solution (h, u) is unique. Therefore, convergence is guaranteed for sufficiently small γ steps in (11.24) and γ_i steps in the inner power allocation update of Algorithm 7.

Intuitively, convergence is also guaranteed if transmission power inner iteration is executed at a sufficiently faster time scale than congestion control and multipliers update. Then, after a link weight update, link capacities fast converge to the adequate solution, and congestion control convergence is guaranteed under standard assumptions of Section 10.4, since after convergence this loop sees a constant capacity in the links.

11.5.1 Implementation Example

We illustrate the convergence properties of the cross-layer scheme proposed for a 6-node wireless network like the one in Fig. 9.10. In the congestion control side, one demand is considered

Algorithm 17 Cross-layer congestion control and power allocation for (11.21)

1: Set $t = 0$, initialize $\pi(0)$ to any value $\pi(0) \geq 0$.
2: *Power allocation*: Inner iterations of Algorithm 7 for multipliers π, to iteratively adjust the power.
3: *Congestion control*: Each demand adjusts its rate to π using (11.23).
4: *Multipliers update*: At times $t = t_1(e), t_2(e), ...$, link e updates π_e using (11.24).

between each node pair, with an α-fair utility function with $\alpha = 1$. Congestion control operates at a time scale one order of magnitude slower than power adjustment. Each demand asynchronously adjusts its rate using the most updated multipliers obtained, and waits a random time between 5 and 15 units until the next update. Also, each link adjusts its multiplier π_e using the local capacity and traffic information and waits a random interval between 5 and 15 time units to the next update. In every update, link multipliers are signaled to the nodes. Signaling messages are assumed to arrive with a random delay between 15 and 45 time units, but 5% of messages are lost. The γ step in the congestion control iteration is $\gamma = 0.0001$.

In the power adjustment side, the algorithm operates in an asynchronous form. Each link uses the most updated π_e value locally known in every moment. The algorithm parameters and wireless channel conditions are equal to the ones described in Section 9.6.1, including the noisy gradients but not the heavy-ball term. The only difference is a reduction of the γ parameter to $\gamma = 1$ (instead of $\gamma = 10$).

Figure 11.3 illustrates the results. As we can see, convergence is smoothly achieved. The reader can check using the Net2Plan code provided, how convergence is reached for a wide range of γ steps in the inner and outer algorithm, and also when time-scales of power allocation and congestion control are similar. In particular, for the same γ steps, speeding up the congestion control one order of magnitude (so both subproblems operate asynchronously, but at the same time-scale) creates no diverging feedback, and just speeds up the convergence.

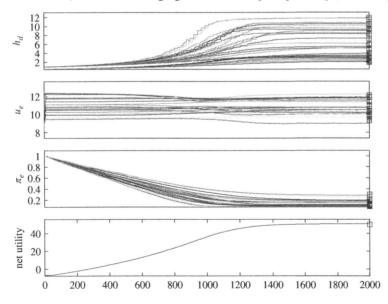

Figure 11.3 Evolution of Algorithm 17. Optimum values are plotted as squares in the right-hand side

11.6   Multidomain Routing

In this section, we use a primal decomposition approach in the routing optimization in a network composed of a set C of *clusters* or *domains*. Each domain $c \in C$ can correspond to an autonomous system in the Internet or any set of nodes controlled by an independent network carrier. Each node n belongs to exactly one domain, the institution managing it. We denote as $c(n)$ to the cluster (carrier or domain) a node belongs to, and $\mathcal{N}_c$ to the set of nodes composing a cluster c. The set of all network nodes is $\mathcal{N} = \bigcup_{c \in C} \mathcal{N}_c$ and the set of all network links is $\mathcal{E}$.

Given a cluster c, $\mathcal{E}_c^i$ is the set of internal or intradomain links of the cluster: that is, those links starting and ending inside it. Naturally, these are not the only links in the network since domains are not isolated but connected by interdomain links, which start in a node of a domain and end in a node of another domain. Domains connected by one or more links are said to be neighbors. We denote $\mathcal{E}_c^f$ as the interdomain links incoming or outgoing to/from cluster c, and $\mathcal{E}^f = \bigcup_{c \in C} \mathcal{E}_c^f$ to the set of all interdomain or frontier links in the network.

The network offered traffic is composed of a traffic matrix $h_{nt}, n, t \in \mathcal{N}$ containing the volume of traffic originated in node n, targeted to node t. When n and t are in different domains, its exchanged traffic must traverse interdomain links, and we call it interdomain traffic. In contrast, intradomain traffic can be routed only through the domain internal links, although this is not mandatory. We denote h_{ct} as the aggregated sum of traffic originated in cluster c targeted to t: $h_{ct} = \sum_{n \in \mathcal{N}_c} h_{nt}$.

Optimizing the routing in an interdomain network is a challenging task, since carriers owning the domains are reluctant to share sensible information like (i) the internal topology of intradomain links, (ii) its intradomain routing, and (iii) fine details of the offered traffic from/to their nodes.

The primal decomposition strategy that follows, targets the destination-based routing (x_{te} formulation) that minimizes the average number of hops in the complete network. Domains engaged in this optimization collaborate on exchanging a limited amount of coordination data in an iterative process, but the shared data does not include any of the three previous types of sensitive information.

We consider the traffic traversing the *interdomain* links as master variables to the problem, and we denote them as:

$$m_{te}, t \in \mathcal{N}, e \in \mathcal{E}^f = \{\text{Traffic targeted to } \quad t \quad \text{traversing interdomain link} \quad e\}$$

Once the master variables are fixed, domains become isolated entities that can optimize their network independent from each other. Each cluster c does that solving the problem:

$$\min \quad \sum_{t, e \in \mathcal{E}_c^i} x_{te} + M u_c \quad \text{subject to:} \tag{11.25a}$$

$$\sum_{e \in \delta^+(n)} x_{te} - \sum_{e \in \delta^-(n)} x_{te} = h_{nt}, \quad \forall t, n \in \mathcal{N}_c, n \neq t \tag{11.25b}$$

$$\sum_t x_{te} \leq u_e + u_c, \quad \forall e \in \mathcal{E}_c^i \tag{11.25c}$$

$$u_c \geq 0, x_{te} \geq 0, \quad \forall t \in \mathcal{N}, e \in \mathcal{E}_c^i \tag{11.25d}$$

$$v_{te}^c : x_{te} = m_{te}, \quad \forall t, e \in \mathcal{E}_c^f \tag{11.25e}$$

The objective function sums two contributions: (i) the amount of traffic carried in the links (proportional to the average number of hops of the routing) and (ii) a penalization factor Mu_c, where u_c is a variable with the maximum amount of oversubscription traffic in any internal link, as enforced by constraint (1.1c). Oversubscription in the links is allowed (although strongly penalized by a large M factor), so that routing subproblems for a domain are always feasible, whatever the master variables are. Recall that this is a requirement for the primal decomposition to work.

Constraints (11.25b) are the flow conservation constraints, *involving only the domain nodes.* Note that constraints (11.25e) fix the x_{te} values for the frontier links to constants given by associated m_{te} values, and v_{te} are the multipliers of these constraints.

We denote $C_c^*(m)$ as the optimum cost of the routing problem in cluster c, given the master variables m. We denote $v_{te}(m)$ as the multipliers of constraints (11.25e). The master program solving the original problem is given by:

$$\min_m \ \sum_c C_c^*(m) + \sum_{e \in \mathcal{E}_f} \sum_t m_{te} \quad \text{subject to:} \tag{11.26a}$$

$$\sum_{e \in \delta^+(c)} m_{te} - \sum_{e \in \delta^-(c)} m_{te} \begin{cases} = h_{ct}, \text{ if } \ t \notin \mathcal{N}_c \\ - \sum_{c' \neq c} h_{c't}, \text{ otherwise} \end{cases}, \quad \forall t \in \mathcal{N}, c \in C \tag{11.26b}$$

$$\sum_t m_{te} \leq u_e, \forall e \in \mathcal{E}^f \tag{11.26c}$$

$$m_{te} \geq 0, \forall t \in \mathcal{N}, e \in \mathcal{E}^f \tag{11.26d}$$

The objective function sums the contributions of the internal costs in each cluster (intradomain carried traffics and oversubscription cost), plus the carried traffic in the interdomain links. (11.26b) are flow conservation constraints *applied seeing a domain as a single node*, and $\delta^+(c)$ and $\delta^-(c)$ are the outgoing and incoming interdomain links for a cluster c, respectively:

- If a target t is not in a domain c, the outgoing minus incoming traffic in c to destination t is h_{ct}, what the domain generates targeted to t.
- If a target t is in a domain c, the incoming minus outgoing traffic in c to destination t sums the total amount of traffic to t generated by external sources: $\sum_{c' \neq c} h_{c't}$.

The primal decomposition rationale lies in solving the master problem using an iterative projected gradient approach. A gradient of the objective function can be obtained by exploiting the fact that $v_{te}^c(m)$ multipliers produced by cluster c under master variables m, are negative subgradients of the function $C_c^*(m)$. Then, the subgradient of the objective function f of the master is:

$$\frac{\partial f}{\partial m_{te}}(m) = -v_{te}^{c(a(e))} - v_{te}^{c(b(e))} + 1, \quad \forall t \in \mathcal{N}, e \in \mathcal{E}^f \tag{11.27}$$

And the master subgradient iteration using a constant γ step is given by:

$$m(k+1) = P_{\mathcal{M}}(m(k) - \gamma \nabla f(m(k))) \tag{11.28}$$

where $P_{\mathcal{M}}(m)$ is the projection of m into the set of constraints (11.26b–d). Convergence of the subgradient iteration with constant step length is guaranteed to the proximity of the optimum, and this holds even if gradients are computed using outdated information, for instance, when v_{te}^c multipliers collected are outdated, and each domain updates its routing asynchronously and independently from others. This inspires Algorithm 18.

Algorithm 18 Distributed multidomain routing

1: *Master iteration*: At times $t = t_1, t_2, \ldots$, master unit:
2:     Collects the multipliers v_{te}^c from the domains.
3:     Computes the new m_{te} variables solving (11.28).
4:     Signals each domain with the m_{te} variables of its frontier links.
5: *Domain iteration*: At times $t = t_1(c), t_2(c), \ldots$, each domain c:
6:     Collects the m_{te} values for its frontier links.
7:     Recomputes its routing tables according to (11.25).

11.6.1 Implementation Example

We test Algorithm 18 in a network of three clusters, as shown in Fig. 11.4. Topology and traffic chosen are that of the Abilene reference topology and traffic matrix, included in the Net2Plan release as `abilene_N12_E30_withTrafficAndClusters3.n2p`. The total offered traffic sums 75 units and link capacities are $u_e = 20$.

Each carrier asynchronously and independently updates its routing solving (11.25), using the current m_{te} master variables just for its frontier links. Then, it signals the obtained v_{te} multipliers to a central unit implementing the master iteration, and waits a random time between 0.5 and 1.5 time units until the next routing update.

The master iteration is implemented asynchronously with respect to each carrier, using the most updated v_{te} information from each of them. After completing the iteration, updated m_{te} master variables are available, and the central unit waits a random time between 0.5 and 1.5 time units until the next master iteration.

When each carrier recomputes its routing, the x_{te} values obtained from (11.25) are used to configure the routing tables of their nodes. As is common in hop-by-hop routing (e.g., IP), these tables state the *fraction* of the incoming traffic to be forwarded to each output link, and

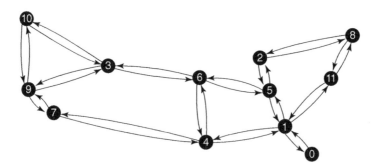

Figure 11.4 Topology example. Cluster 1, nodes $\mathcal{N}_1 = \{3, 7, 9, 10\}$, Cluster 2, nodes $\mathcal{N}_2 = \{0, 1, 4, 6\}$, Cluster 3, nodes $\mathcal{N}_3 = \{2, 5, 8, 11\}$. Link capacity $u_e = 20$, offered traffic in file `abilene_N12_E30_withTrafficAndClusters3.n2p`

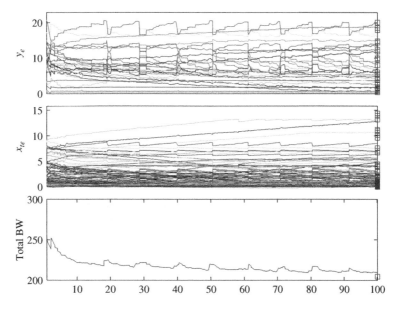

Figure 11.5 Evolution of Algorithm 18, $\gamma = 0.1$. Plots from upper to lower are the traffic in the links, x_{te} values, and total bandwidth consumed in the links (target to minimize). Optimum values are plotted as squares on the right-hand side

not the absolute values. Actually, the x_{te} values observed in the carrier links can be somewhat different to that coming from (11.25). The reason is that the traffic in the frontier links can also be different to m_{te} values, since carriers operate asynchronously, and some carriers may be routing the traffic using an outdated m_{te} information.

Figure 11.5 illustrates the convergence of the algorithm for a fixed step $\gamma = 0.1$. Note that since the master iteration involves a subgradient projection, a fixed step produces convergence to the vicinity of the optimum. Once this is achieved, a somewhat oscillatory behavior can appear, as witnessed by y_e evolution (traffic in the links). A diminishing step rule like $\gamma = 1/t$ would be needed to guarantee exact optimality. The interested reader can check this using the Net2Plan implementation provided.

11.7 Dual Decomposition in Non-Convex Problems

In this section, we present a case study where a dual decomposition is applied to a $\mathcal{NP}$-hard network problem to reduce its computational complexity. As explained in Section 11.2.2, the resulting algorithm has no convergence guarantees to the optimum, since the original problem instance may have a duality gap.

We focus on a network $\mathcal{G}(\mathcal{N}, \mathcal{E})$ with offered traffic given by the set of demands $\mathcal{D}$. The link capacities are constrained to be integer multiples of a basic module of capacity U and the routing is constrained to be non-bifurcated. We are interested in obtaining the routing and capacity allocation that minimizes the network cost, given by the number of capacity modules to install. Flow-link formulation (11.29) models the problem:

$$\min_{x,n} \sum_e n_e \quad \text{subject to:} \tag{11.29a}$$

$$\sum_{e \in \delta^+(n)} \hat{x}_{de} - \sum_{e \in \delta^-(n)} \hat{x}_{de} = \begin{cases} 1, \text{ if } \quad n = a(d) \\ -1, \text{ if } \quad n = b(d) \, , \quad \forall d \in D, n \in \mathcal{N} \\ 0, \text{ otherwise} \end{cases} \tag{11.29b}$$

$$\pi_e : \sum_d h_d \hat{x}_{de} \leq U n_e, \forall e \in \mathcal{E}$$

$$\hat{x}_{de} \in \{0, 1\}, \quad n_e \in \{0, 1, \cdots U_{\max}\}, \quad \forall d \in D, e \in \mathcal{E} \tag{11.29c}$$

where n_e is the integer number of capacity modules installed in link e, $\hat{x}_{de}$ is 1 if the traffic of demand d traverses link e, and 0 otherwise. $U_{\max}$ is the maximum number of capacity modules a link can host.

Recall that integrality of the decision variables makes the problem non-convex, and strong duality is not guaranteed. Dualizing link capacity constraints (11.29c), using π_e as multipliers, the relaxed problem becomes:

$$\min_{(11.29bd)} \sum_e n_e + \sum_e \pi_e \left(\sum_d h_d \hat{x}_{de} - U n_e \right) \tag{11.30}$$

$$= \min_{(11.28bd)} \sum_e n_e (1 - \pi_e U) + \sum_d h_d \sum_e \pi_e (\hat{x}_{de}) \tag{11.31}$$

The number of modules $n_e(\pi)$ in the minimizer for multipliers π can be computed easily:

$$n_e(\pi) = \begin{cases} 0 \text{ if } \quad \pi_e U \leq 1 \\ U_{\max} \text{otherwise} \end{cases}, \quad \forall e \in \mathcal{E} \tag{11.32}$$

The computation of the optimum routing $x(\pi)$ of the relaxed problem can be separately solved for each demand. Moreover, each demand routing can be computed efficiently, since the solution minimizing (11.31) routes the 100% of the demand traffic through the shortest path between its end nodes according to link weights π_e. If more than one shortest path exists, any can be arbitrarily chosen (e.g., in our tests, the one with the shortest length in km).

Note that the solution $n_e(\pi), x(\pi)$ may violate the relaxed link capacity constraints and thus be unfeasible. Every iteration, we can easily compute an associated feasible solution by reusing the routing $x(\pi)$, but recomputing the number of modules per link as the minimum number needed to carry the link traffic $y_e = \sum_d h_d \hat{x}_{de}$:

$$n_e(\pi)^f = \left\lceil \frac{y_e}{U} \right\rceil \tag{11.33}$$

Since strong duality is not guaranteed to hold, the dual function may be non differentiable. The multipliers iteration, for a general γ_k step in the subgradient projection, is given by:

$$\pi_e(k + 1) = \left[\pi_e(k) + \gamma_k \left(\sum_d x_{de}(\pi(k)) - U n_e(\pi(k)) \right) \right]_0, \quad e \in \mathcal{E} \tag{11.34}$$

Algorithm 19 describes the pseudocode of the approach. In every iteration k, the minimizers associated to $\pi(k)$ are computed. Then, an associated feasible solution is extracted with (11.33). Finally, the multipliers are updated using (11.34) and the loop is repeated until a stop condition is met. Note that the algorithm should keep track of the best feasible solution found historically, and return it when it is terminated.

Algorithm 19 Non-convex capacity and routing optimization for (11.29)

1: *Initialization*: Set $k = 0$, initialize $\pi_e(0) = 1, \forall e \in \mathcal{E}$
2: *Minimizer computation*: Obtain $n_e(\pi(k))$ (11.32) and $x(\pi(k))$ (shortest paths).
3: *Feasible solution*: Compute feasible capacities $n_e(\pi)^f$ with (11.33).
4: *Weight update*: Update the π weights using (11.34).

11.7.1   Implementation Example

Algorithm 19 is tested in the network shown in Fig. 11.4 (without any node clustering). Its topology and traffic correspond to the Abilene reference topology and traffic matrix, included in the Net2Plan release as `abilene_N12_E30_withTrafficAndClusters3.n2p`. The total offered traffic sums 75 units, and modules have capacity one ($U = 1$). The maximum number of acceptable modules in a link $U_{\max}$ is set to 75 (enough to host all the total traffic).

The problem instance chosen could be solved to optimality in several tens of seconds using a CPLEX solver interfaced from JOM (optimum cost is 210 capacity units), while GLPK solver could not find a feasible solution when stopped after a 5 minute run.

Figure 11.6 plots the evolution of Algorithm 19 for a decreasing step length $\gamma_k = 0.05/k$. The upper graph shows the convergence of the multipliers. The lower graph draws two lines: (i) the evolution of the cost of the feasible solutions produced, and (ii) the dual cost in each iteration, a lower bound.

We can clearly see how the dual algorithm searches for the multipliers that maximize the dual function, reaching a maximum dual cost of ≈ 202.4 units. As the theory states, any dual cost is a lower bound to the optimum cost, and the maximum of the dual function (202.4 units) is the best among them. Since we know the optimum for this problem instance (210), we can determine that its duality gap is of ≈ 7.6 units: $(210 - 202.4 = 7.6)$. The best solution found by Algorithm 19 is 215.

The reader can use the Net2Plan implementation provided to see how the duality gap tends to decrease as the capacity modules U become smaller. The reasoning behind is that it results in a finer granularity for choosing the link capacity, which approximates a problem without capacity integrality constraints.

11.8   Notes and Sources

Decomposition techniques have formed a part of general optimization theory since their beginning. The publication of the Dantzig–Wolf decomposition principle in 1961 [4], is considered the start of the extensive evolution of large-scale mathematical optimization that followed. Multiple contributions exist, too many to cite them all. Some references exposing basic decomposition principles are [2, 5, 6] and [7]. Problem decomposition is a hot topic today, fueled by the quest of more and more efficient parallelizable algorithms, for example in the big data processing. The reader is referred to specialized journals and conferences in the topic.

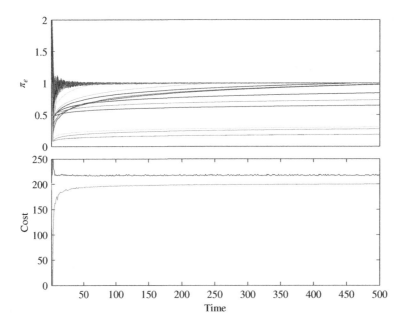

Figure 11.6 Evolution of Algorithm 19, decreasing step $\gamma = 0.05/k$. Plots from upper to lower are link multipliers and primal and dual costs. Optimum cost is 210

The interest in this chapter lays on the fundamental decomposition principles that provide an insight for understanding the interactions between protocols. The path breaking contribution in this line is the work *Layering as optimization decomposition: a mathematical theory of network architectures* [1] that puts together multiple previous contributions under a common view: interactions among network layers can be explained as the coordination of subproblems in a decomposition of a global network algorithm. This is an invaluable resource for the reader interested in the topic that links to more than a hundred of related references.

The organization of the decomposition methods into primal, dual, and problem reformulations is similar to that in [7] and [8]. The case study on cross-layer optimization of congestion control and QoS-aware capacity allocation is adapted from [8]. A dual approach for the cross-layer congestion control and backpressure routing was initially addressed in [9] and also appears in [1]. The cross-layer optimization of congestion control and transmission power in wireless networks is first studied in [3]. For didactic purposes, the algorithm in Section 11.5 is a simplified version of the work in [3], also tailored to reuse the primal power allocation algorithm in Section 9.6.

The idea of an inter-domain cooperative traffic engineering among ISPs in the Internet, supported by a dynamic control that enforces stable, efficient, and predictable interactions is discussed in works like [10, 11]. Domain collaboration schemes based on dual decompositions are presented in [12, 13]. The primal approach for coordination of network domains in Section 11.6 is original, and not published elsewhere.

The dual algorithm for modular capacities and non-bifurcated routing optimization is a simple application of the dual approach with didactic purposes. There is a growing and rich literature studying the duality gaps appearing in Lagrange relaxations and how to characterize strong relaxations that produce tight lower bounds. This analysis is out of the scope of this book, the interested reader is referred to sources like [14, 15].

11.9 Exercises

11.1 In the optimization problem (11.35)

$$\min_{x,y} f(x) + g(y) \quad \text{subject to:} \tag{11.35a}$$

$$x \in \mathcal{X}, y \in \mathcal{Y} \tag{11.35b}$$

$$Fx + Gy \leq h \tag{11.35c}$$

Devise a dual algorithm for the problem dualizing the complicating constraint (11.35c). Comment on the algorithm convergence.

11.2 Repeat Exercise 11.1 using a primal approach. *Hint*: Use an auxiliary variable z and replace (11.35c) by $Fx \leq z, Gy \leq h - z$.

11.3 In the optimization problem (11.36)

$$\min_{\{x_i\},y} \sum_i f_i(x_i, y) \quad \text{subject to:} \tag{11.36a}$$

$$y \in \mathcal{Y}, \quad x_i \in \mathcal{X}_i \quad \forall i \tag{11.36b}$$

$$\sum_i g_i(x_i) \leq b \tag{11.36c}$$

$$h_i(x_i) \leq y, \quad \forall i \tag{11.36d}$$

This problem has a coupling variable y and a complicating constraint (11.36c). Apply a primal decomposition of the problem to get rid of coupling variable y. Then, solve the resulting problem using a dual approach relaxing the coupling constraints. Comment on the resulting algorithm and its convergence.

11.4 Repeat Exercise 11.3 using first a dual approach relaxing constraint (11.36c) and then a primal approach on coupling variable y to solve the relaxed problem. Comment on the resulting algorithm and its convergence.

11.5 Repeat Exercise 11.3 using first a dual approach relaxing constraint (11.36c) and then a dual approach using auxiliary variables y_i and consistency constraints. Comment on the resulting algorithm and its convergence.

11.6 Apply KKT optimality conditions to find a closed expression for the euclidean projection of a point $(a_1, a_2) \in \mathbb{R}^2$ into the $\mathbb{R}^2$ subset $\mathcal{X} = \{x_1 \geq 0, x_2 \geq 0, x_1 + x_2 = c\}$.

11.7 [8] Devise a dual algorithm for solving the joint congestion control and QoS capacity allocation problem (11.11) dualizing the link capacity constraints (11.11bc). Comment on a possible distributed implementation of the scheme. Implement the algorithm in Net2Plan, using as a template the implementation available for Algorithm 15.

11.8   Devise an algorithm for solving the joint congestion control and routing problem (11.15), with a modified objective function:

$$\max_{h,x} \sum_d U_d(h_d) - \epsilon \sum_{de} x_{de}$$

where $\epsilon > 0$ is expected to be a small number. Apply first a primal decomposition, with h_d being the coupling variables, and then solve the routing subproblem using a dual decomposition. Comment on a possible distributed implementation of the scheme. Implement the algorithm in Net2Plan using as a template the implementation available for Algorithm 16.

11.9   Devise an algorithm for solving the joint congestion control and transmission power optimization problem (11.21), with a modified objective function:

$$\max_{h,x} \sum_d U_d(h_d) - \epsilon \sum_e e^{\tilde{p}_e}$$

where $\epsilon > 0$ is expected to be a small number. Apply first a primal decomposition, with h_d being the coupling variables, and then solve the power allocation subproblem using a dual decomposition. Comment on a possible distributed implementation of the scheme. Implement the algorithm in Net2Plan, using as a template the implementation available for Algorithm 17.

11.10 [13] In the interdomain routing problem of Section 11.6, let us assume that each domain c has an utility function $U_c(x^c)$ where x^c represents the destination-link routing variables of its internal and frontier links. The global optimization problem to solve is:

$$\min_{x \geq 0} \sum_c U(x^c) \quad \text{subject to:} \tag{11.37a}$$

$$\sum_{e \in \delta^+(n)} x_{te} - \sum_{e \in \delta^-(n)} x_{te} = h_{nt}, \forall t, n \in \mathcal{N}_c, n \neq t, \forall c \in C \tag{11.37b}$$

$$\sum_t x_{te} \leq u_e, \quad \forall c \in C, e \in \mathcal{E}_c^i \tag{11.37c}$$

$$\sum_t x_{te} \leq u_e, \quad \forall e \in \mathcal{E}^f \tag{11.37d}$$

See that the problem is coupled by flow conservation constraints (11.37b) in frontier nodes, and link capacity constraints (11.37d) in frontier links. For each frontier link $e \in \mathcal{E}^f$ between clusters c_1 and c_2 create the destination-link auxiliary variables: $x_{te}^{c_1}$ and $x_{te}^{c_2}$ for all destinations t, as local variables to each domain of the routing in these links. Create an equivalent problem to (11.37) where each domain c solves an independent routing problem using only local variables and the consistency constraints:

$$x_{te}^{c_1} = x_{te}^{c_2}, \forall t, c_1, c_2 \in C, \text{neighbor domains}, e \in E_{c_1}^f \cap E_{c_2}^f$$

couple the routing problem between neighbor links. Devise a dual algorithm that solves (11.37) relaxing the consistency constraints and coordinating the domain through these multipliers. Elaborate on the conditions for algorithm convergence. Implement the algorithm in Net2Plan, using as a template the implementation available for Algorithm 18.

11.11 We focus on the multilayer network design problem (7.4) described in Section 7.4. Find a dual algorithm targeted to an offline centralized execution, that decouples upper and lower layer problems by relaxing the upper layer link capacity constraints (7.4c). Comment on the resulting algorithm and its convergence in the case when z_c variables are restricted to be integers and when they are not.

11.12 Repeat Exercise 11.11 applying a primal algorithm, using z_c as the complicating variables.

References

[1] M. Chiang, S. H. Low, J. C. Doyle *et al.*, "Layering as optimization decomposition: A mathematical theory of network architectures," *Proceedings of the IEEE*, vol. 95, no. 1, pp. 255–312, 2007.

[2] D. P. Bertsekas and J. N. Tsitsiklis, *Parallel and Distributed Computation: Numerical Methods*. Athena Scientific, 1997.

[3] M. Chiang, "Balancing transport and physical layers in wireless multihop networks: Jointly optimal congestion control and power control," *Selected Areas in Communications, IEEE Journal on*, vol. 23, no. 1, pp. 104–116, 2005.

[4] G. Danzig and P. Wolfe, "The decomposition algorithm for linear programming," *Econometrica*, vol. 4, pp. 767–778, 1961.

[5] D. P. Bertsekas, *Nonlinear Programming*. Bertsekas: Athena Scientific, 1999.

[6] L. Lasdon, *Optimization theory for large systems*, ser. Dover books on Mathematics. Dover Publications, 2002.

[7] S. Boyd, L. Xiao, A. Mutapcic, and J. Mattingley, "Notes on decomposition methods," *Notes for EE364B, Stanford University*, 2007.

[8] D. P. Palomar and M. Chiang, "A tutorial on decomposition methods for network utility maximization," *Selected Areas in Communications, IEEE Journal on*, vol. 24, no. 8, pp. 1439–1451, 2006.

[9] X. Lin and N. B. Shroff, "Joint rate control and scheduling in multihop wireless networks," in *Decision and Control, 2004. CDC. 43rd IEEE Conference on*, vol. 2. IEEE, 2004, pp. 1484–1489.

[10] J. Winick, S. Jamin, and J. Rexford, "Traffic engineering between neighboring domains," Available: http://www.research.att.com/~jrex/papers/interAS.pdf, July 2002.

[11] R. Mahajan, D. Wetherall, and T. Anderson, "Negotiation-based routing between neighboring ISPs," in *Proceedings of the 2nd conference on Symposium on Networked Systems Design & Implementation-Volume 2*. USENIX Association, 2005, pp. 29–42.

[12] A. Tomaszewski, M. Pióro, and M. Mycek, "A distributed scheme for optimization of interdomain routing between collaborating domains," *Annals of Telecommunications-Annales des Télécommunications*, vol. 63, no. 11-12, pp. 631–638, 2008.

[13] G. Shrimali, A. Akella, and A. Mutapcic, "Cooperative interdomain traffic engineering using Nash bargaining and decomposition," *IEEE/ACM Transactions on Networking (TON)*, vol. 18, no. 2, pp. 341–352, 2010.

[14] L. Wolsey, *Integer Programming*, ser. Wiley Series in Discrete Mathematics and Optimization. New York, NY, USA: John Wiley Inc., 1998.

[15] L. A. Wolsey and G. L. Nemhauser, *Integer and combinatorial optimization*. Hoboken, NJ, USA: John Wiley & Sons, Inc., 2014.

12

Heuristic Algorithms

12.1 Introduction

This chapter is devoted to the development of offline algorithms for non-convex $\mathcal{NP}$-hard network problems, typically executed in centralized servers, and without major running time requisites (e.g, from minutes to hours). These problems typically appear in contexts like:

- *Capacity planning*: Planning departments elaborate upgrade plans for the link and node capacities or the placement of new links/nodes for the upcoming year, to cope with a fore-casted traffic growth. These tasks typically receive the name of *capacity planning*.
- *Greenfield network planning*: A greenfield plan involves designing a network from scratch, for example to plan a network deployment in a region where the operator is currently absent.
- *Brownfield network planning*: Brownfield planning tasks refer to a redesign of a large portion of an existing network, reusing some or all of the legacy equipment in place, for example, in a migration plan to a new network technology.
- *Online network optimization*: The increasing introduction of Software Defined Networking (SDN) instruments in the network, permits automating multiple tasks that traditionally required manual intervention. For instance, the unified collection of traffic measurements, routing, and topology information from diverse databases, can now be managed by so-called SDN controllers. Then, the updated network state is available to network optimization engines, which can periodically: (i) redesign the traffic routing or the link capacities (using on-demand capacity services) and (ii) issue the required reconfiguration orders to the SDN controller, which automatically conveys them to the nodes. The online network optimization (sometimes called *in-operation planning* [1]) consists of repeating the process a limited number of times per day to adapt the network to slow-changing traffic conditions.

12.1.1 What Complexity Theory Tells Us that We cannot Do

It is the case that the majority of the optimization problems involved in previous design tasks are non-convex and $\mathcal{NP}$-hard. For instance, node and/or link placement problems, traffic routing under integral or non-bifurcated constraints, non trivial OSPF/ECMP routing variants, is

Optimization of Computer Networks – Modeling and Algorithms: A Hands-On Approach,
First Edition. Pablo Pavón Mariño.
© 2016 John Wiley & Sons, Ltd. Published 2016 by John Wiley & Sons, Ltd.
Companion Website: www.wiley.com/go/PavonMarinoSol16

the case. As stated in Appendix C, this means that there are not known polynomial algorithms solving them and it is conjectured that such algorithms do not exist. Moreover, in Appendix C we see that some problems could be not only hard to solve, but also hard to *approximate*. We presented the concept of *polynomial approximation algorithms*, as those that guarantee finding an ϵ-approximation in polynomial time, where the ϵ value can be chosen beforehand. For instance, in a minimization problem, a 0.1-approximation algorithm ($\epsilon = 0.1$) guarantees producing a solution with, at most, 10% extra cost than the optimal.

In this respect, we saw that:

- $\mathcal{NPO}$-complete optimization problems are not approximable (assuming $\mathcal{P} \neq \mathcal{NP}$), meaning that there are no ϵ-approximation algorithms for any ϵ[1]. Examples are solving general integer linear programs (ILPs) and finding the minimum cost ring in a network or its maximum clique (subset of nodes fully connected among them).
- $\mathcal{APX}$ problems are those for which there are polynomial ϵ-approximations for *some* ϵ. $\mathcal{APX}$-complete problems are those that (assuming $\mathcal{P} \neq \mathcal{NP}$) have no polynomial approximations for small ϵ values. Examples are the node location problem, integral routing, or minimum cost multicast tree problem.
- Problems in $\mathcal{PTAS}$ that are not in $\mathcal{FPTAS}$ have an approximation for any $\epsilon > 0$, but its running time grows worse than a polynomial with respect to the approximation quality $1/\epsilon$. Then, it can still be intractable finding fine approximations.

The reader is referred to Appendix C for deeper explanations.

12.1.2   Our Options

All in all, complexity theory leaves a dark panorama on the limits to what can be done to solve large instances of common non-approximable network problems. In practice, the two main alternatives are:

- *Formulate and solve.* If the problem can be reasonably formulated using integer linear programming, integer convex programming, or general nonlinear programs, it is possible to use standard commercial or freeware solvers to attempt finding good suboptimal solutions for them. For instance, current version of Net2Plan/JOM provides access to CPLEX (commercial) and GLPK (GNU) solvers for mixed integer linear programs. These solvers can be configured with a maximum limit time and an acceptable approximation quality. Then, the solver returns the best solution found so far if the time expires before an acceptable solution is found. In the author's experience, the advantage of this approach is that in many medium-size problems the solver is able to find close to optimal solutions in a reasonable time. However, it may very well fail or crash in large-scale problem instances.
- *Devise an ad hoc heuristic.* Heuristics are algorithms that search for approximate solutions to a problem in polynomial time. However, in contrast to approximation algorithms, they do not provide any guarantee beforehand of the quality (ϵ-approximation) of the solution to return.

---

[1] They have polynomial approximation algorithms with ϵ values that grow with problem size.

12.1.3   Organization and Rationale of this Chapter

This chapter provides the guidelines for creating heuristics for hard network planning problems, suitable for an offline centralized execution in a server, and without major time limits. The interest in heuristic design is motivated by the multiplicity of network technologies and particular conditions when planning a network that demand *ad hoc* algorithm developments. A typical situation faced by planning departments and academia sounds like:

> I have found in the literature no algorithm for dimensioning the network in this potentially interesting context X, although some previous works exist for variations of it, and subproblems inside it. We should devise an ad-hoc heuristic, potentially reusing or adapting some of the algorithms already investigated. The heuristic will enable our report assessing X.

Then, a creative process starts to design our heuristic, which should ideally combine two sources:

- A precise knowledge of the problem to solve and the tactics used by other algorithms, addressing variations or subparts of it.
- A background in *heuristic design strategies*. These are techniques for building successful heuristics as skeletons adaptable to any problem that help to smartly explore among the enormous set of potential solutions.

In this chapter we exemplify the heuristic design process by first describing some general rules applicable (Section 12.2), followed by a didactic review of most common strategies for building heuristics: local search, greedy algorithms, and meta-heuristics like simulated annealing (SAN), tabu search (TS), GRASP, ant-colony optimization (ACO), and evolutionary algorithms (EA).

We accompany each of the techniques with an example, where the heuristic philosophy is applied to a particular $\mathcal{NP}$-hard network problem. This will help us exposing design hints and recommendations. In order to didactically emphasize the differences between the heuristics, the same problem is solved with all of them. For this, we selected a popular traffic engineering problem, consisting of finding the OSPF link weights in an IP network to minimize a measure of network congestion. The particular problem version is described in Example 12.1.

OSPF weight setting is known to be a difficult and tricky task, since modifying a single link weight can drastically change the routing in multiple parts of the network. The reader is referred to Section 4.6.5 for details on how the OSPF/ECMP routing in IP networks works. Formulations of the problem using integer linear programs turn to be intractable even for small instances, since many auxiliary variables and constraints are needed to model the Equal-Cost Multi-Path (ECMP) bifurcation rule [2]. Several heuristics have been also presented for the problem (see Section 12.11), aside of the ones that we will present here.

In order to estimate the performances of the algorithms, we compare the solutions provided with a lower bound to network congestion: the optimum congestion if a destination-based routing without ECMP traffic bifurcation restrictions was used. This lower bound can be computed in polynomial time solving a linear program, as described in Example 12.1.

We illustrate each heuristic with numerical examples, for a problem instance based on the topology and offered traffic given by `NSFNet_N14_E42.n2p` file in Net2Plan tool (see

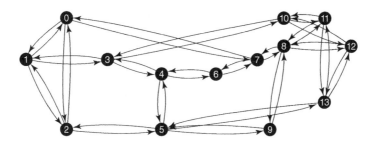

Figure 12.1 NSFNET topology. The network has 21 bidirectional links, all with the same capacity $(u_e = 500$ Gbps). Total offered traffic is as in NSFNet_N14_E42.n2p file, normalized to sum 4 Tbps

Fig. 12.1). The network is composed of 42 links, and link weights $\{w_e, e \in \mathcal{E}\}$ are restricted to integer numbers between 1 and $W_m = 16$.

In the implementation provided, evaluating the network congestion for a set of link weights w needs around 1 ms of computation time. For those heuristics based on the *vicinity* concept, that will be described later in this chapter, we can cut down the computational cost applying the following property.

Proposition 12.1 If a link does not carry traffic of a demand (it is not in a shortest path), it will neither carry traffic of that demand when we increase the link weight, keeping the rest of weights unchanged.

An obvious consequence of previous proposition is that increasing the weight in a link that does not carry traffic, keeping the rest unchanged, does not modify the routing of any demand.

Despite of any code profiling we make, a brute force test of all the possible weight settings would mean evaluating $16^{42} \approx 10^{50}$ solutions, which would need more than 10^{34} years of running time even if we could evaluate each solution in 1 nanosecond.

The problem instance described has a congestion lower bound of $cg_{lb} \approx 0.51$ (with a maximum link utilization ≈ 0.52 and an average link utilization of ≈ 0.4). As we will see, the heuristics presented are all close to it after some minutes of computation. All the algorithms are implemented in the Net2Plan tool and their source codes are available to the reader.

To finalize, Section 12.10 includes a complete and fairly realistic network planning case study, where we develop an ad hoc heuristic for planning a backbone IP over WDM network, under three different recovery schemes: 1+1 lightpath protection, shared lightpath protection, and lightpath restoration. Equipment acquisition and link hiring costs are representative of current technology. A detailed analysis of the results is included to show the reader a realistic network planning application example.

The developed heuristic for the case study is based on combining a GRASP scheme with small size ILPs, with running times in the order of tens of milliseconds (and limited to a a few seconds) in a standard laptop (Intel Core i5, 2.27 GHz, 8 GB RAM). The role of the ILPs is performing an efficient exploration of a reduced solution space, taking benefit of the extremely profiled implementations of sophisticated optimization techniques in ILP solvers. The GRASP scheme is in charge of guiding the search of the ILPs. The combination of an heuristic skeleton repeatedly calling ILP-based routines has shown its strength in multiple practical works (see Section 12.11 for references). This strategy is enabled by libraries like, for example, JOM, that efficiently interface with optimization solvers from programs in general purpose languages, permitting building large formulations in memory in tens of milliseconds.

Example 12.1 Let $\mathcal{N}$ be a set of IP routers and $\mathcal{E}$ the set of links connecting them, with known capacities $\{u_e, e \in \mathcal{E}\}$, and let $\{h_{st}, s, t, \in \mathcal{N}\}$ be the offered traffic matrix. We are interested in optimizing the vector w of OSPF link weights $w = \{w_e, e \in \mathcal{E}\}$. Weights are restricted to be integer numbers between 1 and W_m. The optimization target is minimizing a measure of network congestion given by:

$$cg(w) = 0.9 \max_e \frac{y_e(w)}{u_e} + 0.1 \sum_e \frac{y_e(w)}{u_e}$$

where $y_e(w)$ is the amount of traffic in link e when the OSPF/ECMP routing is applied with weights w. This measure balances two contributions: the maximum link utilization (90%), and the average link utilization (10%).

A lower bound to the network congestion achievable with OSPF routing is provided by the optimum of the destination-based formulation (12.1):

$$\min_{x \geq 0, \rho \geq 0} \ 0.9\rho + 0.1 \sum_e \frac{\sum_t x_{te}}{u_e} \quad \text{subject to:} \tag{12.1a}$$

$$\sum_{e \in \delta^+(n)} x_{te} - \sum_{e \in \delta^-(n)} x_{te} = \begin{cases} h_{nt}, & \text{if } n \neq t \\ -\sum_s h_{st}, & \text{if } n = t \end{cases}, \quad \forall t, n \in \mathcal{N} \tag{12.1b}$$

$$\sum_t x_{te} \leq u_e \rho, \quad \forall e \in \mathcal{E} \tag{12.1c}$$

12.2 Heuristic Design Keys

12.2.1 Heuristic Types

There is a vast diversity of heuristic algorithms, for didactic purposes we classify them in three types:

- *Local search algorithms*: These are iterative algorithms that, in each iteration k, jump from a solution x_k to a *neighbor* solution that improves the current one. The process ends when no improving neighbor solution is found. These methods depend on the particular definition of solution neighborhood or vicinity.
- *Greedy algorithms*: These are *constructive* algorithms. The algorithm starts with an empty solution, for example, no decision variable in the problem is set, and in each iteration some parts of the solution are fixed trying to optimize a myopic or greedy function. The method is constructive, in the sense that there is no "coming back": those parts of the solution already set in previous iterations are not changed and only new pieces are decided. The algorithm ends when a complete solution is created.
- *Metaheuristic algorithms*: These are templates, general schemes or "philosophies" for searching for good problem solutions, which can be adapted in different forms to any optimization problem. Most metaheuristics are inspired in observing how the nature works to solve complex problems.

It is important to emphasize that, in most of the cases, network planning algorithms are designed combining different techniques among those described in this chapter, in a somewhat creative process governed by experience and intuition. The techniques are described separately for didactic purposes.

12.2.2   Intensification versus Diversification

Heuristics target an efficient exploration of a large solution space, riddled with of local optima. This process coordinates two opposite trends that should appear in every heuristic:

- *Intensification*: The heuristic should include techniques that intensify the search on those regions of the solution space that have provided good solutions in previous iterations, guessing that the chances are better to find improving solutions in the future.
- *Diversification*: The heuristic should include techniques that avoid confining the search into a small region in the solution space. For this to happen, it is necessary that the heuristic accepts to iterate among solutions that worsen the current one, with the hope that these iterations end up later in a better solution's region. A common diversification technique consists of, after a number of iterations have passed without improving the *incumbent solution* (best solution found so far), randomly choosing the next solution to explore and continuing the search from it.

Local search algorithms are examples of pure intensification schemes with no diversification. In turn, a pure diversification scheme would randomly pick solutions in each iteration, without any trend to intensify the search near the best solutions found.

Both diversification and intensification techniques should appear in any heuristic. We will see them in different forms in the strategies described in this chapter.

12.2.3   How to Assess the Solution Quality

In contrast to approximation algorithms, heuristics do not provide guarantees beforehand on the quality of the solution to return. Then, to assess it we must compare the returned solution with optimality bounds: lower bounds to the optimal cost in minimization problems or upper bounds to the achievable benefit in maximization problems.

Optimality bounds can be obtained in multiple forms, some of them are:

- *Via the dual problem, fixed multiplier values*: Any dual solution with non-negative values for multipliers of inequality constraints has a dual cost that is an optimality bound. For instance, removing constraints of the type $f(x) \leq 0$ or $h(x) = 0$ is equivalent to computing the dual cost in a relaxed problem where the multipliers of these constraints are set to zero.
- *Via the dual problem, optimum multiplier values (Lagrange relaxation)*: The dual of an optimization problem is a convex problem solvable in polynomial time, even if the original problem is $\mathcal{NP}$-hard. The optimum dual solution provides the best optimality bound achievable using the dual function: the maximum lower bound in minimization problems, the minimum upper bound in maximization problems.

- *Relaxing integrality constraints.* For instance, this converts an ILP into a linear program solvable in polynomial time, whose optimum solution is an optimality bound. An important property regarding to these relaxations is that, in linear programs with integer variables, they are always weaker than Lagrange relaxations (see [3] p. 172). That is, for example in minimization problems, the cost lower bounds achieved with a Lagrange relaxation are always equal or greater than (and thus "better"" or "stronger") than those achieved relaxing integrality constraints and solving the resulting linear program.

Exercise 4.2 in Chapter 4 and Exercise 7.21 in Chapter 7 include examples of easy to compute optimality bounds.

12.2.4   Stop Conditions

There are no good stop conditions for heuristic algorithms, which is a natural consequence of targeting problems that are $\mathcal{NP}$-hard even to approximate. Implementations usually combine some of the following:

- *Approximating an optimality bound*: A solution that is an ϵ-approximation to an optimality bound is also an ϵ-approximation to the true optimum. A common stop condition is halting the algorithm, for example when the cost approximates enough to a cost lower bound precomputed or computed on the fly. However, note that being far from an optimality bound does not mean that our solution is bad: it may be that our bound is inaccurate. Even if the bound is accurate, our heuristic may never come close to it in polynomial time. For these reasons, optimality bounds are auxiliary stop conditions, combined with others.
- *Maximum running time/iterations*: The algorithm stops after some fixed time, or after a maximum number of iterations.
- *Maximum time/iterations without improvement*: The algorithm stops if no improvement in the incumbent solution (best solution found so far) is observed in a given time or number of iterations. Alternatively, instead of stopping the algorithm, it can be restarted using a new randomly chosen initial solution.

12.2.5   Defining the Cost or Fitness Function

Heuristics use the so-called cost function (in minimization problems) and fitness function (in maximization problems) to control the algorithm behavior. This is a function that, given a solution x, returns a real value evaluating its "goodness". The natural value of the cost (fitness) function, is just the objective function to minimize (maximize). However, there are two extra aspects to consider:

- If x is not a feasible solution, its objective function may be not defined. We may be tempted to return in such case an infinite cost. However, to help the heuristic to find better feasible solutions in future iterations, it is important that the cost/fitness function assigns different costs to different unfeasible solutions, for example, distinguishing between those "slightly unfeasible" from those "very unfeasible". As an example, the cost function may be the sum of the objective function plus a penalization term that (i) takes large values, so that feasible solutions are always preferred over even slightly unfeasible solutions, and so that (ii) penalization is higher when unfeasibility is more profound. For instance, it may be proportional to the number of constraints violated, or the extent by which they are not met.

- In some occasions, an exact computation of the objective function is too costly. For instance, it requires an event-driven simulation that can take minutes or hours. In such cases, the cost/fitness function is based on an estimation of the true objective value, which can be efficiently computed, leaving the exact evaluation for just a small set of selected solution candidates.

12.2.6 Coding the Solution

Given a problem solution x, the coding of the solution $c(x)$ refers to the specific form in which it is implemented in the computer. This is often a crucial aspect when designing an heuristic, the following points should be considered:

- *One-to-one representation of the solutions.* There should be a one-to-one relation between the set of feasible problem solutions and the set of possible codifications. This means that every feasible solution x has one and only one coding. In Exercise 12.1 we see an example when this condition is not met, resulting in an artificial and unproductive increase in the size of the solution space to explore.
- *Efficiency to implement heuristic-specific operations.* Some heuristics are based on, given a solution x, enumerating the vicinity set $\mathcal{V}(x)$, the neighbor solutions to x. In such cases, the coding of the solution should make this enumeration computationally inexpensive. In turn, evolutionary algorithms apply special operations as *crossover*, consisting of producing a *child* solution x_c from two different parent solutions x_1, x_2, such that x_c inherits some of the properties in x_1 and x_2. In this case, the coding should permit efficient implementations for producing $c(x_c)$ from $c(x_1)$ and $c(x_2)$.

12.3 Local Search Algorithms

Local search algorithms are iterative schemes targeted to find local optima in complex problems. Algorithm 20 illustrates its pseudocode. In each iteration the algorithm enumerates the solutions y in the *vicinity* of the current solution x ($y \in \mathcal{V}(x)$) and moves to one out of them that improves x. Two alternatives exist:

- *Best-fit*: Every neighbor in $\mathcal{V}(x)$ is evaluated and x is updated to the best among them.
- *First-fit*: Neighbor solutions in $\mathcal{V}(x)$ are evaluated in a particular order, x is updated to the first improving solution found. Then, the remaining neighbor solutions are not evaluated, saving computation time.

In either case, if no improving neighbors exist the algorithm stops returning x, the last and best solution found. Note that the solution returned is a local minimum according to the definition of vicinity adopted, since it is equal or better than their neighbors.

Algorithm 20 Local search (*best-fit*)

1: *Initialization*: Set *x* initial solution
2: **do**
3:     *next* = *x*.
4:     **for each** $y \in \mathcal{V}(x)$
5:         **if** $f(y) < f(x)$ **comment:** *f* is the cost function to minimize.
6:             *next* = *y*   **comment:** Improving solution
7:             **comment:** If *first-fit* then execute **break** here to leave the **for each** loop
8: **while** *next* ≠ *x*   **comment:** End when no improving solution
9: **return** *x*

12.3.1   Design Hints

In this section we provide some hints in the design of local search heuristics:

- *Vicinity size*. Local search algorithms are determined by how we define the set $\mathcal{V}(x)$ of neighbor solutions of *x*. In this respect, a crucial parameter to consider is the vicinity size. A large vicinity is likely to mean better returned solutions at a cost of a higher execution time, since the solution returned is the best among a larger set of neighbors to evaluate. However, this is an *average* behavior and we can find examples when a larger vicinity results in faster and/or worse solutions, since the iterations behave using local information and better solutions in one iteration can lead to worse solutions later on.
- *Connectivity of the solution space*. When defining the solution vicinity we should guarantee that from any initial solution it is possible to reach any other solution in a finite number of steps.
- *Best-fit versus first-fit*. Intuitively, one may imagine that best-fit runs should provide in average better solutions than first-fit, at a cost of higher running times. However, because of the local-based decisions, starting from the same initial solution a first-fit run can be faster or slower than best-fit and/or return a better or worse solution.

Example 12.2 `Offline_fa_ospfWeightLocalSearch` file provides a local search implementation for the OSPF weight setting problem (Example 12.1), where the first-fit versus best-fit option is user-defined. We consider that two solutions are neighbors if all the links but one have the same weight and in that link the weight differs in at most *D* units, where parameter *D* can also be user defined. Then, the maximum number of neighbors *V* of a solution is, for instance:

$$V = 21 \times 2 = 42 \quad \rightarrow \quad \text{if} \quad D = 1$$
$$V = 21 \times 15 = 315 \quad \rightarrow \quad \text{if} \quad D = 15$$

It is easy to check that the solution space is connected under this vicinity definition, since from any weight setting *x* we can reach any other in at most $|\mathcal{E}| \lceil (W_m - 1)/D \rceil$ iterations.

- Figure 12.2 compares the congestion and running time obtained in 100 runs with vicinities defined for $D = 1$ and $D = 15$. We see that in both the first-fit and best-fit executions (Fig. 12.2a and Fig. 12.2b, respectively), small vicinities tend to be less robust, producing both good and bad solutions. Larger vicinities improve in average the solutions returned with also less outliers, at a cost of a significantly higher and more variable computation time.

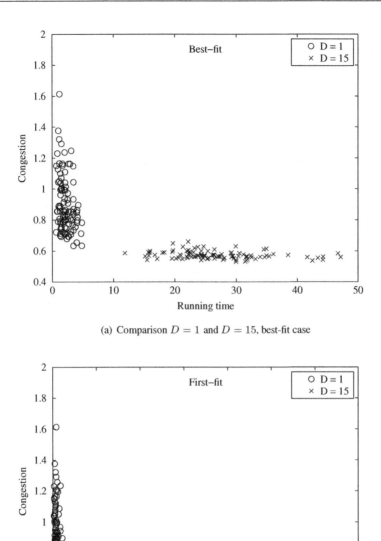

(a) Comparison $D = 1$ and $D = 15$, best-fit case

(b) Comparison $D = 15$ and $D = 15$, first-fit case

Figure 12.2   Local search vicinity size comparison. Congestion lower bound: 0.51

- Figure 12.2 also helps us to compare first-fit versus. best-fit trade-offs. We see that, on average, both provide very similar congestion values and computation times for $D = 15$, while for $D = 1$ the best-fit average computation is twice that of first-fit, without significant congestion improvements.

12.4   Simulated Annealing

Simulated Annealing (SAN) meta-heuristics were first proposed in the 1980s [4] to optimize functions with multiple local optima. It is inspired in emulating the annealing process that brings a fluid into a low-energy state, such as growing a crystal. Annealing consists of melting the fluid and then slowly lowering the temperature, such that molecules arrange into lower energy crystalline structures. A slow temperature decrease, especially in the final the stages, tends to produce better crystals.

SAN algorithms emulate this behavior varying the standard local search algorithms by adding a mechanism that permits *uphill* movements, that is, jumping to a non-improving neighbor solution. The standard SAN pseudocode is shown in Algorithm 21. In each iteration, the algorithm chooses randomly one solution y among the neighbors of the current solution x ($y \in \mathcal{V}(x)$). Then, it applies the so-called Metropolis [5] test to decide whether to jump to y, or stay one more iteration in x:

- Accept y if it is better than x. That is, if $c(x) < c(y)$, where $c(z)$ is the cost of solution z, the objective target to minimize.
- If not, still accept y with probability:

$$P_{acc} = e^{\frac{c(y)-c(x)}{T}}$$

Here T is the *system temperature*, a global variable in the algorithm. The importance of the temperature T is critical in SAN:

- For high T values, non-improving solutions are always accepted:

$$T \to \infty \Rightarrow P_{acc} = e^{\frac{c(y)-c(x)}{T}} \to 1$$

Then, SAN behaves as a random algorithm, which erratically jumps between neighbor solutions whatever their costs are.

- For low T values, non-improving solutions are never accepted:

$$T \to 0 \Rightarrow P_{acc} = e^{\frac{c(y)-c(x)}{T}} \to 0$$

and SAN behaves as a standard first-fit local search and eternally stays in the first local optimum found.

The annealing process is emulated by (i) initializing T to a high value, which is slowly decreased and (ii) every time a temperature decrease occurs, SAN performs a sufficient number of iterations to simulate an equilibrium state for that temperature in the physical crystal equivalence. This is implemented by two nested loops in Algorithm 21, the outer loop reducing the temperature, the inner loop running a usually fixed number of iterations at that temperature (typically between hundreds and thousands).

Algorithm 21 Simulated annealing

1: *Initialization*:
2: Set $T = T(0)$, initial temperature, x initial solution, $x_{best} = x_0$ incumbent solution
3: **do** **comment:** Outer loop
4:   **do** **comment:** Inner loop
5:    $v = $ random chosen neighbor in $\mathcal{V}(x)$
6:    **if** $c(v) < c(x)$ **or** with probability $e^{-\frac{c(v)-c(x)}{T}}$
7:     $x = v$, update x_{best} if needed. **comment:** Jump to v
8:   **while** tempDecreaseCriterion
9:    Decrease T.
10: **while** stopCriterion
11: **return** x_{best}

12.4.1 Design hints

In general, SAN parameters should be tailored empirically, adapted to the particular problem to solve. Here, we provide some hints for that, using the minimum congestion OSPF weight setting problem as an example. We consider that two solutions are neighbors if they differ in the weight of a single link, in any quantity ($D = 15$ case).

- *Initial temperature*: We can fix manually the initial temperature if we (i) estimate the worse case cost difference $c(v) - c(x)$ between two neighbor solutions and (ii) fix a (high) acceptance probability for that case, for example $P_{acc} = 0.5$. For instance, in the OSPF problem example we can fix $T(0)$ to accept with a probability of 0.5, a congestion worsening of 0.25, then:

$$e^{-\frac{0.25}{T(0)}} = 0.5 \Rightarrow T(0) = \frac{-0.25}{\log\ 0.5} \approx 0.36$$

  The maximum temperature can be also computed using an adaptive scheme that initially *increases* the temperature from a low value (*warming stage*), and monitors the percentage of worse solutions that are accepted. Then, the first temperature found that matches a target acceptance threshold (e.g. between 0.5 and 0.9) becomes the initial temperature and the standard *cooling stage* of SAN continues from it.
- *Freezing temperature*: The *freezing* temperature T_f is that which makes the algorithm get stuck in a local optimum. Usually, freezing temperatures are not computed manually, but with adaptive schemes that detect that the system is frozen just observing that the acceptance ratio falls below a threshold (e.g. 0.01). In these cases, the temperature is raised again (e.g. reheated to the initial temperature) to make the algorithm advance.
- *Temperature reduction schedule*. The most popular temperature reduction scheme is the geometric: $T = T \times \alpha$, where $\alpha \in (0, 1)$ is the reduction factor, typical numbers are around $\alpha = 0.9$. Other options are for instance the linear decrease: $T = T - \beta$, for $\beta > 0$.
- *Rapid quenching*: Another technique based on the physical system analogy is the rapid quenching approach. Rapid quenching, quickly reduces the system temperature bringing the algorithm to a local optimum quickly. The system is then reheated to a temperature lower than the initial temperature, and the process is repeated.

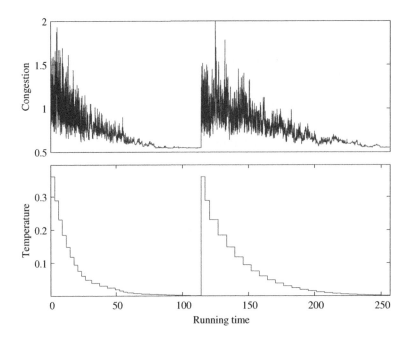

Figure 12.3 Evolution of SAN Algorithm for OSPF weight setting problem, $\alpha = 0.8$, $P_{acc}^{wc} = 0.5$, $P_f = 0.01$, initial solution is chosen randomly. Best congestion objective found ≈ 0.521 (0.51 is a lower bound)

Example 12.3 Offline_fa_ospfWeightSAN provides a SAN implementation for the OSPF weight setting problem. The algorithm fixes the input temperature according to a user-defined worse case acceptance probability, for example of $P_{acc}^{wc} = 0.5$ and applies a geometric decrease factor α. The algorithm monitors the acceptance probabilities and considers that the algorithm is stuck in a local minima if in the inner loop the fraction of moves that change the congestion is below a user-defined fraction P_f. If that happens, the temperature is reset to the initial value and the algorithm continues from there.

Figure 12.3 shows the congestion evolution trace in a 5 minute execution for the same network topology as in Section 12.3.1. This helps to illustrate the typical SAN behavior. High temperatures produce a large variability in the solutions found. As the temperature decreases, the system gets frozen and converges to a local optimum. This is detected by the algorithm that reheated the system around $t = 140$ s. The best congestion objective found is ≈ 0.521 (recall that 0.51 is a lower bound).

12.5 Tabu Search

Tabu search (TS) algorithms were proposed by Fred Glover in the late 1970s and have been applied since then in multiple $\mathcal{NP}$-hard optimization problems.

TS is a variation of the local search scheme that includes techniques to permit jumps to non-improving neighbor iterations. This is accomplished by implementing a so-called *tabu list*, which describes a subset of neighbor solutions that are forbidden. Then, in each iteration we explore all *non-forbidden* neighbor solutions and move to the best among them, which, depending on the tabu list contents, can be non-improving.

Every iteration, the new solution visited or an *attribute* characterizing it is added to the tabu list. The size of the tabu list T (or *tabu tenure*) is typically fixed, so that when the list becomes full after the first iterations, every subsequent step a new solution or attribute is added and the oldest one is removed. Then, the tabu list emulates the behavior of a *short-term memory* that remembers last solutions tried to not repeat them.

We now illustrate the difference between including in the tabu list *full solutions* and *solution attributes*. We focus on a TS method for the minimum congestion OSPF link weight setting problem (Example 12.1), where two solutions are neighbors if their weights are equal in all links but one. Then, in an iteration k a particular link $e(k)$ in the current solution changes its weight from $w_e(k)$ to $w_e(k+1)$. In this context:

- *Full solutions*. Including a full solution in the tabu list means that it cannot be repeated in the near future. Checking if a candidate neighbor w is in the tabu list can be a costful process, since it requires comparing the weights for all the links with all the solutions of the tabu list[2].
- *Solution attributes*. An attribute $a(x)$ of a solution x is a usually short identifier that characterizes the solution x and potentially multiple other solutions. Then, by adding an attribute to the tabu list, all these solutions are simultaneously forbidden. In the OSPF weight setting problem, we can add to the tabu list the link $e(k)$, which has changed its weight in current iteration. Then, once the weight of a link has been changed, it will stay unchanged in the near future. Such type of attributes are called *move attributes* since they characterize a transition between two neighbor solutions.

Algorithm 22 includes the pseudocode of the basic TS procedure described.

Algorithm 22 Tabu search (basic scheme)

1: *Initialization*:
2:    Set x initial solution, $x_{best} = x$
3:    Set $TL = \emptyset$ **comment:** Tabu list initially empty
4: **do**
5:    $x_{bestNeighbor} = \emptyset$.
6:    **for each** $y \in V(x)$, y not tabu
7:       **if** $f(y) < f(x_{bestNeighbor})$
8:          $x_{bestNeighbor} = y$ **comment:** Improving solution
9:    $x = x_{bestNeighbor}$ **comment:** Jump to the best neighbor
10:    $TL = TL \bigcup a(x)$ **comment:** Update the tabu list
11:    **if** $|TL| > T$ **then** remove oldest element in TL
12:    **if** $f(x) < f(x_{best})$ **then** $x_{best} = x$ **comment:** Update incumbent solution
13: **while** stop criterion not met
14: **return** x

---

[2] This process can be sped up if the tabu list is populated with a *hash* number computed from the full solution. This is the strategy followed in [6].

12.5.1   Design Hints

In this section we describe common techniques added to standard basic TS scheme, together with design hints.

- *Tabu tenure.* The size of the tabu list (tabu tenure) is a key design parameter in TS method. Larger tabu lists help to diversify the search, although too large lists may prevent reaching the optimal solution and/or create erratic movements. Typically, the tabu list size is set manually. To tailor it, it is important to consider the fraction of the neighborhood that becomes tabu when the list is full. For instance, in the TS OSPF weight setting algorithm described, a tabu list of size $f|\mathcal{E}|$ links, where $|\mathcal{E}|$ is the number of network links, roughly forbids a fraction $f \in [0, 1]$ of the neighbor solutions. The tabu tenure parameter can also be dynamically adapted. As an example, the Reactive Tabu Search technique [7] suggests increasing the tabu list when cycles in the iterations are detected.
- *Multiple tabu lists.* Multiple tabu lists can be combined in a problem, such that a solution is considered tabu if it is forbidden by any of them. Typically, multiple tabu lists appear as a combination of a per-solution list with a move-attribute list.
- *Medium-term memory.* Medium-term memory structures (or recency memory) can be used to *intensify* the search in promising regions of the solution space. The idea is extracting common features of elite solutions, and confine the search among those solutions sharing these features. In our example, the recency memory can take the form of a vector r_e, which assigns to each link e the number of *consecutive jumps* in the past in which the weight of e did not change. Then, we confine the search during a certain period by not exploring those solutions that change these weights, which are considered a characteristic of "good solutions".
- *Long-term memory.* Long-term memory structures can be included in TS to encourage *diversification*. Their typical representation is a *frequency memory*, which stores for each solution component or for each possible move the number of times that it appeared since the algorithm started. For instance, in the OSPF weight setting example, the long-term memory can be implemented as a r_{ew} matrix, counting the number of times that link e was assigned weight w. The diversification can be applied periodically or when the algorithm gets stuck. Then, the search is restarted from a new solution that includes the components that less frequently appear, maybe randomly combined with those in elite solutions. Also, a penalty in the objective function can be applied to the most frequently used components.
- *Aspiration criterion.* Aspiration criterion is a form to override the tabu list, provided that the neighbor solution initially forbidden is "good enough". The commonly used aspiration criterion consists of accepting a tabu solution if it improves the incumbent solution. This is the approach also followed in our OSPF weight setting example. Note that applying such criterion increases the iteration running time, since all neighbor solutions are evaluated, including those forbidden by the tabu list.

Example 12.4 `Offline_fa_ospfWeightTabuSearch` implements a TS scheme for the OSPF weight setting problem, following the guidelines described above. Figure 12.4 shows the results of two runs of the algorithm, with and without aspiration criterion, and with a tabu tenure equal to half the number of links. More or less similar results have been obtained with

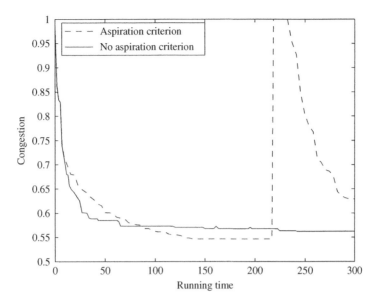

Figure 12.4 Evolution of Tabu Search Algorithm for OSPF weight setting problem. Tabu list size is 21 (50% of the number of links). Randomization occurs after 15 iterations without improving the best solution found since the last randomization event. Best congestion objectives are ≈ 0.546 (aspiration criterion) and ≈ 0.562 (no aspiration criterion). 0.51 is a congestion lower bound

smaller and larger tabu lists, were usually (but not always) the aspiration criterion resulted in better solutions. Best congestion objectives in the plotted case are ≈ 0.546 (aspiration criterion) and ≈ 0.562 (no aspiration criterion). Recall that 0.51 is a congestion lower bound.

The algorithm includes a long-term memory structure, and the current solution is randomized when the algorithm gets stuck. In particular, we keep the best solution found since the last randomization event and the search is restarted after 15 iterations without improving it. In Fig. 12.4 this happened once for the case with aspiration criterion, around $t = 220$ s. In the randomized solution, each link takes the weight in the incumbent solution with a 50% probability, or otherwise one chosen uniformly among the least used weights for that link.

12.6 Greedy Algorithms

Greedy algorithms are typically simple schemes targeted to quickly produce a feasible problem solution, called as subroutines in other heuristics. The defining characteristics of greedy schemes are:

- *Iterative and constructive.* The method is iterative and in every iteration a solution component is fixed, and never changed later, so the algorithm ends when all the components have been added.
- *Myopic decisions.* The decision on how to proceed in each iteration is taken using a locally optimal decision that maximizes an immediate benefit (using a *greedy* or *myopic* function).

Algorithm 23 helps us to illustrate the greedy schemes more formally. In a greedy algorithm a problem solution x is represented as a subset of C, the set of candidate components. Initially, x is empty. In each iteration, one component $c \in C(x)$ is added to it, where $C(x) \subset C$ is the subset of possible components that could be added in this iteration, according to the current state of x. The component c aggregated is one that optimizes (in this case, maximizes) a *myopic* or *greedy* function f_x, where the subindex reflects that the function may be different in different iterations.

Algorithm 23 Greedy algorithm

1: *Initialization*: $x = \emptyset$
2: **do**
3:     Find set $C(x)$ of candidate components.
4:     $x = x \bigcup \{\arg \max_{c \in C(x)} f_x(c)\}$ **comment:** Greedy selection
5: **while** x is not a complete solution
6: **return** x

Then, the defining elements of a greedy algorithm are (i) how we represent a solution as an aggregation of elements in a set C, (ii) how sets $C(x)$ are created and (iii) the greedy functions f_x used.

Example 12.5 File `Offline_fa_ospfWeightGreedy` implements a greedy scheme for the OSPF link weight optimization problem in Example 12.1. Set C is given by all the possible link-weight pairs:

$$C = \{(e, i), \forall e \in \mathcal{E}, i = 1, \cdots, W_m\}$$

The algorithm visits each link in an arbitrary order. In the iteration associated to link e, the set of candidate components is $C(x) = \{(e, 1), \cdots, (e, W_m)\}$. The weight i chosen is the one that minimizes the congestion considering: (i) the weight assigned in x for the already visited links and (ii) an arbitrary weight w_e^0 in the non-visited links.

12.7   GRASP

Greedy Randomized Adaptive Search Procedures (GRASP) were proposed at the end of the 1980s by Thomas Feo and Mauricio Resende [8, 9]. They consist of repeating an arbitrary number of times, until any standard stopping criterion is met, a GRASP iteration, consisting of two consecutive phases (see Algorithm 24):

- *Greedy randomized phase*: A solution x is produced using a *randomized* version of a greedy scheme. This means, a greedy scheme where some steps or decisions are taken randomly, such that different solutions are produced every GRASP iteration.
- *Local search*: A local search heuristic starts using x as initial solution, the one produced in the previous greedy step.

Algorithm 24 GRASP algorithm

1: **do**
2:     x = solution created using a greedy-randomized scheme.
3:     x' = solution resulting from a local search starting in x.
4:     Update x_{best}, the incumbent solution.
5: **while** stopping criterion is not met
6: **return** x_{best}

There are multiple procedures for creating randomized greedy algorithms. An approach described in [10] consists of modifying the standard greedy behavior in line 4 of Algorithm 23. Instead of adding to x the candidate component $c \in C(x)$, which optimizes the myopic function f_x, we use f_x to rank the components in $C(x)$. From it, we create a so-called restricted candidate list (RCL), with the elite components providing the best myopic results. Then, one component chosen randomly out of the RCL is added to x. Note that the larger the size of the RCL, the more random the process becomes. If the RCL is limited to just one element, the best according to the myopic function, we have the standard greedy approach. If all the components are added to the RCL, the greedy choice is totally random.

Example 12.6 File `Offline_fa_ospfWeightGRASP` implements a GRASP scheme for the OSPF link weight optimization problem in Example 12.1. The local search procedure is similar to the one described in Section 12.3.1. The greedy algorithm is randomized by (i) randomizing the order in which the links are visited, (ii) randomizing the set w^0 of initial link weights, and (iii) applying the RCL concept, where the RCL size is controlled by a user-defined parameter $\alpha \in [0, 1]$. In particular, if we define v_{min} as the best and v_{max} as the worse congestion measures among the 16 candidate weights in an iteration, the components in the RCL are given by:

$$RCL = \{c \in C(x) \text{ such that } f_x(c) \leq v_{min} + \alpha(v_{min} - v_{max})\}$$

$\alpha = 0$ restricts the RCL to the local optimum and in $\alpha = 1$ the greedy function has no effect, since x_i is chosen randomly in $C(x)$. Figure 12.5 plots the results of the algorithm in our case study for a parameter $\alpha = 0.5$. Best congestion objective found is ≈ 0.547 (recall that 0.51 is a lower bound).

12.8 Ant Colony Optimization

Ant Colony Optimization (ACO) meta-heuristic was proposed by Marco Dorigo in 1992 [11], inspired in the behavior of ants. It is a representative example of a series of relatively recent metaheuristics based on swarm intelligence: the social behavior of insects and other animals, that enables them to cooperatively solve problems in nature.

ACO takes inspiration in the ants' foraging behavior, in particular in how ants find the shortest path between the food and their nest. In the natural world, ants wander randomly in their quest for food. When an ant finds some, it returns to the nest while laying down a *pheromone* trail. The ants in the proximity of the trail can sense it with a probability proportional to the concentration of pheromones and then stop traveling at random and follow that trail. More ants in a trail means more pheromones in it, with more chances to attract more ants. This creates a

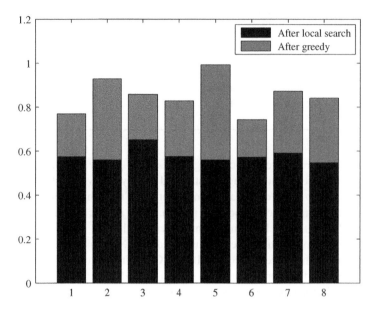

Figure 12.5 Results from GRASP algorithm for OSPF weight setting problem in a 5 minute run (8 GRASP iterations). RCL size corresponds to $\alpha = 0.5$. In each stack, the upper solution comes from the greedy step, the lower solution is returned by the local search procedure. Best congestion objective found is ≈ 0.547 (0.51 is a lower bound)

positive reinforcement. Simultaneously, pheromone trails tend to evaporate, thus reducing its attractive strength. Pheromone evaporation plays a double role: without evaporation the path chosen by the first ants could attract the rest without a sufficient exploration of other options, but a too strong evaporation can quickly eliminate the trail of poor solutions that diversify the search.

The key idea, is that all this process is biased towards shortest paths. Initially, multiple ants can create several paths from the food to the nest. However, since the ant needs less time to march shorter paths, the pheromone density becomes higher there. Then, shorter paths tend to attract more and more ants than longer ones, until eventually all the ants traverse the shortest path.

The goal of the ACO metaheuristic is to mimic this ant behavior, for the solution of arbitrary problems. The scheme guidelines are described in Algorithm 25:

- *Ant algorithm.* In each ACO iteration, a number A of artificial ants are created and their movement from the food to the nest emulated. In its march, each ant creates one problem solution, implemented as a greedy randomized scheme where pheromone information shows bias toward the random decisions. Let C be the solution components and assume that each component $c \in C$ is assigned an amount $\tau_c \geq 0$ of pheromones. Each ant starts with an empty solution x and adds one component in every greedy iteration, representing in our analogy that it traverses this component in its walk. Given a current solution x, $C(x)$ is the set of candidate components that could be added, meaning the possible next choices the ant can take in its position. Then, each possible choice $c \in C(x)$ has assigned two values:

- A non-negative amount of pheromones τ_c, a measure of how often c appeared in good solutions in the past.
- A non-negative value b_c provided by a myopic function f_x, measuring the immediate benefit (the higher the better) of choosing c.

The probability p_c of choosing component c (so called *transition probability*) should then be positively biased towards higher values of τ_c and b_c. This can be done with:

$$p_c = \frac{\gamma \tau_c + (1 - \gamma) b_c}{M} \tag{12.2}$$

where M is a constant to make the transition probabilities sum one ($\sum_{c \in C(x)} p_c = 1$), and $\gamma \in [0, 1]$ is a factor to balance the importance of the pheromones versus the greedy information. In the limit, when $\gamma = 1$ only the pheromones affect the decisions, while with $\gamma = 0$ they are not considered at all and the essence of the ACO meta-heuristic is lost[3].

- *Reinforcement*: After all the ants produce a solution, the pheromones of traversed components are reinforced. The reinforcement of a component should be more intense if it results in better solutions. Let us denote $b(x)$ to the benefit associated to a full solution x, the target to maximize. Then, a typical reinforcement scheme is:

> **for each** $a = 1, \ldots, A$
>   **for each** $c \in x_a$
> $\tau_c \leftarrow \tau_c + b(x_a)$

A common alternative is reinforcing with just those ants producing the best solutions.

- *Evaporation*: Pheromone evaporation uniformly decreases the pheromone values in all the components. This is to avoid an unlimited growth of pheromones and a useful form of *forgetting* poor choices made in the past. This is also needed to forget good but not optimal choices made in the first ACO iterations, which can produce a far too rapid convergence to a sub-optimal region. The usual evaporation rule is:

> **for each** $c \in C$
> $\tau_c \leftarrow \tau_c (1 - \rho)$

where $\rho \in (0, 1]$ is the evaporation rate. When $\rho \to 1$, the pheromone information is soon forgotten, while $\rho \to 0$ make the pheromones laid by good and bad decisions persist longer.

- *Optional local search*: It is possible to perform actions that boost the ACO performance. The main example of such operations is introducing a local search phase starting in the solution produced by each ant, such that pheromone reinforcement is applied on the resulting solution.

---

[3] In many works p_c is computed with the formula $p_c = \frac{\tau_c^\alpha \times b_c^\beta}{M'}$, using parameters α and β to tailor the importance of pheromones and greedy information, respectively. However, this introduces a counter-intuitive effect when tuning α and β: (i) if $\tau_c > 1$ ($b_c > 1$), higher α (β) increases the importance of the pheromones (greedy information), while (ii) if $\tau_c < 1$ ($b_c < 1$) the opposite effect occurs. In contrast, in (12.2) only one parameter has to be adjusted and higher γ always amplifies the importance of the pheromones and reduces that from the greedy information.

Algorithm 25 ACO algorithm

1: *Initialization*:
2:     A = number of ants
3:     τ_c = pheromone of each component $c \in C$
4: **do**
5:     **for each** $a = 1, \ldots, A$
6:         x_a = ant solution created using a greedy-randomized scheme, biased by τ.
7:         Update the incumbent solution x_{best}
8:     *Reinforcement*: increase τ_c of traversed components, more if part of better solutions.
9:     *Evaporation*: reduce τ for all components (used or not by ants).
10: **while** stopping criterion is not met
11: **return** x_{best}

12.8.1 Design Hints

Tailoring ACO parameters needs experimentation with the particular problem to solve. We provide below some hints to guide this process (more details e.g., in [12] and references in [13]):

- *Number of ants*: In each ACO iteration, A ants choose a solution using the same pheromones values τ_c. More ants mean a more intense search in the region of solutions stochastically biased by τ, at a cost of a higher per iteration running time. Typical A values range from one ant to several hundreds. Different studies recommend increasing the number of ants as the algorithm progresses (e.g., 1 ant every 10 iterations), so that intensification occurs once a good region of solutions is found.
- *Algorithm convergence*: The convergence of ACO algorithms occurs when a large majority or all of the ants choose a similar solution, which happens, for example, when pheromone values are such that ant decisions become almost deterministic. This situation can be detected using several methods. For instance, the entropy of the transition probabilities $p_c, c \in C(x)$ measures the uncertainty in the decision process of an ant in a particular greedy-randomized iteration:

$$E_x = - \sum_{c \in C(x)} p_c \log_2 p_c$$

The minimum uncertainty ($E_x = 0$) occurs when $p_c = 1$ for some c and the maximum ($E_x = \log_2 |C(x)|$) when all p_c are equal. By averaging the entropy values of all the decisions of all the ants, we can have a view of the algorithm convergence.
- *Evaporation rate*: Evaporation rate is typically chosen using experimentation, with reported values as small as $\rho = 0.01$ or as high as $\rho = 0.99$. In general, higher evaporation rates make poor decisions be soon forgotten and make the algorithm converge faster.
- *Greedy versus pheromone balance*: The $\gamma \in [0, 1]$ factor in (12.2) weighting the importance of the pheromones over the greedy information should also be chosen experimentally. Higher emphasis in the greedy information can make the process less random, and enforce a too rapid convergence. Some works propose to help the diversification during the initial ACO iterations by reducing the importance of the greedy information and increasing it in the last iterations.

Example 12.7 File `Offline_fa_ospfWeightACO` implements an ACO scheme for the OSPF link weight optimization problem in Example 12.1. An ant greedy-randomized solution is obtained similarly to the GRASP algorithm in Example 12.6. The solution components are all the possible link-weight pairs (e, i), each one is assigned a pheromone quantity τ_{ei}. An ant traverses the links in a random order. The probability p_{ei} of assigning weight i to link e in an iteration is:

$$p_{ei} = \frac{\gamma \tau_{ei} + (1 - \gamma)b_{ei}}{\sum_{i=1}^{W_m} \gamma \tau_{ei} + (1 - \gamma)b_{ei}}$$

where b_{ei} is the inverse of the congestion evaluation for the candidate link-weight, obtained as in Example 12.5. A user-defined fraction f of the best ranked ants contribute to the pheromone reinforcement.

> **for each** a index of best-ranked ant
>> **for each** $c \in x_a$
>>> $\tau_c \leftarrow \tau_c + b(x_a)$

we use as a benefit $b(x)$ of a solution x the inverse of its congestion metric. Pheromone evaporation rate is a user-defined parameter ρ.

Figure 12.6 plots the results of a 5 minute run of the algorithm in our case study, for a system with 100 ants ($A = 100$), initial pheromone values $\tau_{ei} = 1$, and parameters $\gamma = 1$ (only

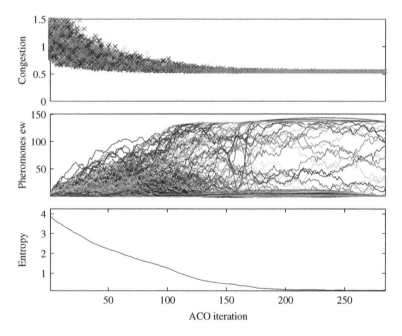

Figure 12.6 Results from the ACO algorithm for a OSPF weight setting problem in a 5 minute run (270 ACO iterations), $A = 100$, $\gamma = 1$, $f = 0.5$, $\rho = 0.4$, initial pheromone values $\tau_c = 1$. Best solution found has a congestion metric $cg = 0.541$ (0.51 is a lower bound)

pheromone information is used), $f = 0.5$ (best half of the ants reinforce the pheromones), and evaporation factor $\rho = 0.4$. As can be seen, bad solutions tend to disappear from the algorithm as iterations evolve and uncertainty in the ant movements tends to decrease. This is witnessed by the average entropy values of the p_{ei} distributions, and how multiple pheromone quantities approximate to zero values.

Best congestion found in this experiment is ≈ 0.541 (recall that 0.51 is a lower bound). The reader can observe, using the provided implementation, how algorithm convergence occurs significantly sooner producing a worse solution when $\rho = 0.9$, or it becomes much slower when $\rho = 0.1$.

12.9 Evolutionary Algorithms

Evolutionary algorithms (EA) are metaheuristics inspired in the behavior of biological systems, in terms of how species evolve along generations, and natural selection enforces the survival of the fittest. EA schemes were first used in the 1960s and since then have shown their validity in problems appearing in very diverse disciplines. When applied to problems where solutions are coded using strings or arrays of numbers, EAs are typically referred to as *genetic algorithms*.

The higher layer pseudocode of EAs is shown in Algorithm 26. Initially, a *population* P of diverse problem solutions is created using a pure random or a greedy-randomized procedure. Then, a sequence of EA iterations follows until any stop criterion is met. Each EA iteration emulates a generation, using the so-called evolutionary operators:

- *Parent selection*: A number of couples of solutions out of the population P are selected.
- *Crossover*: An offspring is created as each couple creates a child solution that inherits some characteristic from both progenitors.
- *Mutation*: Offspring solutions can be slightly and randomly changed to emulate mutations in the crossover process.
- *Selection*: Some elements in the new population (the union of the previous population and the offspring) survive to the next generation while others disappear.

Algorithm 26 Evolutionary algorithm

1: *Initialization*:
2:    P = initial population
3: **do**
4:    *Parents* = selected set of couples (x, y) of solutions in P
5:    *Offspring* = apply the *crossover* operation to every pair in *Parents*
6:    *Offspring* = apply the *mutation* operation to every solution in *Offspring*
7:    P = apply *selection* operation to $P \bigcup$ *Offspring*
8:    Update the incumbent solution x_{best}
9: **while** stopping criterion is not met
10: **return** x_{best}

12.9.1 Design Hints

As with the rest of heuristics, the form in which previous general rules are implemented is problem dependent. We provide some guidelines:

- *Coding the solution*: In the context of EAs, the form in which a solution is coded is usually called a chromosome. The coding should facilitate the implementation of the *crossover* operation, the most characteristic in the EA process, in which two parent solutions are somehow mixed to produce a new solution, inheriting features from both.
- *Initial population*: Initial population should be sufficiently large and diverse, or otherwise the algorithm will narrow the search to a small region of the solution space. Typically, initial population has several thousands of solutions, created using a greedy-randomized algorithm.
- *Parent selection*: Parent selection should be a random process, where the probability p_x that a solution x becomes a parent is related to its *fitness function*. This is the function that evaluates the "goodness" of a solution, typically based on the maximization version of the problem objective function. Then, better solutions have better chances to reproduce and get mixed with other solutions by the crossover operator. It is a design decision letting or not a parent to be selected in more than one couple in a generation. Two classical selection methods are:
  - *Roulette wheel selection*: Each time a parent is chosen, the probability of choosing a particular solution x is proportional to its fitness.
  - k-Tournament selection: Each time a parent is chosen, k solutions are randomly extracted from P, and the best (highest fitness) among them is selected.
- *Offspring size*: The number of couples created in a generation determines the size of the offspring. This size is usually fixed as a fraction (e.g, between 10 and 50%) of the population size $|P|$. Smaller sizes tend to limit diversification, since worse solutions usually survive during a low number of generations, and with smaller numbers of offspring their probability of engaging in a crossover drops.
- *Crossover*: The crossover process may have a random nature, so if the two different parents have several children, they are likely to be different solutions. Also, it should guarantee that child solutions inherit characteristics from both parents. Classical methods are:
  - Crossover for discrete vectors: When a solution is coded as a vector of discrete numbers, the 1 point, N points, or uniform crossovers are often used (see Fig. 12.7). They are based on splitting the parent solutions in one, N or all possible points, so the child solution inherits each part from one parent chosen randomly, but potentially biased by its fitness its fitness.
  - Crossover for real vectors: When solutions are vectors of real numbers, the crossover of two solutions x and y can be computed as its average $(x + y)/2$, or an average weighted by the solution fitness.
  - Crossover for orderings: In some occasions, a solution is coded as a vector where their coordinates are dependent on each other. Then, inheriting a coordinate from a parent may condition the crossover in other coordinates. The typical example is when a solution is a particular ordering of numbers $1, \cdots, N$, so the same number cannot appear twice in a solution. This happens, for example when a solution is the order in which nodes are visited in a TSP problem instance. In this case, we can modify the 1-point crossover as follows. First, the left solution part is inherited from a parent. Then, the already used numbers are removed from the other parent and the resulting solution appended to the child.

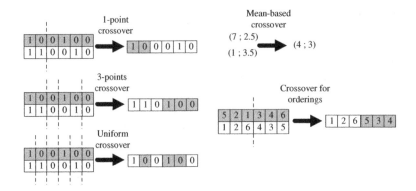

Figure 12.7   Crossover examples in evolutionary algorithms

- *Mutation*: The mutation operator consists of changing randomly and slightly a child solution after the crossover to reflect natural mutation in biological systems. For example, when a solution is coded as an array of numbers, mutation can consist in randomly change one vector coordinate also chosen randomly. The mutation role is increasing diversification.
- *Survival of the fittest*: This operator emulates the natural selection process, where the best elements among the previous population and the offspring pass to the next generation. Often, the population size M is a fixed parameter so in every generation the number of solutions dropped equals the size of the offspring. For instance, 80% of the next generation population can be the best solutions (highest fitness) among the population plus offspring, while the other 20% is chosen randomly among the rest.

Example 12.8 File `Offline_fa_ospfWeightEA` implements an EA scheme for the OSPF link weight optimization problem in Example 12.1. The population size M and offspring size L are user-defined parameters. Initial population is chosen in a pure random form. In parent selection, a used-defined fraction f of parents are the best solutions among the population and the fraction $1 - f$ is chosen randomly (thus, a parent can be selected more than once). Couples are formed randomly. In the crossover, the child solution inherits each link from one parent chosen randomly. Then, mutation randomly changes one link weight of the child solution. Finally, the best M solutions among the previous population and offspring, survive to the next generation.

Figure 12.8 plots the algorithm evolution in our case study. The best congestion found is ≈ 0.545, recall that 0.51 is a lower bound. We see how, as time evolves, worse solutions tend to be eliminated from the population while new better solutions appear and survive. The method converges in the sense that the diversity of the solutions in the population decreases, as witnessed by the reduction of the average entropy in the weight assignment among the population ($\bar{E}$), computed as:

$$\bar{E} = -\frac{\sum_e \sum_w P_{ew} \log_2 P_{ew}}{|\mathcal{E}|}$$

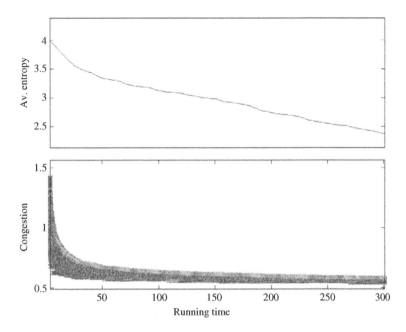

Figure 12.8 Results from an Evolutionary Algorithm for OSPF weight setting problem in a 5 minutes run (356 generations), $M = 1000, M = 200, f = 0.1$. Best solution found has a congestion metric $cg = 0.545$ (0.51 is a lower bound)

where p_{ew} is the fraction of the solutions in the population that assign weight w to link e. Note, however, that in this example, more time would be needed to further reduce the entropy. The reader can check how reducing the offspring size, the entropy values achieved after a 5 minute run are in general lower.

12.10   Case Study: Greenfield Plan with Recovery Schemes Comparison

This section includes a case study with a network greenfield planning example with a double aim:

- Exemplify a realistic network planning task that requires the development of ad hoc heuristics.
- Illustrate a different strategy for building network algorithms where an heuristic scheme is used to guide the search and limit the complexity of integer linear formulations.

12.10.1   Case Study Description

We take the role of a network operator, planning a new optical backbone network deployment. The locations of the core nodes are given and in our example correspond to that of the NSFNET network in Fig. 12.1, and in the NSFNet_N14_E42.n2p file. Distances are normalized so that most distant nodes are separated 4500 km. Nodes should be connected using optical fibers

rented to a dark fiber provider[4] and the fiber topology to deploy is part of the planning decisions to optimize. Wavelength Division Multiplexing (WDM) technology is used, such that
each fiber link is able to carry a maximum of $W = 80$ different channels, each at a different
wavelength.

The input traffic to carry is given by a number of end-to-end optical connections between
each node pair, called lightpaths. Each lightpath can carry 100 Gbps, and is originated in a
transponder at the ingress node, traverses a sequence of fibers, and ends in other transponder card at the egress node. Intermediate nodes with respect to a lightpath do not look into it.
Instead, an Optical Add Drop Multiplexer (OADM) located at each node switches the lightpaths optically. Each lightpath occupies a wavelength in each traversed fiber. Regeneration
equipment can be located along the lightpath route to remove the effects of optical signal
degradation and change the wavelength of a lightpath if needed. However, lightpath propagation delay cannot exceed 50 ms (a realistic latency constraint in backbone design). Considering
a standard propagation speed in the fiber of 200,000 km/s, this limits lightpath lengths to
10,000 km.

The number of lightpaths $h_{nn'}$ to establish between node n and n' is known. In our example,
the traffic is symmetric ($h_{nn'} = h_{n'n}$) and given by:

$$h_{n'n} = h_{nn'} = \left\lceil \frac{10 \times \max\{TM_{nn'}, TM_{n'n}\}}{100} \right\rceil$$

Where $TM_{nn'}$ is the traffic $n \to n'$ in Gbps in NSFNet_N14_E42.n2p file. The resulting
total traffic contains 558 lightpaths. We make the following considerations with respect to the
network costs:

- Each fiber link costs $400 per km and year. This includes the leasing payments to the dark
  fiber provider and the annual prorated cost of the optical amplification and regeneration
  equipment needed. To ease the network management, the topology of fiber links should be
  symmetric: the number $0, 1, 2, \cdots$ of fiber links deployed from a node n to n', is the same to
  the ones from n' to n.
- One OADM is placed in each location. The acquisition cost of the OADM is of $20 K for
  OADMs of degree two (at most, two output fibers) and $40 K for OADMs of a degree higher
  than two. The per-year cost of the OADM is given by the acquisition cost prorated in 5 years
  of expected operation.
- Each lightpath starts and ends in a transponder. Transponders are bidirectional and to ease
  the network management lightpath routes are constrained to also be symmetric. The acquisition cost of a bidirectional transponder is $15 K, which also includes the cost of the router
  line card connected to it. The per-year cost is given by the acquisition cost prorated in 5 years
  of expected operation.

We consider that the network links are vulnerable to fiber cuts[5] and the network design
should add some resiliency for it. In particular, we assume that all the links deployed between

---

[4] Long distance fiber links are typically laid by civil engineering, railway, or highway building companies, and then
leased to network operators. The product is called *dark fiber* when the optical amplification, compensation, transmission, and reception equipment is placed by the network operator, which has the full control of the link.
[5] Some studies [14] estimate that every 1000 km of fiber link can suffer an average of ≈ 1.7 accidental cuts per year,
with repairing time in the order of hours/days.

two nodes in both directions are placed in the same duct, so that if cut, all these links fail simultaneously. As a design constraint, the network must be tolerant to a single duct cut.

In this scenario, the target of the case study is assessing three different network designs, based on three different potential recovery strategies to adopt in the network:

- *1+1 lightpath dedicated protection*: Each lightpath is backed up by a duct-disjoint lightpath between the same nodes. Then, under any duct cut at least one out of the two survives. The recovery is fast and the network management simple, but the number of lightpaths deployed is twice the number of lightpaths needed.
- *Lightpath shared protection*: A set of lightpaths between each node pair (n, n') is established so that under any single duct failure, at least $h_{nn'}$ lightpaths survive. Then, the number of lightpaths deployed can be higher than the number of lightpaths needed, but in general less than twice this number. As a drawback, the recovery is more complex than 1+1 case, since it requires the higher layer to deviate the traffic from the failed lightpaths to the surviving ones. If lightpaths are IP links, this is automated by the OSPF/ECMP rule of the routers.
- *Lightpath restoration*: The number of lightpaths deployed equals the number of lightpaths needed. However, lightpaths are not static, and the network can react rerouting lightpaths to adapt to duct failures. This dynamicity requires a more complex management system, and sophisticated OADMs that are able to reroute added, bypassed and dropped lightpaths[6].

12.10.2   Algorithm Description

Our assessment study requires implementing heuristic algorithms that obtain the fiber links to deploy, the lightpaths, and their routes, such that the total network cost is minimized, in each of the three scenarios considered. In this section, we describe the algorithm developed, available in the `Offline_wdm_physVirtTopDesign` file, suitable to solve the three cases.

The algorithm is based on a GRASP scheme, where GRASP iterations are executed until a maximum running time is met. The two stages of each GRASP iteration are described next, the pseudocode is shown in Algorithm 27:

- *Greedy-randomized algorithm*: One solution z is created using a greedy-randomized method. This stage starts with an empty network where no links nor lightpaths are deployed. Then, node pairs (n, n') are visited in a random order. For each node pair, we call an ILP formulation to deploy the bidirectional lightpaths between them, so the accumulated network cost is minimized. The method is constructive so the lightpaths and links deployed in a greedy step are not changed and affect the decisions taken later. For instance, the ILP search will naturally prefer reusing scarcely occupied links deployed in previous iterations instead of deploying new ones. The ILP to optimize the lightpaths in a node pair is different in the 1+1 case and the other two cases, as will be described later.
- *Local search*: A first-fit local search starts from the solution z created in the greedy stage. The vicinity of a solution is composed of those designs where all the lightpaths follow the same route, but those between two nodes. The enumeration of all the neighbors means listing all the node pairs in an arbitrary order, and then reusing the same ILP as in he previous stage to search for an improving reorganization of the node pair lightpaths.

---

[6] OADMs with this functionality are called colorless-directionless OADMs. They usually add some internal blocking and design constraints (e.g., see [15]) not considered in this example.

Algorithm 27 GRASP iteration, case study algorithm.

1: **comment:** Greedy-randomized stage
2: Order the node pairs randomly, $z = $ empty network
3: **for each** (n, n')
4:     $z = $ ILP-based search of best deployment of new lightpaths in z between (n, n')
5: **comment:** Local search stage starting from z
6: **do**
7:     Order node pairs randomly.
8:     **for each** (n, n')
9:         $z' = $ ILP-based search of best reoptimization of lightpaths in z between (n, n').
10:        **if** $f(z') < f(z)$ **comment:** f is the cost function to minimize.
11:            $z \leftarrow z'$, **break**   **comment:** Improving solution, first-fit
12: **while** z has been improved   **comment:** End when no improving solution
13: **return** z

12.10.2.1 1+1 ILP Formulation

ILP program (12.3) describes the optimization of the primary and backup lightpath routes between a node pair (n, n'), so that the network cost is minimized. We denote $\mathcal{E}$ as the set of all potential ducts where dark fibers can be leased, a full mesh of ducts between all node pairs. We use p_e to denote the number $0, 1, 2, \cdots$ of fiber links to hire in duct e. Both fiber links and ducts are bidirectional. Let D denote the set of requested lightpaths to establish between n and n'. For each lightpath demand d, we compute the set $\mathcal{P}_d$ of admissible paths for it. $\mathcal{P}$ is the union of all the paths for all the lightpaths and $\mathcal{P}_e$ the subset of $\mathcal{P}$ traversing duct e. Decision variables to the problem are:

- $x_p, p \in \mathcal{P}$: 1 if path p carries a lightpath (backup or primary) for demand $d(p)$, 0 otherwise.
- $p_e, e \in \mathcal{E}$: Number of fibers (bidirectional) in duct e (bidirectional).
- $d_n^3, n \in \mathcal{N}$: 1 if node n needs a more expensive OADM, since its degree is higher than two. 0 otherwise.

Also, we denote as c_e the per-year cost of two fibers (one in each direction) in duct e, c_t the cost of two transponders occupied by a bidirectional lightpath and c_O the extra per-year cost associated to a high-degree OADM with respect to a degree two OADM. Finally, y_e is the number of occupied channels in the fibers of duct e, because of already deployed lightpaths.

$$\min \sum_e c_e p_e + c_T \sum_p x_p + c_O \sum_n d_n^3 \quad \text{subject to:} \tag{12.3a}$$

$$\sum_{p \in \mathcal{P}_d} x_p = 2, \quad \forall d \in D \tag{12.3b}$$

$$\sum_{p \in \mathcal{P}_d \cap \mathcal{P}_e} x_p \leq 1, \quad \forall d \in D, \forall e \in \mathcal{E} \tag{12.3c}$$

$$\sum_{p \in \mathcal{P}_e} x_p \leq W p_e - y_e, \quad \forall e \in \mathcal{E} \tag{12.3d}$$

$$\sum_{e \in \delta^+(n)} \leq 2 + M d_n^3, \quad \forall n \in \mathcal{N} \tag{12.3e}$$

The objective function minimizes the network cost. First constraint (12.3b) means that all lightpaths occupy two routes, a primary and a backup. Constraint (12.3c) makes them duct-disjoint, and (12.3d) dimensions the number of fibers needed in each duct according to host the previously and currently deployed lightpaths. Finally, (12.3e) establishes that more expensive OADMs are needed when the node degree exceeds two (M is any large number).

12.10.2.2 Shared Protection Formulation

Program (12.4) optimizes the lightpaths to establish between two nodes n and n' in the shared protection case. $\mathcal{P}_{nn'}$ is the set of bidirectional routes admissible for the lightpaths, and $h_{nn'}$ the number of bidirectional lightpaths requested. Set S contains all the possible $1 + |\mathcal{E}|$ failure states in which the network can be: the state where all ducts are active, and the states when all are active but one. P_s is the set of paths that survive in failure state s. Finally, decision variables x_p now represent the number of lightpaths in the route p (which can be higher than one).

$$\min \sum_e c_e p_e + c_p \sum_p x_p + c \sum_n d_n^3 \quad \text{subject to:} \tag{12.4a}$$

$$\sum_{p \in P_{nn'} \cap P_s} x_p \geq h_{nn'}, \quad s \in S \tag{12.4b}$$

$$\sum_{p \in P_e} x_p \leq W p_e - y_e, \quad \forall e \in \mathcal{E} \tag{12.4c}$$

$$\sum_{e \in \delta^+(n)} \leq 2 + d_n^3, \quad \forall n \in \mathcal{N} \tag{12.4d}$$

Constraint (12.4b) means that at least $h_{nn'}$ lightpaths should survive in any failure state, (12.4c) dimensions the number of fiber links in each duct according to the previously and currently deployed lightpaths, and (12.4d) sets the OADM needs.

12.10.3   Combining Heuristics and ILPs

The combination of ILP formulations and heuristics can be a useful strategy for building planning algorithms, for those problem sizes where formulating and solving the full problem is outside the CPU and memory limits. This happens in the example described.

In our algorithm, a GRASP heuristic skeleton is used to guide the search. The role of the ILPs is being a routine that efficiently explores a reduced solution space, where only the lightpaths between two nodes can be changed. Then, we take benefit of the efficient and sophisticated optimization techniques that solvers implement in their exploration. For instance, the number of different forms in which 10 lightpaths can be routed in 200 candidate paths is[7] $\binom{10+200-1}{10} \approx 3 \times 10^{16}$. Evaluating the objective function for all of them and checking their

---

[7] A lighptath-path allocation is represented by a chain of 209 letters: 200-1 letters s (representing a separator between one path and the next), mixed with 10 letters L (lightpath). The number of different chains is the given expression.

survivability under all failure scenarios would take an unreasonable amount of time in standard implementations. However, ILPs often find optimum solutions to such problems in tens of milliseconds. The reason behind this is their extremely profiled internal implementation that includes branch-and-bound, branch-and-cut, and branch-and-price techniques that efficiently find and rule out suboptimal parts of the solution space[8].

Combining ILP solving and heuristics in the same algorithm is enabled by libraries like for example JOM, that interfaces with CPLEX and GLPK solvers from Java. In our tests, the average time needed by JOM to construct the ILP model was in the order of tens of miliseconds, approximately the same as the *average* time taken then by the solver to find the optimum solutions. As theory predicts, worst-case running times of the ILP solvers can be quite long, since ILP solving is an $\mathcal{NPO}$-complete problem. To cope with this, the solver was configured through JOM with a maximum running time of 2 seconds, so that the best solution found so far is returned in the (very infrequent) cases when this time is exceeded before the ILP optimum is reached.

12.10.4   Results

This section reports some of the results obtained in our example, after running the heuristic algorithm devised for one hour in the 1+1 and shared protection case. The lightpath restoration plan is extracted from the shared protection solution, by just keeping the same deployed links and reducing the lightpaths to exactly that in $h_{nn'}$ matrices. Note that since deployed links in the shared protection case are enough to carry $h_{nn'}$ lightpaths in any failure state $s \in S$, the lightpath restoration is also feasible. Finally, for the sake of comparison, we use the shared protection heuristic to solve the case (unrealistic in the backbone context) where no network recovery is planned. For this, the shared protection algorithm is run with the set S containing just the no-failure state.

In our tests, the set of admissible paths for each lightpath is composed of all the paths traversing a maximum of four ducts, and with a maximum length of 10,000 km, with a maximum of 200 candidate paths per demand (sufficient to include at least all the valid routes traversing three ducts). Computing facilities used are a standard laptop with an Intel i5 CPU (2.27 GHz) with 8 GB of RAM, interfacing the CPLEX v12.2 solver from JOM library.

Table 12.1 plots the results in the four cases tested, that suffices to show the main trends[9]. In the cost side, first observation is that dark fiber leasing costs dominate with respect to transponders and optical switching costs. The lower cost solutions are lightpath restoration and shared protection, while 1+1 protection is the most expensive. The gross of the cost difference lies in the four extra links needed in the 1+1 case. The differences in the number of lightpaths is quite relevant among the three, however, their impact in the cost is reduced in this scenario where link costs dominate. Note, however, that a similar study in a metro network, with link distances in the order of hundreds of km instead of thousands, would make the transponder cost differences impact the final outcome, since link costs would have a much lower weight.

Regarding the algorithm evolution, we see that several thousands ILPs were solved in each run, with an average running time of hundreds of milliseconds. The time needed to build the model with JOM and the time needed to solve it dominate with respect to the rest of the tasks.

---

[8] The interested reader in ILP and mixed ILP solution techniques is referred to specialized literature like [16]

[9] Professional planning reports usually include cost assessments along a 5–20 years period, considering different annual growth traffic profiles.

Table 12.1   Case study results

Description	1+1 protection	Shared protection	Restoration	No recovery
Total cost (K$/year)	29,506.5	28,808.7	27,710.7	15,358.7
Link costs (K$/year)	26,046.5	25,928.7	25,928.7	13,612.7
Transponder costs (K$/year)	3348.0	2772.0	1674.0	1674.0
OADM costs (K$/year)	112	108	108	72
Num. links	62	58	58	32
Num. transponders (bid)	1116	924	558	558
Num. degree 2 OADMs	0	1	1	1
Num. degree >2 OADMs	14	13	13	13
Num. GRASP iterations	11	21	21	92
Num. solver calls	1981	9421	9421	16,698
Av. solver time (s)	0.241	0.194	0.194	0.052
Av. JOM modeling time (s)	1.55	0.182	0.182	0.159

Both are in the same order of magnitude, excepting the 1+1 case where JOM modeling was more time expensive. This is because the number of variables and constraints is higher, as one demand was needed per each lightpath request (set $\mathcal{D}$), instead of an aggregated demand for each node pair.

12.11   Notes and Sources

The interested reader in the techniques for solving integer linear programs and the tightness of Lagrange relaxation optimality bounds in those programs is referred to specialized literature like [3] and [16].

Heuristic and metaheuristic problem solving is a relatively recent discipline, with applications in every knowledge field. Multiple excellent sources exist covering metaheuristics in detail such as in the books [17–19].

Applications to network $\mathcal{NP}$-hard problems are widespread, especially for planning problems in any network technology. The interested reader in a particular problem variant is referred to its specialized literature. In [2] we found a significant amount of examples.

The optimization of link weights in IP/OSPF-like routing protocols is a classical traffic engineering problem. A comprehensive compilation of results can be found in Chapter 7 of [2], a complete planning study involving OSPF link weight setting in a real network is described in [20].

The reader interested in planning problems for IP over WDM networks is referred to surveying sources like [21–24], or [25]. More examples in this field of algorithms combining heuristic skeletons and ILP routines are [26] or [27]. Algorithms considering the internal blocking of colorless-directionless OADMs in the planning decisions can be found in [28, 29].

12.12   Exercises

12.1   Let $\mathcal{E}$ be a set of links and $y_e > 0$ the known traffic carried in each. We have K modules of capacity U to distribute among the network links. The design target is minimizing a blocking function $f(u)$, which assigns to each possible capacity assignment in the links $u = \{u_e, e \in \mathcal{E}\}$, an estimation of the resulting worst-case connection blocking. For the two following forms of coding a solution:

- As a vector of $|\mathcal{E}|$ coordinates, indicating the number of modules in each link.
- As a vector of K coordinates, indicating the identifier of the link that each module is assigned to.

Show in each case if the coding is a one-to-one representation of the solution space.

12.2 In the first solution representation of Exercise 12.1, we assume that two solutions are neighbors if all the links but two have the same number of modules and in those two links one module was moved from one to the other. For the case $|\mathcal{E}| = 20, K = 20$, (i) compute the maximum and minimum number of neighbors of a solution x and (ii) compute the size of the solution space: the number of different problem solutions.

12.3 In the IP/OSPF link weight setting problem described in Example 12.1, for a network of $\mathcal{E}$ links, we assume that two solutions w and w' are neighbors if $w_e = w'_e$ in all the links $e \in \mathcal{E}$, excepting at most k links. Compute the size of the vicinity set for different values of k. In a network of $|\mathcal{E}| = 42$ links, compute the maximum number k for which the number of neighbors of a solution does not exceed one million.

12.4 Let $\mathcal{N}$ be a set of locations of access nodes. Each access node location can host zero or one core nodes. An access node $i \in \mathcal{N}$ must be connected to one core node. The cost c_{ij} of connecting an access node i to a core node in site j and the cost c_j of placing a core node in site j are known. Since there are no specific constraints to access node connections, in the minimum cost placement each access node is linked to its closest core node. A problem solution can be coded with a binary vector of $|\mathcal{N}|$ coordinates, indicating whether or not a site n has a core node. Two solutions x_1, x_2 are neighbors if they have the same placement in all the sites but in k. Indicate the maximum size of the vicinity for different values of k. Implement in Net2Plan a local search algorithm for solving this problem, when c_{ij} equals the distance between i and j nodes. Input parameters to the algorithm should include parameter k and the constant cost M of a core node.

12.5 Modify algorithm in Exercise 12.4 for the case when each core node is constrained to be connected to a maximum of K access nodes. Implement the algorithm in Net2Plan using a metaheuristic among the ones shown in this chapter, or a combination of them. Comment on the tailoring of the heuristic specific parameters.

12.6 Let $\mathcal{G}(\mathcal{N}, \mathcal{E})$ be a given network, u_e the capacity in link e, D a set of offered demands, h_d the demand d offered traffic and P_d the admissible paths for the traffic of demand d. We are interested in finding link-disjoint primary and backup admissible paths for each demand d, such that network survivability A^H (computed as in Section 3.6.3) is maximized, for the set of SRGs defined in the network. Implement in Net2Plan an algorithm to solve this problem, using a metaheuristic among the ones shown in this chapter, or a combination of them. Comment on the tailoring of the heuristic specific parameters.

12.7 Let $\mathcal{N}$ be a set of IP routers running OSPF routing protocol, and c_{ij} the leasing cost of a bidirectional 10 Gbps link between nodes i and j. Offered traffic is given by a traffic matrix $\{h_{ij}, i, j \in \mathcal{N}, i \neq j\}$. We are interested in finding the minimum cost set of links to hire in the network that carry all the traffic. An arbitrary number of links can be hired between each node pair, and all network links have the same OSPF weight. Implement in Net2Plan an algorithm to solve this problem, using a metaheuristic among the ones shown in this chapter, or a combination of them. Comment on the tailoring of the heuristic specific parameters.

12.8 Repeat Exercise 12.7 assuming that OSPF weights are restricted to be integers between one an W_m (an input parameter), jointly optimized together with the network links.

12.9 Let $\mathcal{N}$ be a set of nodes and c_{ij} the cost of connecting nodes i and j with a bidirectional link. Compute the number of different bidirectional rings in the network. Find a one-to-one coding of the solution space. Let v be a vector specifying the order in which the nodes are visited in the ring. We assume that two solutions v and v' are neighbors if they are equal but in k coordinates where their nodes are permuted in any form. Compute the number of neighbors of a solution for different k values.

12.10 Devise a metaheuristic for finding the minimum cost bidirectional ring in a network, where link cost is proportional to its distance using any of the schemes described in this chapter, or a combination on them. Comment on the tailoring of the heuristic specific parameters.

12.11 In the ACO metaheuristic, let us assume that the transition probabilities p_c are computed as:

$$p_c = \frac{\tau_c^{\alpha} \times b_c^{\beta}}{M'}$$

where τ_c is the pheromone in component c, b_c its greedy benefit, $\alpha \geq 0$ and $\beta \geq 0$ two algorithm parameters weighting the importance of pheromones and greedy information in the decision, and M' a constant that makes $\sum_{c \in C(x)} p_c = 1$ in each ant choice.

- If $\tau_c < 1$, a higher α factor means a higher impact of τ_c in the decision?
- If $\tau_c > 1$, a higher α factor means a higher impact of τ_c in the decision?

References

[1] L. Velasco, A. Castro, D. King, O. Gerstel, R. Casellas, and V. Lopez, "In-operation network planning," *Communications Magazine, IEEE*, vol. 52, no. 1, pp. 52–60, 2014.

[2] M. Pioro and D. Medhi, *Routing, Flow, and Capacity Design in Communication and Computer Networks*. Morgan Kaufmann Publishers, 2004.

[3] L. Wolsey, *Integer Programming, ser. Wiley Series in Discrete Mathematics and Optimization*. New York, NY, USA: John Wiley & Sons, Inc., 1998.

[4] S. Kirkpatrick, C. D. Gelatt, M. P. Vecchi *et al.*, "Optimization by simulated annealing," *Science*, vol. 220, no. 4598, pp. 671–680, 1983.

[5] N. Metropolis, A. W. Rosenbluth, M. N. Rosenbluth, A. H. Teller, and E. Teller, "Equation of state calculations by fast computing machines," *The Journal of Chemical Physics*, vol. 21, no. 6, pp. 1087–1092, 1953.

[6] B. Fortz and M. Thorup, "Internet traffic engineering by optimizing OSPF weights," in *INFOCOM 2000. Nineteenth Annual Joint Conference of the IEEE Computer and Communications Societies. Proceedings. IEEE*, vol. 2. IEEE, 2000, pp. 519–528.

[7] R. Battiti and G. Tecchiolli, "The reactive tabu search," *ORSA Journal on Computing*, vol. 6, no. 2, pp. 126–140, 1994.

[8] T. A. Feo and M. G. Resende, "A probabilistic heuristic for a computationally difficult set covering problem," *Operations Research Letters*, vol. 8, no. 2, pp. 67–71, 1989.

[9] T. A. Feo and M. G. Resende , "Greedy randomized adaptive search procedures," *Journal of Global Optimization*, vol. 6, no. 2, pp. 109–133, 1995.

[10] J. P. Hart and A. W. Shogan, "Semi-greedy heuristics: An empirical study," *Operations Research Letters*, vol. 6, no. 3, pp. 107–114, 1987.

[11] M. Dorigo, "Optimization, learning and natural algorithms," *Ph. D. Thesis, Politecnico di Milano, Italy*, 1992.

[12] T. Stützle, M. López-Ibánez, P. Pellegrini, M. Maur, M. M. de Oca, M. Birattari, and M. Dorigo, "Parameter adaptation in ant colony optimization," in *Autonomous Search*. Springer, 2012, pp. 191–215.

[13] Ant colony optimization Available online at: www.aco-metaheuristic.org/.

[14] Y. T'Joens, G. Ester, and M. Vandenhoute, "Resilient optical and Sonet/SDH-based ip networks" in *Proceedings of the Second International Workshop on the Design of Reliable Communication Networks (DRCN 2000)*, 2000, pp. 255–260.

[15] P. Pavon-Marino and M. Bueno-Delgado, "Dimensioning the add/drop contention factor of directionless roads," *Lightwave Technology, Journal of*, vol. 29, no. 21, pp. 3265–3274, 2011.

[16] L. A. Wolsey and G. L. Nemhauser, *Integer and Combinatorial Optimization*. Hoboken, NJ, USA: John Wiley & Sons, Inc., 2014.

[17] F. Glover and G. A. Kochenberger, *Handbook of Metaheuristics*. Heidelberg, Germany: Springer Science & Business Media, 2003.

[18] E.-G. Talbi, *Metaheuristics: From Design to Implementation*.Hoboken, NJ, USA: John Wiley & Sons, Inc., 2009, vol. 74.

[19] Z. Michalewicz and D. B. Fogel, *How to Solve It: Modern Heuristics*. Heidelberg, Germany: Springer Science & Business Media, 2013.

[20] A. Nucci and K. Papagiannaki, *Design, Measurement and Management of Large-Scale IP Networks: Bridging the Gap Between Theory and Practice*. Cambridge, UK: Cambridge University Press, 2009.

[21] C. S. R. Murthy and M. Gurusamy, *WDM Optical Networks: Concepts, Design, and Algorithms*. Prentice Hall, 2002.

[22] B. Mukherjee, *Optical WDM Networks*. Heidelberg, Germany: Springer Science & Business Media, 2006.

[23] A. Somani, *Survivability and Traffic Grooming in WDM Optical Networks*. Cambridge University Press, 2006.

[24] R. Dutta, A. E. Kamal, and G. N. Rouskas, *Traffic Grooming for Optical Networks: Foundations, Techniques and Frontiers*. HeidelBerg, Germany: Springer Science & Business Media, 2008.

[25] J. M. Simmons, *Optical Network Design and Planning*. Springer, 2014.

[26] P. Pavon-Mariño, S. Azodolmolky, R. Aparicio-Pardo, B. Garcia-Manrubia, Y. Pointurier, M. Angelou, J. Sole-Pareta, J. Garcia-Haro, and I. Tomkos, "Offline impairment aware RWA algorithms for cross-layer planning of optical networks," *Journal of Lightwave Technology*, vol. 27, no. 12, pp. 1763–1775, 2009.

[27] B. Garcia-Manrubia, P. Pavon-Marino, R. Aparicio-Pardo, M. Klinkowski, and D. Careglio, "Offline impairment-aware RWA and regenerator placement in translucent optical networks," *Lightwave Technology, Journal of*, vol. 29, no. 3, pp. 265–277, 2011.

[28] P. Pavon-Marino, M.-V. Bueno-Delgado *et al.*, "Add/drop contention-aware RWA with directionless roadms: The offline lightpath restoration case," *Journal of Optical Communications and Networking*, vol. 4, no. 9, pp. 671–680, 2012.

[29] P. Pavon-Marino, M.-V. Bueno-Delgado, and J.-L. Izquierdo-Zaragoza, "Evaluating internal blocking in non-contentionless flex-grid roads [invited]," *Journal of Optical Communications and Networking*, vol. 7, no. 3, pp. A474–A481, 2015.

Appendix A

Convex Sets. Convex Functions

A.1 Convex Sets

Let x^1 and x^2 be two different points in $\mathbb{R}^n$. The *line* crossing x^1 and x^2 is given by the set of points x of the form $x = \alpha x^1 + (1 - \alpha)x^2$, where α ranges all the values in $\mathbb{R}$. In turn, the *segment* between x^1 and x^2 is the set of points $x = \alpha x^1 + (1 - \alpha)x^2$, where α is now restricted to be $\alpha \in [0, 1]$.

A set $\mathcal{X} \in \mathbb{R}^n$ is said to be a *convex set*, if and only if for any two points $x^1, x^2 \in \mathcal{X}$, the segment between them is also contained in $\mathcal{X}$. Figure A.1 helps us to provide a graphical intuition of convex sets, using examples in $\mathbb{R}^2$.

Let $\mathcal{X} = \{x^1, \cdots, x^p\}$ be any arbitrary set of points. We say that y is a *convex combination* of points in $\mathcal{X}$, when it is possible to find a set of coefficients $\alpha_i \geq 0, i = 1, \cdots, p, \sum_{i=1}^{p} \alpha_i = 1$, such that $y = \sum_{x^i \in \mathcal{X}} \alpha_i x^i$.

Extending the convex combination to an infinite set of summands, we can define the *convex hull* of a convex or non-convex set $\mathcal{X}$, denoted $conv(\mathcal{X})$, as the set of all the possible convex combinations of points in $\mathcal{X}$:

$$conv(\mathcal{X}) = \left\{ y = \sum_{x^i \in \mathcal{X}} \alpha^i x^i, \quad \forall \alpha^i \geq 0, \sum_i \alpha^i = 1 \right\}$$

The convex hull $conv(\mathcal{X})$ can be equivalently defined as the smaller convex set that contains $\mathcal{X}$, and thus a set is convex if and only if it is equal to its convex hull. Figure A.2 illustrates the concept of convex hull, using $\mathbb{R}^2$ examples. For instance, the convex hull of a set of two points, is the segment between them, and the convex hull of three not-aligned points, is the triangle among them, including the interior area and the border.

The following operations over sets, preserve the convexity:

- *Intersection of convex sets.* The intersection of a finite or infinite family of convex sets, is a convex set. On the contrary, the union of convex sets is usually non-convex (see Figure A.3 for a graphical illustration).

Optimization of Computer Networks – Modeling and Algorithms: A Hands-On Approach,
First Edition. Pablo Pavón Mariño.
© 2016 John Wiley & Sons, Ltd. Published 2016 by John Wiley & Sons, Ltd.
Companion Website: www.wiley.com/go/PavonMarinoSol16

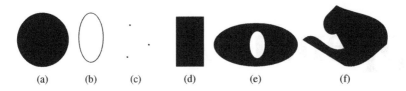

Figure A.1   Convex and non-convex set examples in $\mathbb{R}^2$. Dark points belong to the set, white do not. (a) and (d) are convex sets. The rest are not, since it is possible to find segments with the end points in the set, but that have points not belonging to the set

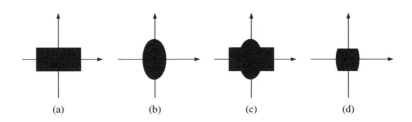

Figure A.2   Convex hull examples in $\mathbb{R}^2$. Dark points belong to the set $\mathcal{X}$, gray points and black points are in $conv(\mathcal{X})$

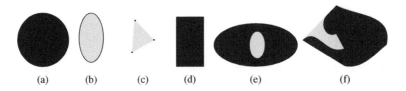

Figure A.3   (a) Set $\mathcal{X}_1$, (b) set $\mathcal{X}_2$, (c) set $\mathcal{X}_1 \bigcup \mathcal{X}_2$ (not convex), (d) $\mathcal{X}_1 \bigcap \mathcal{X}_2$ (convex)

- *Affine transformation of a convex set.* Let $\mathcal{X} \subset \mathbb{R}^n$ be a convex set. The transformation of $\mathcal{X}$ by a linear function:

$$f(\mathcal{X}) = \{Ax + b, \quad \forall x \in \mathcal{X}\}$$

is also convex, for any matrix $A \in \mathbb{R}^{n \times m}$ and vector $b \in \mathbb{R}^m$ defining the transformation. This means that the set $b + \mathcal{X}$ that translates each point in convex set $\mathcal{X}$ by a constant direction b is convex. Also, the projection of a convex set into a subset of its coordinates is convex: if $\mathcal{X} \subset \mathbb{R}^{n+m}$ is convex, then:

$$\{x_1 \in \mathbb{R}^n : (x_1, x_2) \in \mathcal{X} \text{ for at least one } x_2\}$$

is convex.

A.2 Convex and Concave Functions

A function $f : \mathcal{X} \subset \mathbb{R}^n \to \mathbb{R}$ is convex, if its domain $\mathcal{X}$ is a convex set, and for any two points $x, y \in \mathcal{X}$, and all $\alpha \in [0, 1]$, we have that

$$f(\alpha x + (1 - \alpha)y) \leq \alpha f(x) + (1 - \alpha)f(y), \quad \forall x, y \in \mathcal{X} \tag{A.1}$$

and we say that f is strictly convex when the inequality (A.1) holds strictly for $\alpha \in (0, 1)$ whenever $x \neq y$. Figure A.4 helps us to illustrate convex functions. Given two points x, y, right hand-side of (A.1) is the segment in the graph between points $(x, f(x))$ and $(y, f(y))$, which according to (A.1) should be above the graph of f.

Similarly, a function $f : \mathcal{X} \subset \mathbb{R}^n \to \mathbb{R}$ is concave, if its domain $\mathcal{X}$ is a convex set, and for any $x, y \in \mathcal{X}, \alpha \in [0, 1]$:

$$f(\alpha x + (1 - \alpha)y) \geq \alpha f(x) + (1 - \alpha)f(y), \quad \forall x, y \in \mathcal{X} \tag{A.2}$$

Strict concavity holds if previous inequality is strict whenever $x \neq y$, $\alpha \in (0, 1)$. Figure A.5 shows a concave function. It is easy to verify that a function f is (strictly) convex if and only if $-f$ is (strictly) concave.

A.2.1 Convexity in Differentiable Functions

If $f : \mathcal{X} \subset \mathbb{R}^n \to \mathbb{R}$ is a differentiable function, the gradient of f in point $x = (x_1, \cdots, x_n)$ is given by:

$$\nabla f(x) = \left(\frac{\partial f}{\partial x_1}(x), \cdots, \frac{\partial f}{\partial x_n}(x) \right) \in \mathbb{R}^n$$

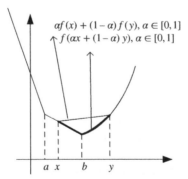

Figure A.4 Convex function $f : \mathbb{R} \to \mathbb{R}$. For any two points x, y, the segment $\alpha f(x) + (1 - \alpha)f(y), \alpha \in [0, 1]$ is above the function graph $f(\alpha x + (1 - \alpha)y), \alpha \in [0, 1]$. The function is not strictly convex, since for some points, for example a, b, the segment between them is not strictly above the graph

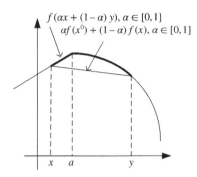

Figure A.5 Concave function $f : \mathbb{R} \to \mathbb{R}$. For any two points x, y, the segment $\alpha f(x) + (1 - \alpha)f(y), \alpha \in [0, 1]$ is below the function graph $f(\alpha x + (1 - \alpha)y), \alpha \in [0, 1]$. The function is not strictly concave, since for some points, for example x, a, the segment between them is not strictly below the graph

The convexity and concavity of f in its convex domain $\mathcal{X}$ can be checked using the gradient of f:

$$f \text{ is convex} \Leftrightarrow f(y) \geq f(x) + (y - x)^T \nabla f(x), \quad \forall x, y \in \mathcal{X}$$

$$f \text{ is concave} \Leftrightarrow f(y) \leq f(x) + (y - x)^T \nabla f(x), \quad \forall x, y \in \mathcal{X}$$

Strict convexity and concavity occurs if previous inequalities are strict for all $x \neq y$. Previous expression means that a differentiable function is convex (concave), if its linear expansion approximation centered in any point x (given by $f(x) + (y - x)^T \nabla f(x)$), is a global subestimator (overestimator) of the function. Figure A.6 illustrates this in real functions.

If f is twice differentiable, the second order Taylor approximation of f around point x is:

$$f(y) \approx f(x) + (y - x)^T \nabla f(x) + \frac{1}{2}(y - x)^T \nabla^2 f(x)(y - x)$$

where $\nabla^2 f(x)$ is the second-derivative or *hessian matrix*. This is an $n \times n$ matrix, with (i, j) coordinate given by $\partial^2 f/\partial x_i \partial x_j$. In the common case, when second partial derivatives are continuous, hessian matrix is symmetric. In the Taylor approximation, we see that the second order term $(y - x)^T \nabla^2 f(x)(y - x)$ represents the *curvature* of f in x: how much f separates from the linear estimation $f(x) + (y - x)^T \nabla f(x)$, when we move from x to y. Then:

- If $(y - x)^T \nabla^2 f(x)(y - x) \geq 0$ for all $x, y \in \mathcal{X}$, the function is always above the linear estimation and thus is convex. This happens if and only if the hessian matrix $\nabla^2 f(x)$ is semidefinite positive (s.d.p.)[1] for all x. If it is definite positive (d.p.), $(y - x)^T \nabla^2 f(x)(y - x) > 0$ for every $x \neq y$ and the function is strictly convex.
- If $(y - x)^T \nabla^2 f(x)(y - x) \leq 0$ for all $x, y \in \mathcal{X}$, the function is always below the linear estimation, and thus is concave. This happens if and only if the hessian matrix $\nabla^2 f(x)$ is

---

[1] A matrix A is definite positive (negative) if all its eigenvalues are strictly greater (lower) than zero, which happens if and only if $x^T A x > 0$ ($x^T A x < 0$) for every $x \neq 0$. Positive and negative semidefiniteness permits some eigenvalues to be zero. A matrix is not defined when it has strictly positive and strictly negative eigenvalues, and thus $x^T A x$ is strictly positive for some x and strictly negative for others.

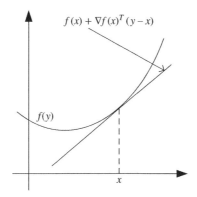

(a) Convex differentiable function, linear subestimator

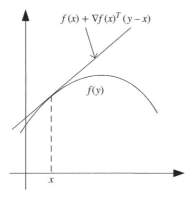

(b) Concave differentiable function, linear overestimator

Figure A.6 Convex and concave differentiable functions

semidefinite negative (s.d.n.) for all x. If it is definite negative (d.n.), the function is strictly concave.

- If $(y-x)^T \nabla^2 f(x)(y-x)$ is strictly positive when we move in some directions $(y-x)$, and strictly negative when we move in other directions, f is neither convex nor concave. This is equivalent to having a hessian matrix $\nabla^2 f(x)$ that is not defined with both strictly positive and strictly negative eigenvalues. If we move in the direction provided by an eigenvector, f is locally curved by an amount given by its associated eigenvalue.

Example A.1 The function $f(x_1) = x_1^2$ is strictly convex with respect to x_1 in all $\mathbb{R}$ ($f''(x_1) = 2$). However, $f(x_1, x_2) = x_1^2$ is convex but not strictly convex with respect to (x_1, x_2), since its hessian equals:

$$\nabla^2 f(x_1, x_2) = \begin{pmatrix} 2 & 0 \\ 0 & 0 \end{pmatrix}$$

which has 2 and 0 as its eigenvalues.

A.2.2 Strong Convexity/Concavity

A twice differentiable function f is strongly convex (concave)[2] when it is strictly convex (concave), and there is a finite number $a > 0$ ($a < 0$), such that all the eigenvalues of $\nabla^2 f(x)$ are strictly greater (lower) than a.

In other words, both strict and strong convexity force the function curvature to be always strictly positive. However, there can be strictly convex functions with curvatures approaching zero in some points. The strong convexity rule out these cases and requires the curvature to be separated from zero at least a constant quantity a. The strongly concave case is analogous.

Example A.2 The function $f(x) = e^x$ is strictly convex in its domain $\mathbb{R}$, since $f''(x) = e^x > 0, \forall x$. However, it is not strongly convex in $\mathbb{R}$, since when $x \to \infty$, the curvature $f''(x)$ can be arbitrarily close to zero.

A.2.3 Convexity in Non-Differentiable Functions

Let $f : \mathcal{X} \subset \mathbb{R}^n \to \mathbb{R}$ be a convex function, not necessarily differentiable in all its convex domain $\mathcal{X}$. Subgradients are a generalization of gradients, suitable also for points where f is not differentiable. We say that a vector $g \in \mathbb{R}^n$ is a subgradient of f in $x \in \mathcal{X}$ when:

$$f(y) \geq f(x) + (y - x)^T g, \quad \forall y \in \mathcal{X} \tag{A.3}$$

That means that, if g is a subgradient of f in x, the linear approximation (first order Taylor approximation) of the function given by $f(x) + (y - x)^T g$, is a global subestimator of f.

When a convex function f is not differentiable in a point x, it can have many subgradients. The set of those subgradients is called the subdifferential of f in x, and is denoted as $\partial f(x)$. When f is differentiable in x, $\nabla f(x)$ is its only subgradient:

$$f \text{ convex and differentiable in } x \Rightarrow \partial f(x) = \{\nabla f(x)\}$$

$$f \text{ convex and } \partial f(x) = \{\nabla f(x)\} \Rightarrow f \text{ differentiable in } x \text{ and } \nabla f(x) = g$$

Figure A.7a helps us to illustrate the concept of subgradient of a convex function. The function is differentiable in x^1 and in this point there is only one line crossing x^1 and that is always below f: the tangent of f in x of slope given by $\nabla f(x^1)$. Then, $\nabla f(x^1)$ is the only subgradient in x^1. In turn, the function is not differentiable in x^0, and there is an infinite number of tangent lines which subestimate f. Their slopes are subgradients of f in x^0. In particular, in Fig. A.7a, all the slopes in the interval $[g_1, g_4]$ are subgradients in x^0: $\partial f(x^0) = [g_1, g_4]$. It can be shown that if g and g' are two subgradients of f in x, all the convex combination of vectors $\alpha g + (1 - \alpha)g', \alpha \in [0, 1]$, are also subgradients. That is, the set of subgradients (subdifferential of f) is always a convex set.

The concept of subgradient can be adapted to concave functions. If $f : \mathcal{X} \subset \mathbb{R}^n \to \mathbb{R}$ is a concave function, a vector $g \in \mathbb{R}^n$ is a subgradient of f in $x \in \mathcal{X}$ when:

$$f(y) \leq f(x) + (y - x)^T g, \quad \forall y \in \mathcal{X} \tag{A.4}$$

---

[2] A definition of strong convexity and concavity exists for non-differentiable functions, not included in this appendix since we mostly restrict to the differentiable case in this book.

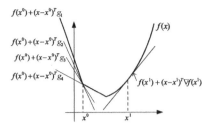

(a) Subgradients in convex functions

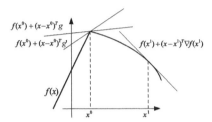

(b) Subgradients in concave functions

Figure A.7 Subgradients in convex and concave functions

This means that in concave functions, subgradients define global overestimators[3]. Figure A.7b shows an example. It can be easily shown that g is a subgradient of a concave function f, if and only if $-g$ is a subgradient of convex function $-f$.

Below are some properties of subgradients:

- If g is a subgradient of f in x, then αg is a subgradient of αf in x for any $\alpha \geq 0$.
- If g_i are subgradients of f_i in x, $i = 1, \cdots, m$, then $\sum_i g_i$ is a subgradient of $\sum_i f_i$ in x.
- Let $f(x) = \max_{i=1,\cdots,m} f_i(x)$. Let k be the index of a function f_i where the maximum is reached in point x. Then, any subgradient of f_k in x is a subgradient of f in x.

A.2.4 Determining the Curvature of a Function

Given a function f, the determination of its convexity or concavity can be quite laborious by direct application of convexity/concavity definitions (A.1, A.2), or computation and definiteness analysis of its hessian matrix. In this section we provide a set of rules that can simplify the process in most of the cases.

First, Table A.1 illustrates the convexity and concavity properties of some basic functions. Proofs are omitted, the reader is referred to Chapter 3 of [1] for details.

---

[3] For this reason, some texts refer to the subgradients of concave functions as supergradients. In this book, we prefer applying the same term subgradient to both convex and concave functions.

Table A.1   Some basic convex functions

Function	Curvature properties
$a^T x + c$	CVX and CVE $\forall x \in \mathbb{R}^n$
$x^T A x + b^T x + c$	CVX (CVE) $\forall x \in \mathbb{R}^n$ if A s.d.p (s.d.n). S-CVX (S-CVE) if A d.p (d.n)
$x^a, a > 1$	S-CVX $\forall x \in \mathbb{R}, x \geq 0$
$x^a, a \in (0, 1)$	S-CVE $\forall x \in \mathbb{R}, x \geq 0$
$x^a, a < 0$	S-CVX $\forall x \in \mathbb{R}, x > 0$
$\|x\|^a, a \geq 1$	CVX $\forall x \in \mathbb{R}$, S-CVX if $a > 1$
e^x	S-CVX $\forall x \in \mathbb{R}$
$\log\ x$	S-CVE $\forall x \in \mathbb{R}, x > 0$
$x \log\ x$	S-CVX $\forall x \in \mathbb{R}, x > 0$
$\|x\|$	CVX $\forall x \in \mathbb{R}^n$
$\|x\|_2^2$	S-CVX $\forall x \in \mathbb{R}^n$
$\max\{x_1 \cdots x_n\}$	CVX $\forall x = (x_1, \cdots, x_n) \in \mathbb{R}^n$
$\frac{x^2}{y}$	CVX $\forall x, y \in \mathbb{R}, y > 0$
$\left(\prod_{i=1}^n x_i\right)^{1/n}$	CVE $\forall x : x_i \geq 0 \forall i = 1, \cdots, n$

CVX: convex, CVE: concave, S-CVX: strictly convex, S-CVE: strictly concave.

Next, we provide a set of function transformations and the resulting function curvature. The application of these properties often permits determining the curvature of complex functions:

- *Non-negative scaling*: If f is a convex function, then αf is convex for every $\alpha \geq 0$. If f is strictly convex, and $\alpha > 0$, then αf is also strictly convex.
- *Non-negative weighted sums*: Let $f_1, \cdots, f_m$ be a set of convex functions in the same domain $\mathcal{X}$. Then, the function $f_1 + \cdots + f_m$ is convex in $\mathcal{X}$. Moreover, if *at least one* function f_i is strictly convex, then f is also strictly convex.
- *Composition with affine transformation.* Let $f : \mathcal{X} : \mathbb{R}^n \to \mathbb{R}$, A an $n \times m$ matrix, and $b \in \mathbb{R}^n$. Define $g : \mathbb{R}^m \to \mathbb{R}$ as:

$$g(x) = f(Ax + b),$$

where the domain of g is $\{x : Ax + b \in \mathcal{X}\}$. Then, if f is convex (concave), so is g. Moreover, if transformation $y = Ax + b$ is one-to-one (bijective), which occurs if $n = m$ and A has full-rank, then if f is strictly convex (concave), so is g.
- *Pointwise maximum.* Let $f_1, \cdots, f_m$ be convex functions defined in the same domain $\mathcal{X}$. The pointwise maximum function:

$$f(x) = \max\{f_1(x), \cdots, f_m(x)\}, \quad \forall x \in \mathcal{X}$$

is convex. If all of the f_i functions are strictly convex, then so is f. Figure A.8a helps us to graphically illustrate the pointwise maximum property. The result can be generalized to the pointwise supremum of an infinite set of convex functions. In this case, strict convexity of all the functions f_i does not guarantee strict convexity of the pointwise supremum.

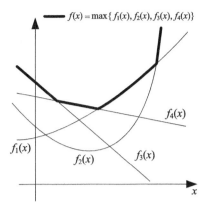

(a) Pointwise maximum of convex functions

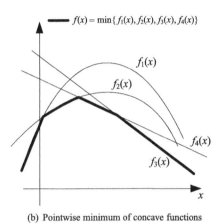

(b) Pointwise minimum of concave functions

Figure A.8 Pointwise maximum and minimum properties for convex and concave functions

- *Pointwise minimum.* Let $f_1, \cdots, f_m$ be concave functions defined in the same domain $\mathcal{X}$. The pointwise minimum function:

$$f(x) = \min\{f_1(x), \cdots, f_m(x)\}, \quad \forall x \in \mathcal{X}$$

is concave. Figure A.8b helps us to graphically illustrate this. The result can be generalized to the pointwise infimum of an infinite number of functions. If the number of functions is finite, and all of the them are strictly concave, then so is f.
- *Composition of functions.* We analyze the curvature of the composite function

$$f(y) = g(h(y)) = (g \circ h)(y)$$

h is a vectorial function $h : \mathcal{X} \subset \mathbb{R}^n \to \mathbb{R}^k$, so that $h(y) = (h_1(y), \cdots, h_k(y))$. g is a real function $g : Im(h) \subset \mathbb{R}^k \to \mathbb{R}$, where $Im(h)$ is the image set of h, and thus f is well defined. In

Table A.2 Convexity of the composite function

g	$h_i, i = 1, \cdots, k$	$f = g(h(y))$
CVX INC	CVX	CVX
CVX DEC	CVX	–
CVE INC	CVX	–
CVE DEC	CVX	CVE
CVX INC	CVE	–
CVX DEC	CVE	CVX
CVE INC	CVE	CVE
CVE DEC	CVE	–

CVX: convex, CVE: concave, INC: increasing, DEC: decreasing, –: any.

some occasions, it is possible to derive the convexity or concavity of function $f : \mathcal{X} \subset \mathbb{R}^n \to \mathbb{R}$ depending on if g and h_i are convex/concave and increasing/decreasing functions. This is shown in Table A.2.

The results in Table A.2 are applicable for any dimensions of the multivariate transformations h, and for differentiable and non-differentiable functions. Still, we can use the chain-rule expression of the second derivative of single variable functions ($k = n = 1$):

$$f(y)'' = (g(h(y)))'' = g''(h(y))(h'(y))^2 + g'(h(y))h''(y) \tag{A.5}$$

as a mnemotecnic rule to produce Table A.2: when the sign of f'', can be derived from the signs of g', g'', h', and h'', we can automatically determine the curvature of f. Strict convexity and concavity in the general case can also be derived from (A.5). For instance, if g is convex and strictly increasing, and all h_i are strictly convex, then so is f.

- *Partial minimization*. Let $f(x, y)$ be a jointly convex function in (x, y) variables. If $\mathcal{X}$ is a compact convex non-empty set, then the function:

$$g(x) = \min_{y \in \mathcal{X}} f(x, y)$$

is convex in its domain, given by the projection of $\mathcal{X}$ into the x-coordinates:

$$\textbf{dom } g = \{x : (x, y) \in \textbf{dom } f \text{ for at least one } y \in \mathcal{X}\}$$

Example A.3 The function $f(x_1, x_2) = (5x_1 - x_2)^2$ is convex since x^2 is a convex real function, and $(5x_1 - x_2)^2$ is x^2 replacing $5x_1 - x_2$ with x. Since this transformation is not one-to-one, $(5x_1 - x_2)^2$ is not strictly convex with respect to (x_1, x_2) although x^2 is strictly convex with respect to x.

A.2.5 Sub-level Sets

Let $f : \mathcal{X} \subset \mathbb{R}^n \to \mathbb{R}$ be a convex function and c any scalar constant. Then, the set of points $x \in \mathcal{X}$:

$$f(x) \leq c$$

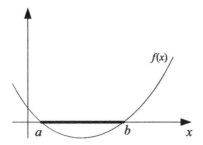

(a) Sub-level set ($c = 0$) of a convex function

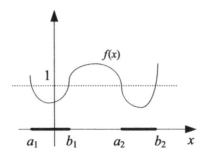

(b) Sub-level set ($c = 1$) of a non-convex function

Figure A.9 Sub-level sets. (a) f is convex and its sub-level set for $f(x) \leq 0$ is the interval $[a, b]$. (b) f is non-convex, and its sub-level set $f(x) \leq 1$ is the set $\{[a_1, b_1] \bigcup [a_2, b_2]\}$, that is not convex

if not empty, is a convex set commonly known as the *sub-level set* of f for constant c. Figure A.9 shows an example of convex sub-level sets of a convex function, and a non-convex sub-level set of a non-convex function.

A.2.6 Epigraphs

The *epigraph* of a function $f : \mathcal{X} \subset \mathbb{R}^n \rightarrow \mathbb{R}$, is defined as:

$$\mathbf{epi} \, f = \{(x, t) : x \in \mathcal{X}, f(x) \leq t\} \subset \mathbb{R}^{n+1}$$

This means that an epigraph is the set of points that are above the graph of f. Then, it holds that a function f is convex if and only if its epigraph is a convex set. Figure A.10 illustrates this.

A.3 Notes and Sources

Although convex optimization has been studied for more than a century, its interest has steadily grown since the 1980s, after recognizing the possibility of efficiently solving convex programs, and the multitude of convex optimization problems appearing in diverse disciplines,

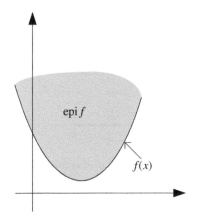

Figure A.10   Epigraph of a convex function

like computer networks. This appendix extracts standard results in convex analysis, which can be found in any recent optimization book. An accessible reference for further reading is [1].

Reference

[1] S. Boyd and L. Vandenberghe, *Convex Optimization*. New York, NY, USA: Cambridge University Press, 2004.

Appendix B

Mathematical Optimization Basics

B.1 Optimization Problems

A mathematical optimization problem, or just an optimization problem, consists of finding a vector $x = (x_1, \ldots, x_n)$ that minimizes an objective function $f : \mathcal{X} \subset \mathbb{R}^n \to \mathbb{R}$, searching only among a set of solutions that satisfy given constraints:

$$\min_x f(x), \quad \text{subject to:} \tag{B.1a}$$

$$g_i(x) \leq 0, \quad i = 1, \ldots, m \tag{B.1b}$$

$$h_j(x) = 0, \quad j = 1, \ldots, p \tag{B.1c}$$

$$x \in S \tag{B.1d}$$

where $x_1, \ldots, x_n$ are the problem decision variables[1]. In network optimization problems, they typically represent quantities like "capacity in a link", "traffic to be carried through a path", or "number of links between two nodes". Function f is called the problem objective function, (B.1b) are the *equality constraints*, (B.1c) the *inequality constraints*, and (B.1d) the *set constraints*.

A solution x which satisfies *all* problem constraints (B.1b–d) is referred to as a feasible solution. The set of all feasible solutions to a problem is called the *feasible set*. All the functions f, g_i, and h_j in the objective and constraints should be well defined in the feasible set. When the problem has no constraints, we refer to it as an *unconstrained optimization* problem, and every vector in $\mathbb{R}^n$ is feasible.

Any minimization problem $\min_{x \in \mathcal{X}} f(x)$, which is a short form of $\min_x f(x)$, subject to $x \in \mathcal{X}$, can be converted into a maximization problem by changing the sign of the objective function. That is, the problem $\min_{x \in \mathcal{X}} f(x)$ is equivalent to $\max_{x \in \mathcal{X}} -f(x)$. Also, note that

---

[1] In this book, we assume that the number of decision variables is finite ($n < \infty$)

Optimization of Computer Networks – Modeling and Algorithms: A Hands-On Approach,
First Edition. Pablo Pavón Mariño.
© 2016 John Wiley & Sons, Ltd. Published 2016 by John Wiley & Sons, Ltd.
Companion Website: www.wiley.com/go/PavonMarinoSol16

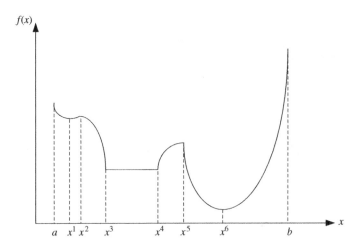

Figure B.1   Example. **dom** $f = \mathcal{X} = [a,b]$. Local maximums are $\{a, x^2, (x^3, x^4), x^5, b\}$. Local minimums are $\{x^1, [x^3, x^4], x^6\}$. Global maximum is b and global minimum x^6

inequalities $g_i(x) \geq 0$ are equivalent to $-g_i(x) \leq 0$. Then, we can study only minimization problems where inequalities are of the form "$\leq$", without loss of generality.

Given an optimization problem $\min_{x \in \mathcal{X}} f(x)$, a vector $x^* \in \mathcal{X}$ is said to be a *global optimum*, when $f(x^*) \leq f(x)$ for all $x \in \mathcal{X}$. $x^* \in \mathcal{X}$ is said to be a *local optimum* when $f(x^*) \leq f(x)$, for every $x \in \mathcal{X} \bigcap B(x^*, \epsilon)$, where $B(x^*, \epsilon)$ is a ball centered in x^* and with a radius $\epsilon > 0$, as small as we wish. In other words, f reaches a minimum in x^*, compared to the feasible points which are in its proximity. Global and local minimums are tagged as *strict*, when the relation $f(x^*) \leq f(x)$ holds strictly when $x \neq x^*$. As an example, Fig. B.1 illustrates the previously defined concepts.

In some occasions (of little practical interest), an optimization problem with feasible solutions, may not have a global optimum. For instance, the problem $\min_{x > 1} x$ has no global optimum since given any point $x > 1$, you can always find one closer to the limit $x = 1$. The intuition would say that the optimum is $x = 1$, but formally $x = 1$ is unfeasible. This is the type of problems that appear when the feasible set is not closed and some boundary points are not feasible. Similarly, the problem $\max_{x > 1} x$ has no global optimum, since for any real number x we can always find a larger one and the limit $x = \infty$ is not a feasible point.

The *Theorem of Weierstrass* provides sufficient conditions to rule out these mathematical technicalities.

Theorem B.1   (Weierstrass) If f is a continuous function, and $\mathcal{X}$ is a non-empty compact set, then the problem $\min_{x \in \mathcal{X}} f(x)$ has at least a global optimum.

In finite dimension spaces like $\mathbb{R}^n$, a set is compact if and only if it is closed on bounded. This is easily achieved in problem (B.1) if, (i) we do not use strict inequalities in the constraint set[2],

---

[2] Actually, general *solver* programs which find numerical solutions to optimization problems, simply treat strict inequalities as if they were non-strict.

(ii) set S in (B.1d) is closed, and (iii) we include *box-constraints* to the decision variables of the problem: $l_i \leq x_i \leq u_i, i = 1, \ldots, n$, where l_i and u_i are finite.

The previous requisites do not pose any loss of generality in practical problems. For example, if a decision variable u_e is the capacity of a link e, we can add the box constraint $0 \leq u_e \leq U_{max}$, where U_{max} is the maximum capacity that any link could have in a realistic situation. If a strict inequality $g(x) < c$ appears, it can just be replaced by $g(x) \leq c - \epsilon$, where $\epsilon > 0$ is any hand-picked constant, so small that it has no influence from an engineering point of view.

B.2   A Classification of Optimization Problems

Optimization problems admit multiple classifications. We follow a classical approach next, and briefly review *linear programs*, *convex programs*, *nonlinear programs*, and *integer programs*. Note that the word *program* here has nothing to do with the concept of a software program, but with the task of creating optimal programs or schedules of activities, one of the initial uses of optimization. Proofs of presented properties will be omitted. Further details can be found in many texts such as [1–4].

B.2.1   Linear Programming

Linear programs are optimization problems (B.1) where all functions f, g_i, h_j are linear[3], and there is no extra set constraints (B.1d). Thus, they are problems of the form:

$$\min_x \quad c_1 x_1 + \ldots + c_n x_n \quad \text{subject to:}$$

$$a_{i1} x_1 + \ldots + a_{in} x_n \leq b_i, \quad \forall i = 1, \ldots, m$$

$$a'_{j1} x_1 + \ldots + a'_{jn} x_n = b'_j, \quad \forall j = 1, \ldots, p$$

This can be expressed in a compact matricial form:

$$\min_x c^T x, \quad \text{subject to: } Ax \leq b, A'x = b'$$

where matrices $A \in \mathbb{R}^{m \times n}, A' \in \mathbb{R}^{p \times n}$, and vectors $c \in \mathbb{R}^n$, $b \in \mathbb{R}^m$, $b' \in \mathbb{R}^p$, are input constants defining the program.

The points satisfying a linear inequality constraint are called a *semispace*. The points satisfying linear equality constraint are called a *hyperplane*. The set of feasible points in a linear program is a *polyhedron*. A semispace $\{x \in \mathbb{R}^n : a^T x \leq b\}$ is a convex set since it is the sub-level set of the convex function $f(x) = a^T x$. This applies also if we reverse the inequality sign, since the function $-a^T x$ is also convex, and thus have convex sub-level sets. Then, a hyperplane $\{x \in \mathbb{R}^n : a^T x = b\}$ is a convex set since it can be expressed as the intersection of two semispaces: $\{x \in \mathbb{R}^n : a^T x \leq b, a^T x \geq b\}$. Finally, the polyhedron is the intersection of semispaces and hyperplanes, and thus is also convex.

---

[3] A function $f : \mathbb{R}^n \to \mathbb{R}$ is linear if it can be expressed as $f(x) = c^T x = c_1 x_1 + \ldots + c_n x_n$, where $c = (c_1, \ldots, c_n)$ is any constant vector.

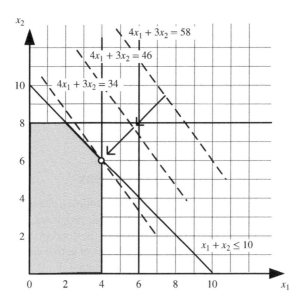

Figure B.2   Linear program example

In the rest of this section, we will use the example formulation (B.2) to give graphical insights and fundamental properties of linear programs.

$$\max_{\{x_1,x_2\}} 4x_1 + 3x_2, \quad \text{subject to:} \tag{B.2a}$$

$$x_1 + x_2 \le 10 \tag{B.2b}$$

$$0 \le x_1 \le 4 \tag{B.2c}$$

$$0 \le x_2 \le 8 \tag{B.2d}$$

Figure B.2 shows the polyhedron of points (x_1, x_2) which are feasible to the problem. The dotted parallel lines in the upper part of Fig. B.2 are solutions with the same benefit $B = 4x_1 + 3x_2$. The optimum solution will then belong to the parallel line which touches the feasible set, and has the highest B value. This is the line $4x_1 + 3x_2 = 34$, and the only one feasible point in the line $(4, 6)$ is the optimum solution to (B.2).

We see that the optimum solution is a *vertex* or *extreme point* of the feasible set. More formally, a point x in a set $\mathcal{X}$ is a vertex of $\mathcal{X}$, if it is not possible to find two different points $y_1, y_2 \in \mathcal{X}$, both different to x, such that x lays in the segment between y_1 and y_2. This definition of a vertex coincides with the geometrical vertex definition in a polyhedron. In Fig. B.2, the vertexes of the feasible set are $\{(0, 0), (0, 8), (2, 8), (4, 6), (4, 0)\}$.

The observation that optimum solution of (B.2) lays in a vertex of the feasible set is not casual, as stated by the following theorem.

Theorem B.2 (Fundamental theorem of linear programming) Let P be a linear program, and $\mathcal{X}$ its feasible set. Then:

- If the problem has optimum solutions, at least one of them is a vertex of $\mathcal{X}$.
- If x^1, x^2 are two optimum solutions, all the points in the segment between them are also optimum.

Theorem B.2 applies only to problems with optimum solutions. These are problems with non-empty feasible sets $\mathcal{X}$, and for which the optimum benefit to maximize cannot be made arbitrarily large (e.g., that would happen if we remove constraints (B.2bc) in our example).

A linear program can have multiple optimum solutions. For instance, if we change the objective function (B.2a) to $\max 3x_1 + 3x_2$, the iso-benefit lines $3x_1 + 3x_2 = B$ become parallel to the boundary of constraint $x_1 + x_2 \leq 10$, and all the points in the segment between $(2, 8)$ and $(4, 6)$ are optimal.

The following proposition is derived from Theorem B.2.

Proposition B.1 Let $\{\min_x c^T x, \text{ subject to: } A_1 x \leq b_1, A_2 x = b_2, x \geq 0\}$, be a linear program where $A_1 \in \mathbb{R}^{m \times n}, A_2 \in \mathbb{R}^{p \times n}$. If the problem has optimum solutions, then at least one of them has at most $m + p$ non-zero coordinates.

This comes from the fact that the vertexes of the polyhedron $A_1 x \leq b_1, A_2 x = b_2, x \geq 0$ are vectors $x = (x_1, \ldots, x_n)$ which have at most $m + p$ non-zero coordinates.

B.2.1.1 Solution Methods

Today, it is possible to find numerical solutions to linear programs of thousands of decision variables and constraints, using standard computing facilities. Even programs several orders of magnitude larger can be efficiently solved if the constraint matrices A and A' are sparse, and have a structure that can be exploited.

The celebrated *simplex method* (George Dantzig, [5]) is the first and for many years best method for solving linear programs. For this and other contributions, George Dantzig (1914–2005) is considered "the father of linear programming". The simplex method checks in each iteration the optimality of a vertex of the feasible set. If the vertex is not the problem optimum, it jumps to an adjacent one that improves the cost to minimize. This procedure is supported by the Fundamental Theorem of Linear Programming: if the problem has optimum solutions, at least one of them is a vertex. The number of different vertexes in a polyhedron can exponentially grow with the number of decision variables and constraints. Simplex method has an exponential worst-case time complexity, since in some pathological problem instances the number of vertexes visited can grow exponentially with problem size. Anyway, in most of the practical cases, the simplex method can efficiently solve large-scale problems.

The first worst-case polynomial method for solving linear programs is the ellipsoid method published in 1980 [6]. The ellipsoid method is slower than simplex in non-pathological instances. Both have been beaten by the so-called *interior point methods* that jump in each iteration between interior points of the feasible set, approaching the optimum solution, a vertex in the boundary. First interior point method was proposed by Narendra Karmarkar in 1984 [7]. Interior point methods have a polynomial worst-case complexity, and have gone through multiple refinements that make them outperform simplex method implementations. JOM library integrated in Net2Plan, used in the exercises, permits interfacing with linear solvers GLPK, IPOPT (open-source) and CPLEX (commercial). GLPK and CPLEX permit choosing between simplex and interior point algorithms for solving linear programs.

B.2.2   Convex Programs

A program $\min_{x \in \mathcal{X}} f(x)$ or $\max_{x \in \mathcal{X}} -f(x)$, is a convex program when f is a convex function, and $\mathcal{X}$ is a convex set.

A sufficient condition for a problem of the form (B.1) to be convex is:

- Objective function f is convex.
- g_i functions in inequality constraints $g_i(x) \leq 0$, are convex.
- h_j functions in equality constraints $h_j(x) = 0$ are affine (linear plus a constant).
- S set in the constraint $x \in S$, is a convex set.

Convex programs have particular properties that provide them of special importance in optimization. We enumerate some of them.

Proposition B.2 If x is a local optimum of a convex problem, then x is also a global optimum.

Proposition B.3 If x^1 and x^2 are optimum solutions of a convex problem, all the points in the segment between them are also optimal.

Proposition B.4 If the objective function f of a convex minimization program is strictly convex, then if a solution exists, it is unique.

Proposition B.2 means that if the problem is convex, any solution that is better than its neighbor solutions (local optimum), is also better than all the feasible solutions (global optimum). Prop B.3 implies that the set of optimum solutions of a convex problem, is a convex set. Finally, Prop. B.4 sets simple sufficient conditions for uniqueness of the global optimum, and will be extensively used throughout the book.

Note that linear programs are subtypes of convex programs. However, while optimum solutions of linear programs are always boundary points of the feasibility set, the optima of a convex problem can be either interior or boundary points. For instance, the convex program

$$\min_x (x_1 - 1)^2 + (x_2 - 1)^2, \text{ subject to: } x_1^2 + x_2^2 \leq 9$$

has a unique optimum solution in the point $(1, 1)$, which is in the interior of the feasibility set (see Fig. B.3a). In turn, the program

$$\min_x (x_1 - 3)^2 + (x_2 - 3)^2, \text{ subject to: } x_1^2 + x_2^2 \leq 9$$

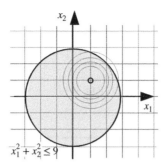

(a) Optimum in the interior of the feasibility set

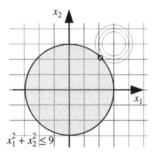

(b) Optimum in the boundary of the feasibility set

Figure B.3   Convex program examples

has $3(\cos\frac{\pi}{4}, \sin\frac{\pi}{4})$ as its unique optimum solution, a boundary point (see Fig. B.3b). Note that in both cases the optimum solution is unique, since the problem is convex with a strictly convex objective function. Strict convexity of the objective function does not guarantee uniqueness of the optimum solution, when the problem is not convex. As an example, the program (see Fig. B.4):

$$\min_x (x_1 - 2)^2 + (x_2 - 2)^2, \text{ subject to: } x_1^2 + x_2^2 \le 9, x_1 x_2 \le 0 \qquad (B.3)$$

has a strictly convex objective function but two optimum solutions $(2,0)$ and $(0,2)$. This is because the problem is not convex, since the feasibility set is not convex, due to constraint $x_1 x_2 \le 0$. Note also that Prop. B.3 is not met, since the points in the segment between $(2,0)$ and $(0,2)$ are not optimum solutions to the problem.

B.2.2.1   Solution Methods

The interior point methods described for linear programs can be generalized for convex programs and convex programs have polynomial worst-case complexity. Actually, they are considered the most difficult problems that can be efficiently solved, where "efficiently" means "in polynomial time".

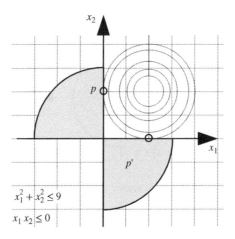

$x_1^2 + x_2^2 \leq 9$

$x_1 x_2 \leq 0$

Figure B.4   Non-convex program example

Interior point methods work quite well in practice, and there is a number of commercial and open-source libraries that solve efficiently generic convex programs. Problems of hundreds of variables and thousands of constraints can be solved in a few seconds using standard computing facilities [8]. The convex models in Part I of this book are solved in Net2Plan algorithms that rely in the open-source IPOPT solver for nonlinear programs to find numerical solutions. Part II of the book provides examples and guidelines to produce algorithms specifically designed to exploit the structure of network problems enabling a distributed implementation.

B.2.3   Nonlinear Programs

Problems of the form:

$$\min_{x} f(x), \quad \text{subject to:} \tag{B.4a}$$

$$g_i(x) \leq 0, \quad i = 1, \ldots, m \tag{B.4b}$$

$$h_j(x) = 0, \quad j = 1, \ldots, p \tag{B.4c}$$

that cannot be classified as convex programs, are classically named *nonlinear programs*, or more specifically *non-convex nonlinear programs*. In this type of problem, none of the properties Prop. B.2, Prop. B.3, or Prop. B.4 have to be met. For instance, in Fig. B.4 associated to problem (B.3) we see an example where the set of optimum solutions of a nonlinear program does not have to be convex and having a strictly convex objective function does not guarantee uniqueness of the global optimum. In turn, the program:

$$\min_{x}(x_1 - 1)^2 + (x_2 - 2)^2, \quad \text{subject to:} \ x_1^2 + x_2^2 \leq 9, x_1 x_2 \leq 0 \tag{B.5}$$

is an example that nonlinear programs can have multiple local minima (Fig. B.5).

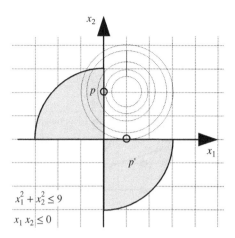

Figure B.5 Example of nonlinear problem (B.5) with two local minimum $p = (1,0)$, with cost 4 and $p' = (0,2)$, with cost 1

B.2.3.1 Solution Methods

There are no efficient methods (worst-case polynomial) to find the global optimum in generic nonlinear programs. Nonlinear solvers like IPOPT that can be interfaced from Net2Plan, contempt with finding a local minimum in differentiable problems. If the global optimum is pursued, we should rely on heuristic approaches (usually ad hoc developed) that provide suboptimal solutions in polynomial time. Chapter 12 in Part II of the book is devoted to this type of algorithms in the network optimization context.

B.2.4 Integer Programs

Optimization problems of the form:

$$\min_{x} f(x), \quad \text{subject to:} \tag{B.6a}$$

$$g_i(x) \leq 0, \quad i = 1, \dots, m \tag{B.6b}$$

$$h_j(x) = 0, \quad j = 1, \dots, p \tag{B.6c}$$

$$x_k \in \mathbb{Z}, \quad \forall k \in \mathcal{I} \tag{B.6d}$$

are called *integer programs* when all of the decision variables are constrained to be integer, and *mixed integer programs* when only some of them are. In (B.6), $\mathcal{I}$ represents the set of integer constrained variables. When a decision variable is restricted to take just the values zero or one, it is called a binary variable. When all the decision variables are binary, the program is referred to as a *binary program*.

The most common case of integer and mixed integer programs occur when functions f, g_i, h_j in (B.6) are all linear. In this case, they receive the name of *Integer Linear Programs*

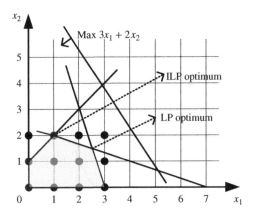

Figure B.6   Integer program example

(ILP), and *Mixed Integer Linear Programs* (MILP). As is shown in Part I of the book, there is a multitude of network optimization problems which are modeled using ILPs and MILPs. Non-bifurcated routing, modular capacities, topology design problems, or multicast delivery are some examples.

B.2.4.1   Solution Methods

Integer decision variables create discontinuities in the feasibility set, that impedes the application in integer programs of the solution methods for convex programs. Actually, even the *linear* integer versions (ILPs and MILPs) are known to be $\mathcal{NPO}$-complete, and thus there is no algorithm that can solve or even approximate its generic version in worst-case polynomial time[4]. Figure B.6 shows the graphical solution of the problem (B.7), an example to illustrate the difficulties in solving integer problems:

$$\max_{x_1, x_2} 3x_1 + 2x_2, \quad \text{subject to:} \tag{B.7a}$$

$$3x_1 + x_2 \leq 9 \tag{B.7b}$$

$$x_1 + 3x_2 \leq 7 \tag{B.7c}$$

$$-x_1 + x_2 \leq 1 \tag{B.7d}$$

$$x_1, x_2 \in \mathbb{Z} \tag{B.7e}$$

The optimum solution is achieved in point $(3, 0)$, with a benefit of nine units. This can be graphically observed, since if we move down the iso-benefit line $3x_1 + 2x_2 = B$, for decreasing values of B, the first line touching a feasible point occurs for $x_1 = 3, x_2 = 0$, and $B = 9$. It is also possible to compute the optimum solution by *enumerating all the feasible points*, since there is a finite number of them, computing each feasible solution benefit, and taking the best.

---

[4] No algorithm is known and the assertion that such algorithm does not exist is true assuming that $\mathcal{P} \neq \mathcal{NP}$.

Table B.1 Brute force enumeration for problem (B.7)

	(x_1, x_2)	Benefit $3x_1 + 2x_2$
1	$(0, 0)$	0
2	$(1, 0)$	3
3	$(2, 0)$	6
4	$(3, 0)$	**9**
5	$(0, 1)$	2
6	$(1, 1)$	5
7	$(2, 1)$	8
8	$(1, 2)$	7

This approach is commonly called *brute force* method. Such enumeration for problem (B.7) is shown in Table B.1. Brute force methods are not valid in practice but in very small problem instances, even if enumerating all the feasible solutions was easy (which is often not true). The impossibility lays on that the number of feasible solutions can grow exponentially with the number of decision variables. For instance, a brute force enumeration of a moderate size problem with 100 binary variables, would require enumerating and evaluating the objective function for $2^{100} \approx 10^{30}$ solutions. If we could process a solution every picosecond (10^{-12} s), it would take around 30,000 million years to end, more than the estimated age of the Universe!

Other technique that we can intuitively presume that can help to solve linear integer problems is (i) finding the optimum of the non-integer linear program (LP) and (ii) use a *rounding technique* to find the optimum integer solution from it. Unfortunately, rounding a non-integer solution to find the integer optimum is a problem that is generally as difficult as the original one. In Fig. B.6, we show that the LP optimum is the point $x_{LP} = (2.5, 1.5)$ and that no rounding produces the integer optimum $(3, 0)$.

Although there is no worst-case polynomial time algorithm for solving generic convex or linear programs with integer constraints, sophisticated techniques in commercial and open-source solvers can effectively solve some medium size instances. Most common techniques are variations of the basic *branch-and-bound* scheme (B&B). In a B&B iteration, the original problem is modified by fixing the value of a subset $\mathcal{I}_1$ of the integer variables, and optimizing the rest without integer constraints using a standard linear or convex solver. This provides a lower bound to the integer modified version. An upper bound can be obtained using a heuristic method to find a feasible solution. Then, if the lower bound obtained is worse than an upper bound of other B&B instances, we can safely assume that the values assigned to $\mathcal{I}_1$ variables, are not part of any optimum solution of the original problem. Different forms of deciding how the $\mathcal{I}_1$ sets are chosen or how the lower and upper bounds are calculated, provide multiple variations to B&B. The interested reader is referred to specialized books in the topic like [9] or [10], for further details in B&B techniques, and other variations like *branch-and-cut* or *branch-and-price*.

ILPs and MILPs models in Part I of the book are solved using Net2Plan in numerous exercises, interfacing with GLPK (free, open-source) and CPLEX (commercial, non disclosed) solvers. In Part II, we address the design of heuristic algorithms to provide (in general suboptimal) solutions of $\mathcal{NP}$-complete integer programs.

B.3 Duality

We consider an optimization problem:

$$\min_{x} f(x), \quad \text{subject to:} \tag{B.8a}$$

$$\pi_i : g_i(x) \le 0, \quad i = 1, \dots, m \tag{B.8b}$$

$$\lambda_j : h_j(x) = 0, \quad j = 1, \dots, p \tag{B.8c}$$

$$x \in S \tag{B.8d}$$

where f, g_i, h_j are continuous functions (not necessarily convex) correctly defined in the problem feasibility set and S is any arbitrary set. Along this section, we call problem (B.8) the *primal* problem, refer to x as the primal variables, and denote as p^* the global optimum cost of (B.8).

B.3.1 Dual Function

We define the *Lagrangian function L* of problem (B.8) as a function $L : S \times \mathbb{R}_+^m \times \mathbb{R}^p \to \mathbb{R}$:

$$L(x, \pi, \lambda) = f(x) + \sum_i \pi_i g_i(x) + \sum_j \lambda_j h_j(x)$$

we refer π_i as the *Lagrange multiplier* or *dual variable* associated to inequality constraint $g_i(x) \le 0$, and λ_j the Lagrange multiplier or dual variable associated to equality constraint $h_j(x) = 0$. Note that inequality multipliers π_i must be non-negative, while equality multipliers λ_j can take values in all $\mathbb{R}$.

The Lagrange function "'moves" to the objective function some of the constraints, summing them multiplied by a real number (the multiplier). This is called *relaxing* or *dualizing* a constraint. We use set S to denote the non relaxed constraints. See that π_i and λ_j symbols are written followed by a colon next to the constraints in the problem definition (B.8). This is a short form of denoting that these constraints are to be dualized and that π_i and λ_j represent their multipliers.

The *dual function* $w(\pi, \lambda)$ of a problem (B.8) returns the *minimum cost of a relaxed problem version*, for particular values of the multipliers ($\pi \ge 0, \lambda$):

$$w(\pi, \lambda) = \min_{x \in S} \left\{ f(x) + \sum_i \pi_i g_i(x) + \sum_j \lambda_j h_j(x) \right\} = \min_{x \in S} L(x, \pi, \lambda)$$

The value $w(\pi, \lambda)$ is usually called the *dual cost* or *relaxed cost* associated to multipliers (π, λ). The domain of the dual function (**dom** w) is composed of those multipliers ($\pi \ge 0$, $\lambda \in \mathbb{R}^p$) such that the minimum $\min_{x \in S} L(x, \pi, \lambda)$ exists. For instance, those multipliers for which $w(\pi, \lambda) = -\infty$, are outside of **dom**$w$. Recall that if S is a non-empty compact set, a finite minimum always exists and **dom** $w = \{(\pi, \lambda) : \pi \ge 0, \lambda \in \mathbb{R}^p\}$.

We can interpret the dual function as follows. Let us assume that a constraint $g_i(x) \le 0$ represents that a solution x cannot use more resources of a type i than available. For instance, that the traffic carried in a particular link cannot exceed its capacity. Then, relaxing the constraint means that now we accept solutions that violate it. However, if a solution x is such that

$g_i(x) > 0$, it is penalized in cost by an amount $\pi_i g_i(x) > 0$, as if we had to buy the extra resources needed at a price of π_i cost units per resource unit. In turn, if a solution uses less resources of type i than available $(g_i(x) < 0)$, it is favored reducing its cost since now $\pi_i g_i(x) < 0$, as if we were able to sell unused resources at a price π_i. A similar interpretation can be made when relaxing equality constraints.

Because of this interpretation, multipliers are often referred as *prices* or *dual prices*. In this context, the dual function for a particular set of multipliers (π, λ) returns the optimum cost achieved if constraints were relaxed and associated resources could be acquired and sold at the prices (π, λ).

Example B.1 In the optimization problem (B.9)

$$\min(x_1 - 1)^2 + (x_2 - 1)^2, \quad \text{subject to:} \tag{B.9a}$$

$$\pi : x_1 + 2x_2 \le 2 \tag{B.9b}$$

the dual function relaxing constraint (B.9b) is given by:

$$w(\pi) = \min_{(x_1, x_2) \in \mathbb{R}^2} \{(x_1 - 1)^2 + (x_2 - 1)^2 + \pi(x_1 + 2x_2 - 2)\}, \quad \forall \pi \ge 0 \tag{B.10}$$

Since we relax all the constraints, the set of non-relaxed constraints is $S = \mathbb{R}^2$, which is not compact. Thus, the dual function will not be formally defined for those π values for which $w(\pi) = -\infty$. There is no such point in this case, so **dom** $w = \{\pi \ge 0\}$. Given a $\pi \ge 0$ value, we can find the points (x_1, x_2) which minimize the Lagrangian, as those points where its gradient

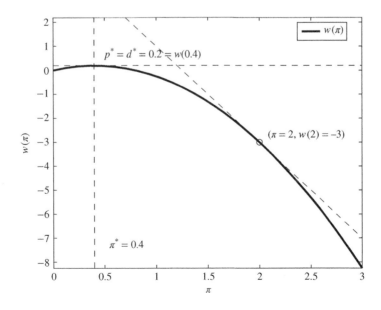

Figure B.7   Dual function of problem (B.9)

vanishes: $\frac{\partial L}{\partial x_1} = \frac{\partial L}{\partial x_2} = 0$. This occurs for $x_1 = 1 - \frac{\pi}{2}, x_2 = 1 - \pi$. Then, substituting these values in (B.10) we have after some manipulations:

$$w(\pi) = -\frac{5}{4}\pi^2 + \pi$$

which is plotted in Fig. B.7.

B.3.1.1 Properties of the Dual Function

We denote $\mathcal{X}^*(\pi, \lambda)$ as the set of all the primal solutions $\tilde{x}$ that are optimal for the relaxed problem with multipliers (π, λ):

$$\mathcal{X}^*(\pi, \lambda) = \left\{ \tilde{x} \in S : \tilde{x} \text{ solves } \min_{x \in S} f(x) + \sum_i \pi_i g_i(x) + \sum_j \lambda_j h_j(x) \right\}$$

Each solution to the relaxed problem $\tilde{x} \in \mathcal{X}^*(\pi, \lambda)$ is called a *minimizer* for multipliers (π, λ). It is important to remark, that a minimizer $\tilde{x}$ must satisfy the non relaxed constraints $\{\tilde{x} \in S\}$, but may violate the relaxed ones. That is, it can happen that $g_i(\tilde{x}) > 0$ and/or $h_j(\tilde{x}) \neq 0$.

If S is a compact non-empty set, the relaxed problem has always at least one solution, and $\mathcal{X}^*(\pi, \lambda)$ is never empty. In the general case, $\mathcal{X}^*(\pi, \lambda)$ can have multiple minimizers. However, if the primal problem is convex with an strictly convex objective function, according to Prop. B.4 the relaxed problem has a unique solution.

Proposition B.5 If in problem (B.8), the objective function f is strictly convex, g_i are convex functions, h_j are linear functions, and S is a non-empty compact and convex set, there is exactly one minimizer for each multiplier vector (π, λ).

Example B.2 All the assumptions of Prop. B.5 hold in Example B.1, but one: set $S = \mathbb{R}^2$ in the example, which is not compact. Still, it was shown in Example B.1 that each multiplier $\pi \geq 0$, has just one minimizer $\mathcal{X}^*(\pi) = \{(1 - \frac{\pi}{2}, 1 - \pi)\}$.

Proposition B.6 (Weak duality) For any problem of the form (B.8), and any multipliers $\pi \geq 0, \lambda \in \mathbb{R}^p$ within the domain of the dual function, it holds that:

$$w(\pi, \lambda) \leq p^* \tag{B.11}$$

Previous important property is called *weak duality*, and holds for any optimization problem of the form (B.8), for example convex or not, with or without integer constraints. It means that relaxing some constraints using whatever multipliers or prices, produces a dual cost that is always lower or equal than the optimum cost of the original problem p^*. Thus, the dual function can be seen as a generator of lower bounds: computing the dual function for any particular multiplier vector produces a *lower bound* to p^*.

Proposition B.7 The dual function $w(\pi, \lambda)$ is concave in its domain, for any optimization problem (B.8).

Example B.3 Following with Example B.1, we see that the dual function $w(\pi)$ is concave in all its domain $\pi \geq 0$: $w(\pi) = -\frac{5}{4}\pi^2 + \pi$, as established by Prop. B.7

Since the dual function is concave, any local maximum is a global maximum, and it has subgradients. The following properties characterize the subgradients of the dual function, and provide sufficient conditions to guarantee that the dual function is differentiable and thus subgradients in each (π, λ) point are unique (the gradients).

Proposition B.8 ([3], p. 423, [2], p. 210) Let $\pi \geq 0$ and λ be multipliers of the problem (B.8), and $\tilde{x} \in \mathcal{X}^*(\pi, \lambda)$ be any minimizer of the relaxed problem $\tilde{x} \in \mathcal{X}^*(\pi, \lambda)$. Then, the vector:

$$(g(\tilde{x}), h(\tilde{x})) = (g_i(\tilde{x}), i = 1, \dots m, h_j(\tilde{x}), j = 1, \dots, p)$$

is a subgradient of the dual function w in (π, λ). Moreover, the set $\partial w(\pi, \lambda)$ of all subgradients of the dual function in (π, λ) is the convex hull of the subgradients generated by the minimizers in $\mathcal{X}^*(\pi, \lambda)$:

$$\partial w(\pi, \lambda) = \text{conv}(\{(g(\tilde{x}), h(\tilde{x})), \forall \tilde{x} \in \mathcal{X}^*(\pi, \lambda)\})$$

Proposition B.8 shows that a subgradient of w can be obtained by just finding a minimizer $\tilde{x}$ for the relaxed problem and evaluating it in each relaxed constraint. Since the minimizers can violate any relaxed constraint, any subgradient coordinate associated to $g_i(\tilde{x})$ or $h_j(\tilde{x})$ can be negative, positive or zero.

Next proposition provides sufficient conditions for the dual function to be differentiable: when the minimizer is unique and the subgradients characterized in Prop. B.8 are actually gradients.

Proposition B.9 ([3], p. 427) Let $w(\pi, \lambda)$ be the dual function of problem (B.8), where f is strictly convex, g_i functions are convex, h_j functions are linear, and S is convex. Then, the dual function is differentiable in all its domain. Recall that if S is compact, the domain of the dual function is all $(\pi \geq 0, \lambda)$.

Example B.4 Following with Example B.1, we see that the dual function $w(\pi) = -\frac{5}{4}\pi^2 + \pi$ is everywhere differentiable, as established by Prop. B.9. Gradients of the dual function are: $\frac{dw}{d\pi} = -\frac{5}{2}\pi + 1$. For instance, in the point $\pi = 2$, we have that $\frac{dw}{d\pi}(2) = -4$. The minimizer for $\pi = 2$ is $x_1 = 1 - \frac{\pi}{2} = 0, x_2 = 1 - \pi = -1$. The relaxed constraint (B.9b) $g(x) \leq 0$, with $g(x) = x_1 + 2x_2 - 2$, when evaluated in point $(0, -1)$ yields $g(0, -1) = -4 = \frac{dw}{d\pi}(2)$, as predicted by Prop. B.8 and Prop. B.9. The line touching the dual function in $(\pi = 2, w(2) = -3)$, with a slope of -4 (gradient) is the only global overestimator of the dual function: the line tangent to the graph in that point (see Fig. B.7).

B.3.1.2 Dual Problem

The *dual problem* associated to primal problem (B.8) with constraints (B.8b,c) relaxed, is defined as the optimization problem:

$$\max_{(\pi,\lambda)\in\text{dom } w} w(\pi, \lambda) = \max_{(\pi,\lambda)\in\text{dom } w} \left\{ \min_{x\in S} f(x) + \sum_i \pi_i g_i(x) + \sum_j \lambda_j h_j(x) \right\} \qquad \text{(B.12)}$$

That means, the dual problem intends to find the multipliers that return the best possible (higher) lower bound: the one that has a dual cost as close as possible to the primal cost p^*. We denote d^* to the optimum dual cost, provided by the solution of (B.12). Note that the weak duality property states that $d^* \leq p^*$.

Recall that if S is compact, **dom** $w = \{(\pi, \lambda) : \pi \geq 0\}$. In this common case, since according to Prop B.7 the dual function is concave, we have that the dual problem involves maximizing a concave function in a very simple convex domain $\{\pi \geq 0, \lambda \in \mathbb{R}^p\}$, and thus is a convex problem which can always be efficiently solved (in polynomial time). Note that this holds even if the primal problem (B.8) is non-convex or has integer constraints.

Example B.5 The dual problem of (B.9) in Example B.1, is:

$$\max_{\pi \geq 0} -\frac{5}{4}\pi^2 + \pi$$

That has a maximum in $\pi^* = \frac{2}{5} = 0.4$. The optimal dual cost is given by: $d^* = w(0.4) = \frac{1}{5} = 0.2$.

In Part II of the book, we will show that the application of gradient algorithms to maximize the dual function, exploiting Prop. B.8 and Prop. B.9, are effective methods to, for example, create distributed algorithms for some network design problems.

B.3.1.3 Strong Duality

A problem of the form (B.8) is said to enjoy the property of *strong duality* when $d^* = p^*$. This means that the maximum of the dual problem has the same cost as the primal optimum. When a problem does not enjoy strong duality, $d^* < p^*$, and the difference $p^* - d^*$ is called the *duality gap*. Also, it can be shown that the dual function is never differentiable in the optimum when the problem has a duality gap.

Proposition B.10 ([2], p. 212) If strong duality does not hold, the dual function is not differentiable in the optimum.

If strong duality holds, the door is open to attempt solving a problem by (i) finding some optimum dual multipliers (π^*, λ^*) for it, (ii) choosing a minimizer $\tilde{x} \in \mathcal{X}^*(\pi^*, \lambda^*)$ that is an optimum solution to the primal problem. This is enabled by the following propositions.

Proposition B.11 ([2], p. 212) Let (B.8) be a problem with strong duality. Let (π^*, λ^*) be a dual optimum solution. Then, at least one $\tilde{x}$ vector among the minimizers of (π^*, λ^*) ($\tilde{x} \in \mathcal{X}^*(\pi^*, \lambda^*)$), is a primal global optimum.

Previous proposition shows up the difficulty that some minimizers $\tilde{x} \in \mathcal{X}^*(\pi^*, \lambda^*)$ may be not a primal optimum. Actually, this happens when those minimizers violate some relaxed constraints and thus are unfeasible[5]. We can rule out this difficulty, in the case when there

---

[5] If a minimizer associated to the optimum multipliers is feasible in a problem with strong duality, then it is a global optimum, since it will be a feasible solution with a cost $d^* = p^*$.

is only one minimizer, since according to Prop. B.11 it must be then feasible and a primal optimum.

Proposition B.12 Let (B.8) be a problem with strong duality, and for which the objective function f is strictly convex, g_i functions are convex, h_j functions are linear, and set S is non-empty, compact, and convex. Then, the unique minimizer associated to the dual optimum solution (π^*, λ^*) is the unique primal global optimum of (B.8).

It is important to remark that under the assumptions of previous proposition, the primal problem has a unique optimum solution x^*, but the dual problem may have multiple optimum multipliers (π^*, λ^*). It is just that any of them has x^* as its unique minimizer.

The following proposition provides a sufficient condition for a problem to enjoy strong duality and is extensively applied throughout the book.

Proposition B.13 ([8]) (Sufficient condition for strong duality). Let (B.13) be a convex problem of the form:

$$\min_x f(x), \quad \text{subject to:} \tag{B.13a}$$

$$g_i(x) \leq 0, \quad i = 1, \ldots, m \tag{B.13b}$$

$$Ax \leq b, A'x = b' \tag{B.13c}$$

$$x \in S \tag{B.13d}$$

where f and g_i functions are convex and well defined, meaning that its domain includes the feasibility set. (B.13c) represents any linear equality and inequality conditions, and S is a closed convex set, not necessarily compact. We assume (Slater condition) that there exists at least one feasible solution $\tilde{x}$, which strictly satisfies convex inequality constraints:

$$\exists \tilde{x} \text{ such that } \{A\tilde{x} \leq b; A'\tilde{x} = b', g_i(\tilde{x}) < 0, \forall i = 1, \ldots, m, \tilde{x} \in \textbf{relint}(S)\} \tag{B.14}$$

Then, problem (B.13) has the property of strong duality.

Notation **relint** (S) stands for the relative interior of S^6. Conditions (B.14) are called the *Slater conditions* and are a variation of those proposed by Morton Slater in [11]. They can be considered more a mathematical technicality than a real limitation from an engineering point of view. In particular, if a convex problem does not satisfy them, we can always modify it by replacing $g_i(x) \leq 0$ constraints by $g_i(x) \leq \epsilon$ with ϵ being such a small positive number that it means no real difference from an engineering point of view. Then, Slater conditions would be met in the modified problem. Finally, note that strong duality always holds in convex problems with linear constraints.

Example B.6 Strong duality properties hold for problem (B.9) in Example B.1, since the objective function is convex (actually strictly convex) and constraints are linear. We saw that

---

[6] A vector $x \in \mathcal{X}$ is in the relative interior of a set $\mathcal{X}$, when there exists a ball $B(x)$ centered in x, such that $B(x) \cap \text{aff}(\mathcal{X}) \subset \mathcal{X}$, where aff $\mathcal{X}$ is the smallest affine set that contains $\mathcal{X}$. Note that if x is in the interior of $\mathcal{X}$, it is also in its relative interior. As an example, the set $\{(x, y, 0) : x \geq 0, y \geq 1\}$ has empty interior, but has $\{(x, y, 0) : x > 0, y > 1\}$ as its relative interior.

the dual optimum is reached for $\pi^* = 0.4$, with optimal dual cost $d^* = w(0.4) = 0.2$. The dual optimum has only one associated minimizer $x_1 = 1 - \frac{\pi^*}{2} = 0.8, x_2 = 1 - \pi^* = 0.6$, which has to be the primal optimum. Effectively, we see that $(0.8, 0.6)$ is feasible, since it satisfies constraint (B.9b) and its primal cost equals the optimal dual cost:

$$p^* = (x_1 - 1)^2 + (x_2 - 1)^2 = 0.2^2 + 0.4^2 = 0.2 = d^*$$

B.4 Optimality Conditions

In this section, we characterize the optimum solutions of constrained optimization problems of the form:

$$\min_{x} f(x), \quad \text{subject to:} \tag{B.15a}$$

$$\pi_i : g_i(x) \leq 0, \quad i = 1, \dots, m \tag{B.15b}$$

$$\lambda_j : h_j(x) = 0, \quad j = 1, \dots, p \tag{B.15c}$$

$$x \in S \tag{B.15d}$$

We study necessary conditions and sufficient conditions for optimality under different assumptions. This family of conditions are usually named Karush–Kuhn and Tucker conditions, or KKT conditions for short, after the works of William Karush in 1939 [12], and Harold Kuhn and Albert Tucker in 1951 [13].

We will treat separately the problems with and without strong duality. When strong duality holds, it is possible to find KKT necessary optimality conditions that are also sufficient in convex problems. In simple cases, these conditions yield to close expressions solving optimization problems. For the rest, they give insight on how the optimum looks like and how it can be reached.

KKT conditions will be applied profusely throughout the book. When applied to the particular case of unconstrained differentiable convex problems, they reduce to the well known statement that the gradient of the objective function should vanish in the optimum (x^* optimum $\Leftrightarrow \nabla f(x^*) = 0$). In problems without strong duality, the KKT optimality conditions do not hold, and the insights on the optimum are weaker.

B.4.1 Optimality Conditions in Problems with Strong Duality

Let (π^*, λ^*) be a maximum of the dual function of a problem (B.15) with strong duality, and x^* a primal optimal solution of (B.15). Then, it holds that:

$$f(x^*) = w(\pi^*, \lambda^*) = \min_{x \in S} \left\{ f(x) + \sum_i \pi_i^* g_i(x) + \sum_j \lambda_j^* h_j(x) \right\}$$

$$\leq f(x^*) + \sum_i \pi_i^* g_i(x^*) + \sum_j \lambda_j^* h_j(x^*)$$

$$\leq f(x^*) \tag{B.16}$$

The first equality in (B.16) is met because of strong duality property. The next equality in the first line is the definition of the dual function. The first inequality holds since the minimum in a set S is lower or equal than the value in a particular point $x^* \in S$. The last inequality holds since $\pi_i^* g_i(x^*) \leq 0$ (as $\pi_i \geq 0$ and $g_i(x^*) \leq 0$) and $h_j(x^*) = 0$. Then, the two inequalities are equalities if and only if the problem has strong duality. This leads to the following proposition.

Proposition B.14 For a problem (B.15), we assume that there exists a primal solution x^*, and the problem dual function has a maximum in $(\pi^* \geq 0, \lambda^*)$. Then, strong duality holds for the problem, if and only if:

- *Lagrange minimization*: x^* is a minimizer of $\min_{x \in S} \{ f(x) + \sum_i \pi_i^* g_i(x) + \sum_j \lambda_j^* h_j(x) \}$, and
- *Complementary slackness*: $\pi_i^* g_i(x^*) = 0, \forall i = 1, \dots, m$.

Complementary slackness conditions appear only for inequality constraints[7]. They mean that, if x^* is an optimal solution and π_i^* an optimum multiplier for a constraint $g_i(x) \leq 0$, then:

- If the multiplier $\pi_i^* > 0$, then the constraint is satisfied with equality in the optimum ($g_i(x^*) = 0$). We say that the constraint is *tight* or *active*.
- If a constraint is *inactive*, or *loose* ($g_i(x^*) < 0$), then its associated multiplier must be zero $\pi_i^* = 0$.

Proposition B.14 is the root of the derivations of the KKT conditions elaborated on in next propositions.

Proposition B.15 (KKT necessary and sufficient conditions) Let (B.15) be a problem with strong duality, and which has at least an optimal solution, with finite cost. Then, x^* is an optimal solution of (B.15) and (π^*, λ^*) is an optimal solution of the dual problem, if and only if all of the following conditions hold:

- *Primal feasibility*: x^* satisfies all constraints in (B.15).
- *Dual feasibility*: $\pi^* \geq 0$.
- *Lagrange minimization*: x^* minimizes $\min_{x \in S} \{ f(x) + \sum_i \pi_i^* g_i(x) + \sum_j \lambda_j^* h_j(x) \}$.
- *Complementary slackness*: $\pi_i^* g_i(x^*) = 0, \forall i = 1, \dots, m$.

If we apply the KKT conditions in a convex differentiable problem where all the constraints are relaxed ($S = \mathbb{R}^n$), we have:

Proposition B.16 (KKT necessary conditions for convex differentiable problems). Let (B.15) be a problem with strong duality, and which has at least an optimal solution, with finite cost. f, g_i are convex and differentiable, h_j is linear, and $S = \mathbb{R}^n$. Then, x^* is an optimal solution of

---

[7] Note that in equality constraints, $h_j(x^*) = 0$ in the optimum and in any feasible solution. Thus it always trivially holds that $\lambda_j^* h_j(x^*) = 0$.

(B.15) and (π^*, λ^*) is an optimal solution of the dual problem, if (i) x^* is feasible, (ii) $\pi \geq 0$, (iii) complementary slackness holds, and:

$$\nabla f(x^*) + \sum_i \pi_i^* \nabla g_i(x^*) + \sum_j \lambda_j^* \nabla h_j(x^*) = 0 \qquad (B.17)$$

If f and/or g_i in Prop. B.16 are not everywhere differentiable the necessary and sufficient optimality conditions hold replacing: (B.17) by:

$$0 \in \partial f(x^*) + \sum_i \pi_i^* \partial g_i(x^*) + \sum_j \lambda_j^* \partial h_j(x^*) \qquad (B.18)$$

When applying Prop. B.18 to unconstrained problems $\{\min f(x)\}$, we have the well known relations:

- If f is differentiable, x^* is an optimum if and only if $\nabla f(x^*) = 0$.
- If f is not everywhere differentiable, x^* is an optimum if and only if $0 \in \partial f(x^*)$.

The following property can be easily derived applying KKT conditions in Prop. B.15, in the case when no constraints are relaxed.

Proposition B.17 In the problem $\min_{x \in S} f(x)$, where f is a continuously differentiable convex function and S is a non-empty, closed convex set, $x \in S$ is the global minimum if and only if:

$$(y - x)^T \nabla f(x) \geq 0, \quad \forall y \in S$$

From Prop. B.4 we know that convex problems with a strictly convex objective function, the optimum (if it exists) is unique. Still, it is possible to have multiple optimum dual solutions that have the same common primal solution x^* as their unique minimizer. The following proposition sets some conditions to guarantee that the dual optimum is unique.

Proposition B.18 (Uniqueness of the primal-dual optimal pair). Under the assumptions of Prop. B.16, if the objective function f is strictly convex, and x^* is the unique global optimum. Then, if the gradients of the active constraints in the optimum are linearly independent (x^* is then called a *regular point*), the optimal dual multipliers (π^*, λ^*) are unique. Note that the active constraints are all equality constraints and inequality constraints for which $g_i(x^*) = 0$.

Example B.7 Problem (B.9) is a convex problem with strong duality. Lagrange function is: $L(x, \pi) = (x_1 - 1)^2 + (x_2 - 1)^2 + \pi(x_1 + 2x_2 - 2)$. Lagrange minimization optimality conditions mean that:

$$\frac{\partial L}{\partial x_1} = 0 \Leftrightarrow 2(x_1 - 1) + \pi = 0$$

$$\frac{\partial L}{\partial x_2} = 0 \Leftrightarrow 2(x_2 - 1) + 2\pi = 0$$

From this we have that in the optimum: $x_2 = 2x_1 - 1$. If in the optimum $\pi = 0$, we have that $x_1 = x_2 = 1$, which is an unfeasible solution. If $\pi > 0$, then problem constraint should be tight, and thus $x_1 + 2x_2 = 2$. This yields to $x_1 = \frac{4}{5}$, $x_2 = \frac{3}{5}$, and $\pi = \frac{2}{5}$, which is the primal dual optimum. Note that the primal optimum is unique, since it is a convex problem with a strictly convex objective. Also, the optimum dual is also unique, since there is only one constraint, and thus conditions of Prop. B.18 are met.

B.4.2 Graphical Interpretation of KKT Conditions

To provide a graphical insight on KKT conditions, we focus on an optimization problem in $\mathbb{R}^2$ of the form:

$$\min_x f(x), \quad \text{subject to:} \tag{B.19a}$$

$$\pi_1 : g_1(x) \le 0 \tag{B.19b}$$

$$\pi_2 : g_2(x) \le 0 \tag{B.19c}$$

$$\pi_3 : g_3(x) \le 0 \tag{B.19d}$$

we assume that f, g_1, g_2, g_3 are differentiable, and such that KKT conditions are necessary and sufficient. We denote as $\mathcal{X}$ the set of feasible solutions of (B.19). x^* and π^* feasible vectors are primal and dual optimal if and only if:

$$\nabla f(x^*) + \pi_1^* \nabla g_1(x^*) + \pi_2^* \nabla g_2(x^*) + \pi_3^* \nabla g_3(x^*) = 0$$

$$\pi_1^* g_1(x^*) = 0, \pi_2^* g_2(x^*) = 0, \pi_3^* g_3(x^*) = 0$$

Then, we explore three possibilities:

- x^* is an interior point of $\mathcal{X}$: Fig. B.8 illustrates this case. In an interior point, the three constraints are loose, and because of complementary slackness $\pi_1^* = \pi_2^* = \pi_3^* = 0$. Then, Lagrange minimization condition becomes:

$$x^* \text{optimum} \Leftrightarrow \nabla f(x^*) = 0$$

  which are the well-known optimality conditions for unconstrained convex problems.
- x^* is a boundary point of $\mathcal{X}$, with one active constraint: Fig. B.9 illustrates the case when $g_2(x^*) = 0$, while $g_1(x^*) < 0$ and $g_2(x^*) < 0$. Then, $\pi_1^* = \pi_3^* = 0$ by complementary slackness and optimality conditions are:

$$x^* \text{optimum} \Leftrightarrow \nabla f(x^*) + \pi_2^* \nabla g_2(x^*) = 0 \Leftrightarrow -\nabla f(x^*) = \pi_2^* \nabla g_2(x^*)$$

  Vector $-\nabla f(x^*)$ is such that if we move from x^* in any direction with an angle of up to $90°$ with it, the objective function decreases (improves). In turn, if we move from x^* in any direction with an angle of up to $90°$ with $\pi_2^* \nabla g_2(x^*)$, the new point becomes unfeasible. Thus, optimality condition means that all directions that locally improve the solution, yield to unfeasible points (these are called *not admissible* directions).

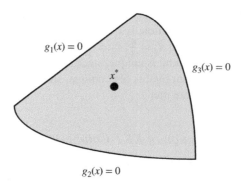

Figure B.8   Optimum point x^* in the interior of the feasibility set

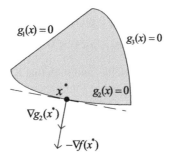

(a) x^* is an optimum point

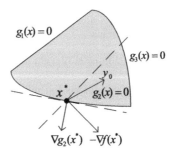

(b) x^* is not optimum. $-\nabla f(x^*)$ not proportional to $\nabla g_2(x^*)$. It is possible to find a direction y_0 that is both improving and admissible.

Figure B.9   Point x^* in the boundary of the feasibility set with one active constraint

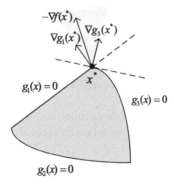

(a) x^* is an optimum point

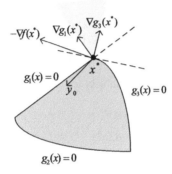

(b) x^* is not optimum. $-\nabla f(x^*)$ is not in the cone generated by $\nabla g_1(x^*)$ and $\nabla g_3(x^*)$. It is possible to find a direction y_0 that is both improving and admissible.

Figure B.10   Point x^* in the boundary of the feasibility set with two active constraints

- x^* is a boundary point of $\mathcal{X}$, with two active constraints: Fig. B.10 illustrates optimality when $g_1(x^*) = 0$, and $g_3(x^*) = 0$, while $g_2(x^*) < 0$ and thus $\pi_2^* = 0$. Lagrange minimization conditions become:

$$x^* \text{optimum} \Leftrightarrow -\nabla f(x^*) = \pi_1^* \nabla g_1(x^*) + \pi_3^* \nabla g_3(x^*)$$

The set of vectors $\pi_1^* \nabla g_1(x^*) + \pi_3^* \nabla g_3(x^*)$, for different values of $\pi_1^* \geq 0, \pi_3^* \geq 0$, is the cone drawn as a light shaded area in Fig. B.10. It is possible to show that if we move in a direction that makes an angle of up to 90° with *at least one* direction in the cone, then we move out from the feasibility set. A point is optimal if vector $-\nabla f(x^*)$ lays on this cone. Again this means that a point is optimal, if all the directions that can locally improve the solution (make an angle of up to 90° with $-\nabla f(x^*)$), are not admissible.

B.4.3   Optimality Conditions in Problems Without Strong Duality

According to Prop. B.14, if strong duality does not hold for a problem, Lagrange minimization and/or complementary slackness conditions do not hold for it either. Then, there is no primal-dual pair that satisfies KKT optimality conditions. That is, even though the problem can have a perfectly defined optimal primal solution and optimal dual solution:

- If the optimal primal x^* is a minimizer of the optimal dual (π^*, λ^*), complementary slackness will not hold ($\sum_i \pi_i^* g_i(x^*) > 0$) and we cannot assert optimality using the lower bound given by d^*, since there is a non-zero duality gap $p^* - d^* = \sum_i \pi_i^* g_i(x^*) > 0$. This is the case in Example B.8.
- It can happen that the primal optimal *is not a minimizer of any dual solution*. This is the case in Example B.9.

Still, the maximization of the dual function can help to find approximate primal solutions. This remark follows from the following proposition.

Proposition B.19 ([2], p. 213) For any $(\pi \geq 0, \lambda)$ multipliers for problem (B.15), not necessarily dual optimal, any minimizer $x^* \in \mathcal{X}^*(\pi, \lambda)$ is a global optimum of the perturbed problem (where the right-hand side of the constraints is modified):

$$\min_x f(x), \quad \text{subject to:}$$

$$g_i(x) \leq g_i(x^*), \quad i = 1, \dots, m$$

$$h_j(x) = h_j(x^*), \quad j = 1, \dots, p$$

$$x \in S$$

This property can be used in different manners. We can search for the optimal dual solution, which is always a convex program. For any minimizer of the dual optimum x^*, it holds that $g_i(x^*)$ is the coordinate of a subgradient and for those constraints with a non-zero multiplier, this subgradient tends to be small. Then, even if the minimizers associated are unfeasible for the original problem, they will be at the global optimum for perturbed problems that can be very similar to the original one (although this is not guaranteed). This can be of use, if in our problem we have some margin to accept unfeasible solutions.

Example B.8 In the integer program (B.20):

$$\min -3x_1 + x_2 - x_3, \quad \text{subject to:} \tag{B.20a}$$

$$\pi : 2x_1 + 3x_2 + 2x_3 \leq 3 \tag{B.20b}$$

$$x \in S = \{x_i \in \{0, 1\}, i = 1, 2, 3\} \tag{B.20c}$$

when constraint (B.20b) is relaxed, we have eight points (x_1, x_2, x_3) in set S. Each point produces a linear function of π: $-3x_1 + x_2 - x_3 + \pi(2x_1 + 3x_2 + 2x_3 - 3)$. The dual function for a multiplier π is the minimum among the eight lines in this particular point. This is shown in Fig. B.11 as a thick black line. The three points associated to the three lines that are minimizers at some π values are annotated next to them. See that the function is concave but not everywhere differentiable. This occurs in linear integer problems. The optimum multiplier is

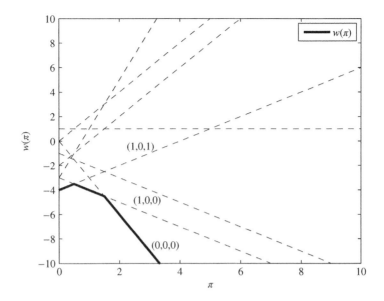

Figure B.11   Example. Dual function of problem (B.20)

$\pi^* = 0.5$, where the dual cost is maximum, $d^* = w(0.5) = -3.5$. The primal optimal solution can be obtained by brute force enumeration and is $(1, 0, 0)$, which has a primal cost of $p^* = -3$. Thus, $d^* < p^*$, the strong duality does not hold for this problem and the duality gap is $d^* - p^* = 0.5$. Note that one of the minimizers associated to the dual optimum is the optimal problem solution.

Example B.9 If we replace inequality constraint of (B.20) by an equality constraint, the dual function of the problem is the same in the $\pi \geq 0$ range and it extends in $\pi \leq 0$. The maximum of the dual function is the same $\pi^* = 0.5$ and also its maximum, $d^* = w(0.5) = -3.5$. The primal optimal solution is $(0, 1, 0)$, which has a primal cost of $p^* = 1$, and the duality gap is $d^* - p^* = 4.5$. Note that no π value has the optimum solution as a minimizer.

B.5   Sensitivity Analysis

In problems with strong duality, the optimal multipliers provide significant insight into how the optimal cost would change if the problem constraints were modified. To study this, we consider a primal problem of the form (B.15) and define its *perturbed problem* (B.21) as:

$$\min_x f(x), \quad \text{subject to:} \tag{B.21a}$$

$$\pi_i : g_i(x) \leq u_i, \quad i = 1, \ldots, m \tag{B.21b}$$

$$\lambda_j : h_j(x) = v_j, \quad j = 1, \ldots, p \tag{B.21c}$$

$$x \in S \tag{B.21d}$$

That is, the perturbed problem replaces the zeros in the right hand-side of the original constraints by constants $u = (u_i, i = 1, \ldots, m)$, $v = (v_j, j = 1, \ldots, p)$. We define the perturbation function of (B.21) and denote it $p^*(u, v)$, as the function that returns the optimal cost of (B.21). Thus, $p^*(0, 0)$ is the optimal cost of the original problem. We make $p^*(u, v) = \infty$ when the problem is unfeasible.

The following properties help us to characterize the perturbation function.

Proposition B.20 ([14], p. 10) If (B.15) is a convex problem, $p^*(u, v)$ is a convex function with respect to (u, v).

Proposition B.21 For any problem (B.15), the perturbation function is non-increasing with respect to perturbations in inequality constraints:

$$u_1 \geq u_2 \Rightarrow p^*(u_1, v) \leq p^*(u_2, v), \quad \forall v$$

Example B.10 The perturbed problem of (B.9) is given by:

$$\min(x_1 - 1)^2 + (x_2 - 1)^2, \quad \text{subject to:} \tag{B.22a}$$

$$\pi : x_1 + 2x_2 \leq 2 + u \tag{B.22b}$$

The dual function is $w(\pi) = -\frac{5}{4}\pi^2 + \pi(1 - u)$, with minimizer: $x_1 = 1 - \frac{\pi}{2}, x_2 = 1 - \pi$. The optimal multiplier is the maximum of the dual function: $\pi^* = \frac{2-2u}{5}$ if $u \leq 1$, and $\pi^* = 0$ if $u > 1$. Then:

$$p^*(u) = \begin{cases} \frac{(1-u)^2}{5}, & \text{if} \quad u \leq 1 \\ 0, & \text{if} \quad u > 1 \end{cases}$$

which is a convex and non-increasing function of u (see Fig. B.12).

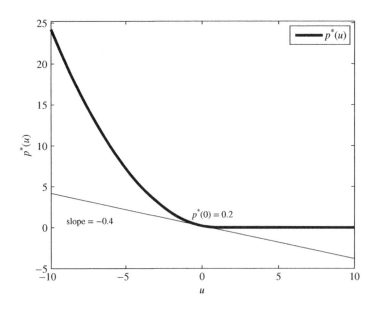

Figure B.12   Perturbation function of problem (B.22)

Proposition B.22 ([8] p. 250) Let be a perturbed problem (B.21) with strong duality, and an optimum solution x^* of finite cost. Then, (π^*, λ^*) are optimum multipliers for the problem, if and only if,

$$p^*(u, v) \geq p^*(0, 0) - \sum_i \pi_i^* u_i - \sum_j \lambda_j^* v_j, \quad \forall u \in \mathbb{R}^m, v \in \mathbb{R}^p \qquad (B.23)$$

this is equivalent to saying that $-(u, v)$ is a subgradient of the perturbation function in point $u = 0, v = 0$.

Example B.11 In the perturbed problem of (B.9), the perturbation function is differentiable. The tangent line to $p^*(u)$ in $u = 0$ is plotted. Its slope is given by minus the optimum multiplier of the unperturbed problem $\pi^* = 0.2$, as computed in Example B.5.

Inequality (B.23) provides us with *optimistic bounds* to the optimum cost when we perturb a problem with strong duality. For instance, if $\pi_i^* > 0$ for an inequality constraint and we tighten it ($u_i < 0$), the optimal cost is guaranteed to worsen in at least $|\pi_i^* u_i|$ units. However, if $\pi_i^* = 0$, losing the constraint ($u_i > 0$) will not improve the cost.

Inequality (B.23) has a global application, meaning that it holds for any perturbation $u \in \mathbb{R}^m, v \in \mathbb{R}^p$. If the dual function of the original problem is differentiable, there is only a subgradient which is a gradient. Then, the perturbation function is also differentiable in $u = 0, v = 0$, and it holds that:

$$\frac{\partial p^*}{\partial u_i} = -\pi_i^*, \quad \forall i = 1, \ldots, m$$

$$\frac{\partial p^*}{\partial v_j} = -\lambda_j^*, \quad \forall j = 1, \ldots, p$$

Then, the multipliers are a linear approximation on how the optimal cost p^* would change if we make small perturbations to the problem (e.g., this occurs in Example B.10).

B.6 Notes and Sources

Many excellent books exist covering and extending the optimization introduction in this chapter, like [1–4]. Convex programming is specifically dealt with in, for example [8], and integer programming in [9] or [10].

Many software optimization packages exist that can be used for finding numerical solutions to optimization problems, too many to cite them all. A non-exhaustive relation of mixed linear integer solvers and/or nonlinear software packages is CPLEX [15], Gurobi [16] Mosek [17], SCIP [18], GLPK [19], COIN-OR [20], or Lindo [21]. In this book we often refer to GLPK (MILP, free, open-source), IPOPT (nonlinear differentiable, free, open-source, part of COIN-OR project), and CPLEX (MILP, commercial) solvers, since these are the ones interfaced by the current version of the JOM modeling library, used in conjunction with Net2Plan in multiple exercises throughout the book. However, this is not in any case a sign of superiority with respect to the many other existing free and commercial optimization packages.

References

[1] D. G. Luenberger and Y. Ye, *Linear and Nonlinear Programming*. Heidelberg, Germany: Springer, 2008, vol. 116.

[2] M. Minoux, *Mathematical Programming: Theory and Algorithms*, ser. Wiley series in discrete mathematics and optimization. New York, NY, USA: John Wiley & Sons, Inc., 1986.

[3] L. Lasdon, *Optimization Theory for Large Systems*, ser. Dover books on Mathematics. Dover Publications, 2002.

[4] D. P. Bertsekas, *Nonlinear Programming*. Bertsekas: Athena Scientific, 1999.

[5] G. B. Dantzig, "Maximization of a linear function of variables subject to linear inequalities," *Activity Analysis of Production and Allocation*, pp. 339–347, 1947.

[6] G. J. Tee, "Khachian's efficient algorithm for linear inequalities and linear programming," *SIGNUM Newsl.*, vol. 15, pp. 13–15, March 1980.

[7] N. Karmarkar, "A new polynomial-time algorithm for linear programming," *Combinatorica*, vol. 4, no. 4, pp. 373–396, 1984.

[8] S. Boyd and L. Vandenberghe, *Convex Optimization*. New York, NY, USA: Cambridge University Press, 2004.

[9] G. L. Nemhauser and L. A. Wolsey, *Integer and combinatorial optimization*. New York, NY, USA: Wiley-Interscience, 1999.

[10] L. Wolsey, *Integer Programming*, ser. Wiley Series in Discrete Mathematics and Optimization. Wiley, 1998.

[11] M. Slater, Lagrange multipliers revisited: a contribution to non-linear programming, 1950.

[12] W. Karush, "Minima of functions of several variables with inequalities as side conditions," Master's thesis, Department of Mathematics, University of Chicago, Chicago, IL, USA, 1939.

[13] H. W. Kuhn and A. W. Tucker, "Nonlinear programming," in *Proceedings of the Second Berkeley Symposium on Mathematical Statistics and Probability*, J. Neyman, Ed. Berkeley: University of California Press, 1951, pp. 481–492.

[14] A. Geoffrion, *Duality in nonlinear programming: a simplified applications-oriented development*, ser. Memorandum (Rand Corporation). Rand Corp., 1970.

[15] Ibm ilog cplex optimization studio.

[16] Gurobi optimization.

[17] Mosek optimization.

[18] T. Achterberg, "Scip: Solving constraint integer programs," *Mathematical Programming Computation*, vol. 1, no. 1, pp. 1–41, 2009.

[19] Gnu linear programming kit.

[20] Computational infrastructure for operations research (COIN-OR).

[21] Lindo optimization.

Appendix C

Complexity Theory

C.1 Introduction

Complexity theory studies the computational complexity of the problems and the algorithms solving them. In this appendix, we are interested in transmitting the main concepts and theoretical limits to algorithm efficiency coming from this theory. and drop some rigor in the description, for the sake of brevity and clarity.

Complexity theory deals with so called *algorithmic problems*, which are those that can be coded in a computer, and for which there is no ambiguity in distinguishing between correct and incorrect answers. This leaves aside, for instance, religious or philosophical problems. An algorithmic problem p is defined by:

- A description of the possible problem input parameters, formally described in any finite alphabet (the symbol set in our computer).
- A function that relates any possible input z_p for problem p (which we call a problem *instance*), with one or more elements in a set of possible outputs: the right answers to that problem instance.

As an example, "Given a positive integer number k, determine if it is prime" is the description of a problem p. The possible inputs are integer positive integer numbers, the outputs are {yes , no}. A problem instance is "determine if 17 is prime", which has a single correct answer "yes".

Among algorithmic problems, we define *optimization problems* as those for which there can be more than one correct solution for an instance that are in some way optimal. In turn, *evaluation problems* are those for which the correct solution is always unique. Evaluation problems for which only two answers can be correct: "yes" or "no" are called *decision problems*.

A similar problem can have an optimization, evaluation, and decision version. For instance, "given a graph and two nodes in them, find the shortest path in number of hops" is an optimization problem: the shortest paths may not be unique. An evaluation version could be "given a graph and two nodes, find the number of hops in a shortest path", which has a unique solution, and "given a graph and two nodes, is there a path between them of less than five hops?" is a decision version of the problem.

Interestingly, complexity theory has shown that the difficulty in solving the optimization, evaluation, and decision versions of a problem is often similar. For this reason, many books concentrate on the analysis to decision problems. We will also do that in this appendix. However, in the last sections we will extend the view to optimization problems to study *approximation algorithms*.

C.2   Deterministic Machines and Deterministic Algorithms

Leaving the details aside[1], *deterministic machines* correspond to current computers available, based on finite memory and a processor with an instruction set that determines without any ambiguity the steps to follow. Programs using such instruction set executed in such computers are *deterministic algorithms*.

We say that a deterministic algorithm A solves a decision problem p when, for any problem instance:

- It always produces an answer.
- And the answer is always correct.

C.2.1   Complexity of a Deterministic Algorithm

There are two basic measures of complexity in deterministic algorithms:

- *Space complexity*, related to the amount of memory the algorithm needs.
- *Time complexity*, related to the time needed for the algorithm to finish.

Both time and space complexity have a formal background we skip in this appendix[2]. In the following, we focus on time complexity for its practical importance and assume that algorithms are executed in machines with sufficient storage space.

Let A be an algorithm solving a decision problem p and z be a particular problem instance. We use $t_A(z)$ to denote the computation time of A for the problem instance z. This computation time depends on:

---

[1] The formal definition of deterministic machines is based on the celebrated *Turing machine*. In a Turing machine, input data is stored in an unbounded *tape* composed of cells linearly arranged, each cell containing a symbol of a finite alphabet. Also, the machine has an external finite *memory* which it can access at any time, storing the *state* of the machine. Initially, the tape has the problem input codified in some form in the cells 0,1, ... The rest of the cells contain a special blank symbol. The machine is initially reading the cell 0. In each execution step, the Turing machine (i) reads the cell and (ii) decides the next action to do depending on the read value and the machine state. The function defining the actions for each read value and state is actually the Turing algorithm or *program*. The possible actions are (all possible at the same time): (i) write a new symbol in the read cell, (ii) modify the machine state, and (iii) move to the left or right neighboring cell. The algorithm ends when it saves in the state a special code for *halting*. It is assumed that at that moment, the answer has been written in the tape in cells 1,2, ... coded in any suitable form.

[2] In the Turing machine definition in the previous note, the space complexity relates to the number of cells the algorithm needs to occupy to run, while the time complexity is related to the number of instructions executed before the algorithm halts.

- The chosen computer where the algorithm is run.
- The chosen programming language.
- How the algorithm is actually implemented.

Complexity theory targets measuring the complexity of any existing and non-existing deterministic algorithm. In this context, the previous three dependencies are arbitrary and have no theoretical interest. The theory gets rid of them by assuming that deterministic machines have a standard instruction set capable of doing the elementary operations appearing in all programming languages: arithmetic, assignment, memory access, conditional jumps, and so on. Then, a *uniform cost model* is assumed, which means that all of these operations take one time unit to be completed. Naturally, this is not true since we know, for example. that some arithmetic operations may take a longer time to complete than others (e.g., divisions versus sums). Still, these differences are unimportant when a rough complexity measure is pursued.

C.2.2 Worst-Case Algorithm Complexity

Complexity theory is interested in a measure of time complexity that depends on the so-called problem instance *size* (denoted by n), but not the particular sample of that size. As an example, in an algorithm A ordering lists, we pursue a value of time complexity $t_A(n)$ that condenses how long does it take to order lists of n elements.

But, how to define, for example $t_A(100)$, the complexity of A for ordering lists of $n = 100$ elements, since it is evident that some lists (e.g., those almost already ordered), can be processed faster than others? Three main alternatives exist:

- *Empirical analysis*: We enumerate some (but not all) lists of 100 elements, probably the most representative ones in our case study, and average the time complexity of A for them.
- *Average-case analysis*: We try to devise the mathematical expectation of the running time for all the lists of size 100.
- *Worst-case analysis*: In this case, $t_A(100)$ is the longest running time among all the lists of 100 elements.

The worst-case analysis is by far the most accepted algorithm complexity measure, since empirical tests are arbitrary, and average-case analysis are commonly impossible to compute. Then, in the sequel we adopt the worst-case analysis, and define $t_A(n)$ as:

$$t_A(n) = \sup \{t_A(z) : z \text{ problem instance of size } n\}$$

C.2.3 Asymptotic Algorithm Complexity

Computing precise worst-case values of $t_A(n)$ for an algorithm is still a tedious and difficult task. Fortunately, complexity theory is not much focused on computing $t_A(n)$ accurately, but in estimating it approximately: the inaccuracies already brought by the uniform cost model makes unproductive investing effort in a precise computation of $t_A(n)$. In addition, complexity theory puts the focus on how $t_A(n)$ evolves when $n \to \infty$. That is, in large-scale problem instances, where the efficiency differences of the algorithms have a higher practical importance.

Three notations of $t_A(n)$ are used to convey the approximate complexity of the algorithms in large-scale instances:

- $\mathcal{O}$ notation: An algorithm A is of complexity $\mathcal{O}(g(n))$, and we denote it $t_A(n) = \mathcal{O}(g(n))$ (this reads t_A is big-O of g) when there exists a constant $c < \infty$ such that:

$$\lim_{n \to \infty} \frac{t_A(n)}{g(n)} \leq c$$

- Ω notation: An algorithm A is of complexity $\Omega(g(n))$ and we denote it $t_A(n) = \Omega(g(n))$ (this reads t_A is big-omega of g) when there exists a constant c such that:

$$\lim_{n \to \infty} \frac{t_A(n)}{g(n)} \geq c$$

In summary, $t_A(n) = \mathcal{O}(g(n))$ means that $g(n)$ is an upper bound to the complexity growth of $t_A(n)$, and $t_A(n) = \Omega(g(n))$ that $g(n)$ is a lower bound (both asymptotically, and up to a multiplying constant).

- Θ-notation: In turn $t_A(n) = \Theta(g(n))$ (t_A is big-theta of g) means that the growth rate is precise up to a multiplying constant:

$$t_A(n) = \Theta(g(n)) \Leftrightarrow \{t_A(n) = \mathcal{O}(g(n)) \quad \text{and} \quad t_A(n) = \Omega(g(n))\}$$

Note that the notation $t_A(n) = \mathcal{O}(g(n))$ (and the same for Ω and Θ) is prone to confusion, since it suggests that $\mathcal{O}(g(n)) = t_A(n)$ also holds, while this reversed relation is not defined.

Example C.1 If $t_A(n)$ is a polynomial of degree $k(t_A(n) = \sum_{i=0}^{k} a_i n^i)$, then $t_A(n) = \mathcal{O}(n^k)$, since the lower order terms of the polynomial become unimportant when $n \to \infty$, and are eliminated by $\mathcal{O}$ notation.

The growth rates more frequently found in the algorithms are:

- Constant running time: $t_A(n) = \mathcal{O}(1)$.
- Logarithmic complexity: $t_A(n) = \mathcal{O}(\log n)$.
- Linear complexity: $t_A(n) = \mathcal{O}(n)$.
- Polynomial complexity (order $k \in \mathbb{N}$): $t_A(n) = \mathcal{O}(n^k)$. For $k = 2$ we read it as quadratic complexity, $k = 3$ cubic complexity.
- Superpolynomial complexity: $t_A(n) = \Omega(n^k)$, for every k. That is, it grows faster than any polynomial.
- Subexponential complexity: $t_A(n) = \mathcal{O}(2^{n^\epsilon})$, for any $\epsilon > 0$. That is, it grows slower than any exponential function.
- Exponential complexity: $t_A(n) = \Omega(2^{n^\epsilon})$ for some $\epsilon > 0$. That is, at least as fast as an exponential function.
- Factorial complexity: $t_A(n) = \mathcal{O}(n!)$. Recall that $n! \approx \sqrt{2\pi n}(\frac{n}{e})^n$, which, for example grows faster that 2^n, since both the base and the exponent grow with n.

When computation times depend on two or more parameters, $\mathcal{O}$, Ω, and Θ notations are still applicable. For instance, the complexity of the famous Dijkstra algorithm for finding the shortest path between two nodes in a network, is $t_A(n, m) = \mathcal{O}(m + n \log n)$, where n and m are the number of network nodes and links, respectively. This means that there is a constant c such that $t_A(n, m)/(m + n \log n) \leq c$ when n and m tend to infinity.

C.2.4 Complexity is a Real Barrier

At this point, it is important to stress the practical differences of having algorithms of different complexity types, in special the differences between polynomial and exponential algorithms. The goal of this section is removing any hope of the type:

"I have an exponential algorithm for this problem, and I guess I will not be able to solve large-scale instances in my home computer. But what if we sum the resources of millions of computers? I think that with enough money, and enough time I can solve any problem instance..."

Table C.1 illustrates why this belief is false, showing the worst-case number of operations for different algorithms and problem sizes n. In the left columns, polynomial and sub-polynomial complexities are shown. First, logarithm complexity numbers reflect that this is an extremely favorable complexity growth. Polynomial growths may be not that nice, for example n^3 results in 10^{12} operations in instances of size $n = 1000$. Still, polynomial algorithms are tagged as "easy" algorithms in complexity theory, while exponential algorithms are tagged as "hard". The reason is made evident when we see the running time growth in exponential and factorial algorithms, for example $n = 1000 \Rightarrow 2^n \approx 10^{300}$. Now, the fact that exponential complexity issue cannot be addressed with more time or computing facilities is clear comparing these numbers with other large quantities like:

- The estimated number of atoms in the Universe $\approx 10^{80}$.
- If a computer executing 10^{12} instructions per second was running since the start of the Universe up until today ($\approx 15,000$ million years), it would execute $\approx 10^{30}$ operations.
- If every atom of the Universe was a computer like the one before, running from the start of the Universe, the overall computation effort would be $\approx 10^{80} \times 10^{30} = 10^{110}$ operations.

Table C.1 Case studies in Chapter 9

n	$\log n$	$\sqrt{n}$	n^2	n^3	2^n	$n!$
10	3.32	3.16	10^2	10^3	10^3	3.6×10^6
100	6.64	10	10^4	10^6	1.27×10^{30}	9.3×10^{157}
1000	9.97	31.62	10^6	10^9	10^{301}	4×10^{2567}
10000	13.29	100	10^8	10^{12}	10^{3010}	2.8×10^{35659}

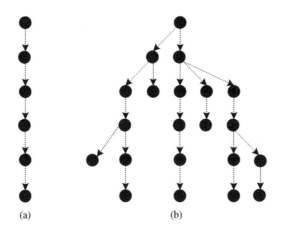

(a)   (b)

Figure C.1   (a) Execution path in a deterministic machine. (b) Tree of execution paths created by a non-deterministic machine

C.3   Non-Deterministic Machines and Non-Deterministic Algorithms

Different textbooks provide diverse but equivalent descriptions of *non-deterministic machines* (NDM), that try to make more understandable its operation. A somewhat informal but simple description is that a NDM is a computer that in each iteration can (i) execute the following instruction as a conventional (deterministic) computer, or (ii) execute a *clone* instruction that creates a finite number of replicas of the computer and its state, but each jumping to a different selected next instruction. After such cloning, each replica can clone again and again in future computation steps, but two computers cannot communicate with each other[3].

The difference between deterministic and non-deterministic machines is illustrated in Fig. C.1. The left-hand figure represents the execution of a deterministic machine, with a single thread that in each step executes the action determined by previous computations. In turn, the cloning feature of NDMs permits emulating the parallel execution of multiple threads, each potentially executing different routines, without any communication among them. We call each of the paths starting in the common root and ending in a leaf as a *computation path*. For instance, Fig. C.1 illustrates a NDM with seven computation paths.

An algorithm for a NDM, including such special *clone* instructions, is referred to as a *non-deterministic algorithm* (NDA). It is important to note that any problem solvable by a deterministic machine can also be solved by a non-deterministic machine, since deterministic algorithms with no clone instructions can be executed in NDMs. Interestingly, this works also the other way around: an NDA can be executed in a deterministic machine by sequentially executing, instead of in parallel, the many (but finite) computation paths. Then, we can state that there is no difference in the set of problems that deterministic and non-deterministic machines can solve, it is just that NDMs could be much faster thanks to their arbitrary parallel processing capabilities.

---

[3] For those familiar with UNIX programming, the *clone* operation emulates a *fork* instruction with an arbitrary number of child processes, without any communication among them nor the father process.

C.3.1   Complexity of a Non-Deterministic Algorithm

Prior to define the complexity of NDAs, we should revisit the definition of the sentence "an algorithm A solves a problem p". Recall that, in the deterministic case, this means that the algorithm always produces the correct answer. However, in the non-deterministic case this definition is relaxed.

Definition C.1 We say that a non-deterministic algorithm A solves a decision problem p, when for those instances when the correct answer is *yes*, at least a computation path ends answering it correctly, while the others may wrongly answer "no". Instead, for those problem instances when the correct answer is *no*, all the computation paths answer "no".

The reader may be shocked on the previous definition, which contradicts the intuitive meaning of "solving a problem". Also, we see that *yes* problem instances are treated differently to *no* instances. We will comment on this asymmetry later. A second important definition follows:

Definition C.2 We define the running time of a NDA A in a problem instance z ($t_A(z)$) as the maximum number of computation steps needed by any path. Then, we say that A runs in polynomial time if the running time of A for an instance of size n is bounded above by a polynomial on n.

As a final remark, we stress that non-deterministic machines and algorithms are mathematical tools used in complexity theory, but are non-realizable concepts: a machine with such unlimited parallel execution capabilities *cannot be built*. However, it can be made equivalent to a non-parallel machine that in every *clone* instruction, instead of making replicas of the computer, *randomly* chooses one of the branching computation paths and continues with it. That is, it possesses an special instruction ? that permits implementing pseudocodes like:

$$\textbf{if ? then } \texttt{program1} \textbf{ else } \texttt{program2}$$

In this context, we say that an NDA solves a problem when the probability that such randomized machine correctly solves a *yes* instance is higher than zero and correctly solves *no* instances with probability one.

C.4 $\mathcal{P}$ and $\mathcal{NP}$ Complexity Classes

We now define $\mathcal{P}$ and $\mathcal{NP}$ as the set of decision problems that can be solved in polynomial time by deterministic and non-deterministic machines, respectively.

Definition C.3 A decision problem p is of class $\mathcal{P}$ if there exists a deterministic algorithm A that solves p, with complexity $t_A(n) = \mathcal{O}(n^k)$, for some k.

Definition C.4 A decision problem p is of class $\mathcal{NP}$ if there exists a non-deterministic algorithm A that solves p, with complexity $t_A(n) = \mathcal{O}(n^k)$, for some k.

We recall that for being in $\mathcal{P}$ the problem must be solvable in polynomial time by standard computers, answering always correctly to *yes* and *no* instances. However, for being in $\mathcal{NP}$

we just need to show that a non-deterministic algorithm exists that answers *yes* at least some times to *yes* instances and never fails to answer correctly *no* instances.

Clearly, $\mathcal{P} \subseteq \mathcal{NP}$, since a polynomial deterministic algorithm solving a problem can also be executed in an NDM. It is conjectured, but still not proved, that the other relation does not hold and that: $\mathcal{NP} \nsubseteq \mathcal{P}$ and thus $\mathcal{P} \neq \mathcal{NP}$. We will come back to this question later, one of the most intriguing dilemmas in computer science.

It is customary to qualify $\mathcal{P}$ problems as those that can be *solved* in polynomial time, while $\mathcal{NP}$ problems are those that can be *verified* in polynomial time. We provide an intuitive explanation of this with an example. Let us focus on the decision problem:

> "Given a network $\mathcal{G}(\mathcal{N}, \mathcal{E})$, with a cost c_e associated to each link $e \in \mathcal{E}$, there exists a unidirectional ring traversing all the nodes, with an aggregated cost of the traversed links lower than 100?"

We denote the previous problem TSP, since it is a version of the classical Traveling Salesman Problem (TSP)[4]. The size of the problem is defined by the number of nodes and links in the network. To date, it is not known any polynomial deterministic algorithm solving it. Then, we cannot assert whether TSP is in $\mathcal{P}$ or not. However, we know that TSP is in $\mathcal{NP}$. For this, let us imagine a NDA working in two consecutive steps, called *guess* and *verify*.

1. *Step 1. Guess.* Using the non-deterministic cloning functionalities, we create a computation path for each possible sequence of $|\mathcal{N}|$ links. This is equivalent as saying that we create a random sequence of $|\mathcal{N}|$ links, and store it in memory.
2. *Step 2. Verify.* The outcome of previous step is verified. It checks if (i) the links are consecutive (one ends where the next starts), (ii) do not traverse a node twice, and (iii) their cost sums to less than 100. If so, the algorithm returns *yes*. If not, it returns *no*.

We see that previous algorithm *solves* TSP in the (awkward) NDM sense:

- If the correct answer is *yes*, then at least one ring of cost less than 100 exists, and one computation path answers *yes*: the one that guesses it.
- If the correct answer is *no*, all guesses yield to computation paths returning *no*.

The guess step can be executed in polynomial time in NDMs (or in exponential time in deterministic machines). The verify step in each computation path does not need special NDM capabilities and actually could be executed in polynomial time by a standard computer. Then, we can safely say that TSP is polynomial in NDMs and thus TSP $\in \mathcal{NP}$.

More generally, a problem is said to be in $\mathcal{NP}$, when the previous guess-verify approach is applicable. Then, it is possible to *guess* candidate solutions (called *certificates*) that could be checked by a polynomial *verifier* algorithm. The name certificate comes since if positively verified, a certificate is a proof that the correct answer is *yes* (although a non-verification does not prove that the correct answer is *no*).

Now we can see the implications of the asymmetry treating differently *yes* and *no* problem instances. Let us imagine the so-called complement of the TSP problem described above, which we denote as co-TSP:

---

[4] The name TSP comes after the classical problem description, where a salesman has to visit a set of $\mathcal{N}$ cities exactly once, starting and ending in the same city, traversing the minimum distance.

> "Given a network $G(\mathcal{N}, \mathcal{E})$, with a cost c_e associated to each link $e \in \mathcal{E}$, all the unidirectional rings in it traversing all the nodes, have an aggregated cost equal or higher than 100?"

the answer to CO-TSP is *no* for those instances when TSP answers *yes* and the other way around. This type of reversed decision problems are called *counterexamples*. The difference lies in that we cannot apply the guess-verify strategy described before. If the correct answer is *no* and we are presented with a candidate ring with a cost higher than 100, we cannot answer *yes*. Intuitively TSP and CO-TSP are essentially different versions and it is not clear if CO-TSP is actually an $\mathcal{NP}$ problem. Actually, those decision problems whose complement is $\mathcal{NP}$ are called co-$\mathcal{NP}$ problems and it is conjectured (although still not proved) that $\mathcal{NP} \neq$ co-$\mathcal{NP}$, and that for instance CO-TSP, is not in $\mathcal{NP}$.

C.5   Polynomial Reductions

Given two decision problems p and q, we are interested in ordering them according to *how complex* they are to solve. That is, determining if p is more or less complex than q, or if their complexities are similar. For this, polynomial complexity will play the role of stating the granularity to consider two complexities similar.

Definition C.5 Given two decision problems p and q, we say that p can be polynomially reduced to q ($p \leq_P q$) if:

1. There exists a polynomial deterministic algorithm $A_{p \to q}$ that transforms every input z_p to p into an input to q, $z_q = A_{p \to q}(z_p)$.
2. With A_q being any algorithm solving q, the combined algorithm that takes a z_p instance, transforms it with $A_{p \to q}$ and then applies A_q to it, solves problem p.

Figure C.2 illustrates this ordering relation. The key is that if $p \leq_P q$, there is an indirect form of solving p using A_q as a subroutine, after a *fast* (polynomial time) transformation of the input using $A_{p \to q}$.

Using the sign $\leq_P$ in the relation suggests on purpose that the complexity of p is somewhat smaller or equal than that of q. Actually this intuition is correct, and $p \leq_P q$ can be read as "p is easier or similar to q". Actually, it holds that if q is polynomial (easy), and $p \leq_P q$, then p is also easy:

$$p \leq_P q \Rightarrow \{ \text{if } q \in \mathcal{P} \text{ then } p \in \mathcal{P} \}$$

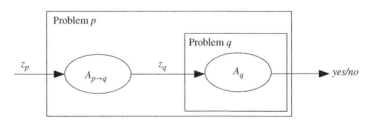

Figure C.2   Polynomial reduction, p reduces to q ($p \leq_P q$)

The proof is straightforward: if q is a polynomial problem and A_q a polynomial algorithm for it, then the indirect algorithm combining $A_{p \to q}$ and A_q solves p in polynomial time.

Definition C.6 If $p \leq_P q$ and $q \leq_P p$ we say that p and q problems are polynomially equivalent and we denote it as $p \equiv_P q$. Then:

$$p \equiv_P q \Rightarrow \{q \in \mathcal{P} \Leftrightarrow p \in \mathcal{P}\}$$

C.5.1 A Polynomial Time Reduction Example

Polynomial time reductions are important to compare the complexity of problems that may look quite different. The following example helps to clarify this concept. Let $\mathcal{G}(\mathcal{N}, \mathcal{E})$ be a graph where links are bidirectional. A subset of nodes $\mathcal{N}' \subset \mathcal{N}$ is:

- An *independent set*, when no two nodes in $\mathcal{N}'$ are connected by a link.
- A *clique*, when all the node pairs in $\mathcal{N}'$ are connected by a link.

The independent set decision problem (`ISet`) reads: given a graph $\mathcal{G}$ and a number k, $\mathcal{G}$ has an independent set of size k?. Equivalently, the clique problem (`Clique`) reads: given a graph $\mathcal{G}$ and a number k, $\mathcal{G}$ has a clique of size k?.

We can show that `ISet` $\leq_P$ `Clique` as follows:

- Given an input graph $\mathcal{G}$, we transform it into the complement graph $\mathcal{G}'$: a link between n_1 and n_2 exists in $\mathcal{G}'$ if and only if it does not exist in $\mathcal{G}$. This transformation is fast (polynomial time).
- Then, we use $\mathcal{G}'$ and k as an input to `Clique`.

The polynomial reduction holds since an independent set in $\mathcal{G}$ is a clique in $\mathcal{G}'$. Figure C.3 illustrates this with an example.

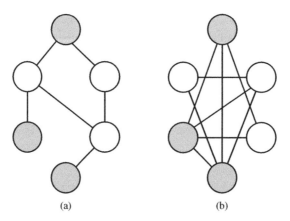

(a) (b)

Figure C.3 Example equivalence between independent sets and cliques. Gray nodes highlight (a) an independent set of $\mathcal{G}$, (b) a clique of $\mathcal{G}'$

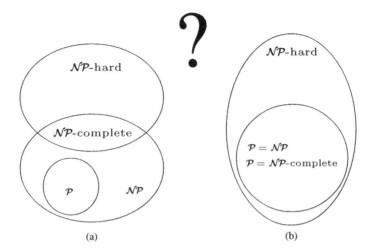

Figure C.4 The $P = \mathcal{N}P$ dilemma. (a) $P \neq \mathcal{N}P$, the widely accepted conjecture, (b) $P = \mathcal{N}P$, then $\mathcal{N}P$-complete problems are all P

C.6 $\mathcal{N}P$-Completeness

Definition C.7 A decision problem p is $\mathcal{N}P$-hard when every $\mathcal{N}P$ problem is easier or similar to p.

$$q \text{ is } \mathcal{N}P\text{-hard} \Leftrightarrow \{\forall p \in \mathcal{N}P, p \leq_P q\}$$

Definition C.8 A decision problem p is $\mathcal{N}P$-complete when it is $\mathcal{N}P$-hard and $p \in \mathcal{N}P$.

That is, $\mathcal{N}P$-hard problems are those which are equally or more difficult than any $\mathcal{N}P$ problem, and thus $\mathcal{N}P$-complete problems are the hardest among $\mathcal{N}P$.

From the definition of $\mathcal{N}P$-completeness, it becomes clear that if a single $\mathcal{N}P$-complete problem q can be solved in polynomial time, then every $\mathcal{N}P$ problem can also be solved in polynomial time, and thus $P = \mathcal{N}P$. This is probably the most famous dilemma in computer science[5]. Figure C.4 illustrates it. The general belief is that $P \neq \mathcal{N}P$ and thus that $\mathcal{N}P$-complete problems cannot be solved in polynomial time. What is absolutely true is that if a problem p is shown to be $\mathcal{N}P$-complete, we know that: (i) no one in the history of computing has ever found a polynomial algorithm for it and (ii) many smart people thinks that such an algorithm does not exist. This is actually useful information to know.

Since 1971, numerous problems in engineering, mathematics science and daily live have been shown to be $\mathcal{N}P$-complete. It was at that moment when A. S. Cook [1] (and later Levin in [2], using a different approach) found the first $\mathcal{N}P$-complete problem. This was the so-called satisfiability (SAT) problem, that reads:

SAT: Let $x_1, \cdots, x_n$ be a set of Boolean variables, that can be either true or false. A *literal*, is a variable x_i or its negation $\bar{x}_i$, and a *clause* is an arbitrary set

---
[5] The dilemma $P \neq \mathcal{N}P$ is stated as a Millennium Problem by the Clay Mathematics Institute, and a correct solution to it is awarded with a million dollar prize.

of literals linked by OR operations[6]. Then, given a set of m clauses, is there any true/false assignment to the n variables, that make all the clauses evaluate to true?

After showing that SAT is $\mathcal{N}\mathcal{P}$-complete, proving that other problem q is $\mathcal{N}\mathcal{P}$-complete does not need to show that $p \leq_p q$, for all $q \in \mathcal{N}\mathcal{P}$. Instead, it is enough to show that SAT is easier or similar than q (SAT$\leq_p q$), or, as the list of $\mathcal{N}\mathcal{P}$-complete problems found grows, it is enough to show that q is harder than one other $\mathcal{N}\mathcal{P}$-complete problem.

C.6.1   An Example Proving $\mathcal{N}\mathcal{P}$-Completeness for a Problem

As an example, we show in this section that ISet decision problem is $\mathcal{N}\mathcal{P}$-complete, by proving that ISet can be used to solve SAT (SAT$\leq_p$ISet)[7]. The transformation of the SAT input (variables and clauses) into a graph where to apply ISet is as follows:

- Build a graph $\mathcal{G}$ with one node per each literal in each clause.
- Fully connect all the nodes of the same clause.
- Add a link between any node pairs of the form $(x_i, \bar{x}_i)$.

SAT is solved by the question: is there an independent set of size m in graph $\mathcal{G}$? The hints to prove that are: (i) any independent set of $\mathcal{G}$ is composed of nodes *in different clauses*, since nodes in the same clause are fully connected among them and (ii) the nodes in the independent set represent literals that do not conflict each other. Then, a literal x_i ($\bar{x}_i$) in the independent set means that $x_i = true$ ($x_i = false$) in the SAT solution. Figure C.5 illustrates this with an example.

Finally, note that since ISet $\leq_p$ Clique, it immediately follows that Clique is also $\mathcal{N}\mathcal{P}$-hard.

C.7   Optimization Problems and Approximation Schemes

To this point, we have focused on decision problems that produce a yes/no answer. Instead, network design problems are *optimization problems*, which target the computation of a solution consisting of a vector of values that are optimal in a certain sense. Studying decision problems was motivated by the fact that, for most of the problems of practical interest, their optimization version is polynomial if and only if the decision version is.

As an example, let us focus on Min-TSP, the optimization version of TSP problem:

- Min-TSP: Given a set of nodes $\mathcal{N}$ and a non-negative integer cost c_{ij} for each possible bidirectional link between each node pair, compute the minimum cost ring connecting all the nodes
- TSP: For the same input as before, there exists a ring of cost less than C?

---

[6] For instance a clause $(x_1$ OR x_3 OR $\bar{x}_4)$ is composed of literals $x_1, x_3, \bar{x}_4$. The clause evaluates true when x_1 is true, x_3 is true, or x_4 is false.
[7] This actually proves that ISet is $\mathcal{N}\mathcal{P}$-hard. Proving that ISet is in $\mathcal{N}\mathcal{P}$ is easy by finding a guess-verify algorithm for it. This is left as an exercise for the reader.

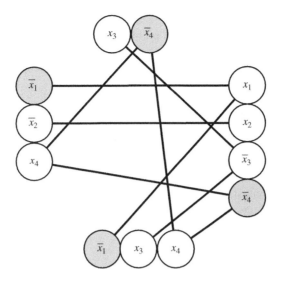

Figure C.5 Example of reduction of SAT to ISet. Representation of the problem: $(\bar{x}_1\,\text{OR}\,\bar{x}_2$ OR $x_4)$AND $(x_3\,\text{OR}\,\bar{x}_4)$AND $(x_1\,\text{OR}\,x_2\,\text{OR}\,\bar{x}_3\,\text{OR}\,\bar{x}_4)$AND $(\bar{x}_1\,\text{OR}\,x_3\,\text{OR}\,x_4)$. Links among the nodes of the same clause are not shown. In gray, an independent set that reflects a SAT solution: x_1 = false, x_4 = false, x_2, x_3

It is trivial that if Min-TSP is polynomial, we can answer TSP just finding the minimum cost ring, and checking if its cost is higher or not than c. In turn, if TSP is polynomial, we can find the minimum cost ring by a two-stage approach, where each of the two consecutive stages are polynomial[8]:

1. We iterate increasing values of c and use TSP to check if there is a ring of such cost. We start with c equal to a ring cost lower bound, for instance, the sum of the $\mathcal{N}$ cheapest links and stop when the first (minimum) c value with a *yes* answer is found.
2. For the minimum c value obtained, we find a minimum solution by removing the network links one by one. After each removal, we check if the resulting graph still has a ring of cost c. If no, we add the link again and continue with the next link. The process ends when there are exactly $|\mathcal{N}|$ links in the graph, which form a minimum cost ring.

C.7.1   The $\mathcal{NPO}$ Class

It is time to more formally define optimization problems. We use subindex z to denote an instance to the problem. In the Min-P optimization version, a problem instance z takes the form $\min f_z(x), x \in \mathcal{X}_z$, where f_z and $\mathcal{X}_z$ are the objective function and feasibility set, respectively. The Max-P version just turns min into max. In the Min-TSP example before, each problem instance z is defined by a particular set $\mathcal{N}$ and c_{ij} values. A feasible solution for z is a set of links that are a ring and the objective function sums the c_{ij} costs of the links in it.

---

[8] More formally, we say that Min-TSP is polynomial-time Turing reducible to TSP.

We are interested in replicating in optimization problems, the role that $\mathcal{NP}$ class had in decision problems. We recall that in $\mathcal{NP}$ problems tentative solutions could be *verified* in polynomial time. We define $\mathcal{NPO}$ class of optimization problems as an analogue to $\mathcal{NP}$, where the feasibility check and objective function evaluation of a tentative solution now play the verification role.

Definition C.9 An optimization problem p belongs to complexity class $\mathcal{NPO}$ (non-deterministic polynomial time optimization problem) if for each instance z, given any tentative solution x, it is possible to (i) check in polynomial time if x is feasible ($x \in \mathcal{X}_z$) and (ii) evaluate in polynomial time the objective function in x ($f_z(x)$).

In the optimization context, we also use letter $\mathcal{P}$ to denote those problems in $\mathcal{NPO}$ that can be solved in polynomial time.

C.7.2   Approximation Algorithms

The theory of $\mathcal{NP}$-completeness has helped to draw limits on what we cannot do in polynomial time in decisions problems. As we have seen, similar limits apply to the optimization problem versions. However, in optimization problems we have a workaround on these limits if we contempt with solutions that are *not optimal*, but are *close enough*. For instance, in the Min-TSP problem, we may be interested in finding rings with a cost close to optimal, which can be computed in *polynomial time*.

Approximation algorithms that run in polynomial time for difficult optimization problems are a useful practical alternative. In this section we will expose different forms of approximation algorithms, and classify optimization problems according to how *approximable* they are in polynomial time. We will see that unless $\mathcal{P} = \mathcal{NP}$, multiple problems of interest in network design can only be polynomially approximable to a certain limit. In other words, finding better approximation than this limit is $\mathcal{NP}$-hard.

Given a problem instance z, we denote f_z^* the optimum value of such an instance and $f_z(x)$ the value of a feasible solution x. We measure the goodness of approximations using the *approximation ratio* concept.

Definition C.10 Let x be a feasible solution of a problem instance z, we define the approximation ratio of x, $r_z(x)$ as:

- $r_z(x) = f_z(x)/f_z^*$ for minimization problems.
- $r_z(x) = f_z^*/f_z(x)$ for maximization problems,

We restrict ourselves to problems where the solution values are always positive ($f_z(x) > 0$) and thus approximation ratios are positive numbers higher or equal to one. In particular, $r_z(x) = 1$ means that the solution x is optimal and higher values of $r_z(x)$ reflect worse approximations.

An equivalent form of writing the approximation ratio is the ϵ notation. When a solution x has an approximation ratio $r_z(x)$, we say that it is an ϵ-approximation, where $\epsilon = r_z(x) - 1$. For instance, if the optimum solution to a Min-TSP instance is 100, a suboptimal solution of cost 120 is a 0.2-approximation.

Definition C.11 Let A be a polynomial time approximation algorithm and $r_z(A)$ denote the approximation ratio of the solution produced by A to a problem instance z. We define the *worst-case approximation ratio* $r_A(n)$ for problem instances of size n, as:

$$r_A(n) = \sup \{r_z(A), z \text{ an instance of size } n\}$$

We are primarily interested in those approximation algorithms whose ratio $r_A(n)$ (or equivalently its $\epsilon = r_A(n) - 1$) is independent on the problem size, in opposition to those whose approximation grows with n, and thus gets worse and worse for large problems. In particular, $\mathcal{APX}$ class contains the problems with constant factor approximations.

Definition C.12 The complexity class $\mathcal{APX}$ contains all optimization problems in $\mathcal{NPO}$ that have an ϵ-approximation algorithm, for at least one ϵ.

As an example, we briefly describe an ϵ-approximation algorithm for the $\mathcal{NP}$-hard problem called `Min-VertexCover`, which given a graph $\mathcal{G}(\mathcal{N}, \mathcal{E})$, computes a minimum size subset of its nodes ($\mathcal{N}' \subset \mathcal{N}$), such that any link in $\mathcal{E}$ has at least one of its ends in $\mathcal{N}'$ (is *covered* by $\mathcal{N}'$). Algorithm 28 illustrates the approximation.

Algorithm 28 `Min-VertexCover` 1-approximation

1: *Initialization*: Set $\mathcal{N}' = \emptyset$
2: **while** $\exists e \in \mathcal{E}$ not covered by $\mathcal{N}'$
3:     Add both e end nodes to $\mathcal{N}'$
4: **return** $\mathcal{N}'$

To show that Algorithm 28 is an 1-approximation, we see that if the resulting cover has k nodes, it is because the original graph $\mathcal{G}$ has at least $k/2$ links without any common node. Then, there cannot be any other covering with less than $k/2$ nodes.

We define now two so-called approximation schemes, which are algorithms for which we can choose any approximation ratio $\epsilon > 0$.

Definition C.13 A polynomial-time approximation scheme (PTAS) for an $\mathcal{NPO}$ problem is an algorithm A that includes as an input the approximation ratio ϵ, and for every $\epsilon > 0$ and every problem instance, A produces an ϵ-approximation with a time complexity that is polynomial with respect to instance size. The complexity class $\mathcal{PTAS}$ contains all $\mathcal{NPO}$ problems for which there is at least a PTAS.

Definition C.14 A fully-polynomial-time approximation scheme (FPTAS) for an $\mathcal{NPO}$ problem is a PTAS with a complexity that is polynomial with respect to both the problem instance size and with respect to the approximation quality $1/\epsilon$. The complexity class $\mathcal{FPTAS}$ contains all $\mathcal{NPO}$ problems for which there is at least a FPTAS.

Differences between PTAS and FPTAS are evident in an example like this. Let A and A' be two PTAS of time complexities given by $t_A(n, \epsilon) = n^2 2^{\frac{1}{\epsilon}}$, and $t_{A'}(n, \epsilon) = \frac{n^2}{\epsilon}$. Second algorithm is a FPTAS while the first is not. If we are pursuing approximating a problem instance of size $n = 100$, with a precision of $\epsilon = 0.001$:

- Approximation using PTAS A would be intractable, since $t_A(n, \epsilon) = n^2 2^{\frac{1}{\epsilon}} \approx 10^{305}$.
- Approximation using FPTAS A' would be tractable, with $t_{A'}(n, \epsilon) = \frac{n^2}{\epsilon} = 10^7$.

From previous definitions, it easily results in:

$$\mathcal{P} \subseteq \mathcal{FPTAS} \subseteq \mathcal{PTAS} \subseteq \mathcal{APX}$$

C.7.3 PTAS Reductions

Using a similar approach as that of the polynomial reductions, but using PTAS instead of polynomial algorithms, it is possible to define a relation of the form: $p \leq_{PTAS} q$.

Definition C.15 A PTAS reduction of optimization problem p to q (denoted $p \leq_{PTAS} q$) consists of a triple of three functions (f, g, α):

- f is polynomial and maps instances of problem p into instances of problem z.
- α is a surjective map, which determines in polynomial time the approximation ratio $\alpha(\epsilon)$ to be used in problem q, in order to have an approximation ratio in p of ϵ, using the PTAS reduction.
- g is polynomial and maps the $\alpha(\epsilon)$ approximated solution coming from q, into an ϵ-approximation of p.

Figure C.6 helps to illustrate the reduction.

C.7.4 $\mathcal{NPO}$-Complete Problems

PTAS reductions can be used to order the difficulty of approximating optimization problems in a similar form to polynomial reductions, which helped to order the difficulty of decision problems. The following properties easily follow:

- If $p \leq_{PTAS} q$, and $q \in \mathcal{PTAS}$, then $p \in \mathcal{PTAS}$. Equivalently, if $p \notin \mathcal{PTAS}$, then $q \notin \mathcal{PTAS}$.
- If $p \leq_{PTAS} q$, and $q \in \mathcal{APX}$, then $p \in \mathcal{APX}$. Equivalently, if $p \notin \mathcal{APX}$, then $q \notin \mathcal{APX}$.

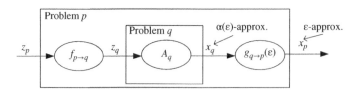

Figure C.6   PTAS reduction of optimization problem p into q, $p \leq_{PTAS} q$

Table C.2   Complexity of some optimization problems of interest in network design.

Name	Description	Complexity
Conv	General convex programs	Polynomial ($\mathcal{P}$)
ILP	General integer linear programs	$\mathcal{NPO}$-complete
BLP	General binary linear programs	$\mathcal{NPO}$-complete
Min-kMST	k-minimum cost spanning trees	Polynomial ($\mathcal{P}$)
Min-kSP	k-minimum cost paths	Polynomial ($\mathcal{P}$)
Min-kSP	k-minimum cost paths	Polynomial ($\mathcal{P}$)
Min-Ste	Min cost multicast tree (Steiner tree)	0.55-approx., $\mathcal{APX}$-complete
Max-Clique	Maximum size clique	$\mathcal{NPO}$-complete
Min-TSP	Min cost ring	$\mathcal{NPO}$-complete
Min-NonBif	Min congestion non-bifurcated routing	2.23-approx., $\mathcal{APX}$-complete
Max-IntegralFlow	Max integral k multicommodity flow on trees	1-approx., $\mathcal{APX}$-complete
Min-NodeLocation	Min cost node location, no connectivity limit	1.4-approx., $\mathcal{APX}$-complete

Then, $p \leq_{PTAS} q$ relation can be read as "p is easier or similar to approximate than q", or that "q is similarly or harder to approximate than p". Using $\leq_{PTAS}$ we can now define the classes $\mathcal{APX}$-complete, $\mathcal{PTAS}$-complete, and $\mathcal{NPO}$-complete, containing the hardest problems among their classes.

- An optimization problem q is $\mathcal{NPO}$-complete, if it belongs to $\mathcal{NPO}$, and $p \leq_{PTAS} q$ for all the problems in $\mathcal{NPO}$.
- An optimization problem q is $\mathcal{APX}$-complete, if it belongs to $\mathcal{APX}$, and $p \leq_{PTAS} q$ for all the problems in $\mathcal{APX}$.
- An optimization problem q is $\mathcal{PTAS}$-complete, if it belongs to $\mathcal{PTAS}$, and $p \leq_{PTAS} q$ for all the problems in $\mathcal{PTAS}$.

C.8   Complexity of Network Design Problems

Many of the problems of interest in network design happen to be hard to solve and hard to approximate. In general, the large majority of non-convex network problems fall into this category. Fruitful literature exists presenting approximations to some versions of these problems, together with, in some occasions, results stating their inapproximability.

Table C.2 includes some examples brought from the collection [3]. Negative (inapproximability) results are true conjecturing that $\mathcal{P} \notin \mathcal{NP}$, as is customary. Full references can be checked in [3], the interested reader is referred there for further descriptions.

C.9   Notes and Sources

Complexity theory is a well established discipline in computer science for which many good sources exist. The concise introduction in this appendix is mostly based on [4, 5] and some explanations in [6] and [7]. A compendium of complexity results of multiple $\mathcal{NP}$ problems is available in [3].

References

[1] S. A. Cook, *The complexity of theorem-proving procedures*, Proceedings of the third annual ACM symposium on Theory of computing, 1971.

[2] L. A. Levin (1973) Russian Academy of Sciences, Branch of Informatics, Computer Equipment and Automatization, *Universal sequential search problems, (Problemy Peredachi Informatsii)* **9**(3), pp. 115–116.

[3] P. Crescenzi, V. Kann, and M. Halldórsson, *"A compendium of NP optimization problems,"* Available online at www.nada.kth.se/~viggo/problemlist/compendium.html, Department of Science, University of La Sapienza, Rome. 1995.

[4] M. R. Garey and D. S. Johnson, *Computers and Intractability: A Guide to NP Completeness.* San Francisco, CA, USA: W. H. Freeman, 1979.

[5] I. Wegener, *Complexity Theory: Exploring the Limits of Efficient Algorithms.* Heidelberg, Germany: Springer Science & Business Media, 2005.

[6] M. Pioro and D. Medhi, *Routing, Flow, and Capacity Design in Communication and Computer Networks.* Morgan Kaufmann Publishers, 2004.

[7] J. Bergstra and M. Burgess, *Handbook of Network and System Administration.* Amsterdam, The Netherlands: Elsevier, 2011.

Appendix D

Net2Plan

D.1 Net2Plan

Net2Plan is an open-source and free to use Java-based software, developed by Pablo Pavón Mariño and (up to version 0.3.1) José Luis Izquierdo Zaragoza, and licensed under the GNU Lesser General Public License (LGPL). Net2Plan has its origins in September 2011, as a resource for network optimization courses at Technical University of Cartagena. Since then, it spread to other Universities, and has been applied in a number of works in the academia and industry. Installing instructions, documentation (including video tutorials), research publications, and teaching materials can be accessed via the website

`www.net2plan.com`

Specific instructions to more easily access the book materials are available at:

`www.net2plan.com/ocn-book`

Net2Plan was designed with the aim to overcome the barriers imposed by existing network planning tools in two forms: (i) users are not limited to execute non-disclosed built-in algorithms, but also can integrate their own algorithms, applicable to any network instance, as Java classes implementing particular interfaces, and (ii) Net2Plan defines a network representation, the so-called network plan, based on abstract concepts such as nodes, links, traffic demands, routes, protection segments, shared-risk groups, and network layers.

Network instances can have an arbitrary number of layers, arranged in arbitrary forms. Technology-specific information can be introduced via user-defined attributes attached to nodes, links, routes, layers, and so on in the network plan. The combination of a technology-agnostic substrate and technology-related attributes provides the required flexibility to model any network technology within Net2Plan, an added value from a didactic point of view. In this respect, current Net2Plan version provides specific libraries to ease the design of IP, wireless and optical networks.

Optimization of Computer Networks – Modeling and Algorithms: A Hands-On Approach,
First Edition. Pablo Pavón Mariño.
© 2016 John Wiley & Sons, Ltd. Published 2016 by John Wiley & Sons, Ltd.
Companion Website: www.wiley.com/go/PavonMarinoSol16

Net2Plan provides both a graphical user interface (GUI) and a command-line interface (CLI). In either mode, Net2Plan includes four tools:

- *Offline network design*: Targeted to execute offline planning algorithms, that receive a network design as an input and modify it in any form (e.g., optimize the routing, the capacities, topology, etc.). Network design algorithms in Part I and heuristic algorithms in Chapter 11 of this book are implemented in this form. Algorithms based on constrained optimization formulations (e.g., ILPs or convex formulations) use the open-source freeware Java Optimization Modeler (JOM)

    www.net2plan.com/jom

    developed by the author to interface from Java to a number of external solvers such as GPLK, CPLEX or IPOPT, that produce a numerical solution. The modeling syntax of JOM is human-readable, and capable of handling arrays of decision variables and constraints of arbitrary dimensions, facilitating the definition and solving of complex models directly from Java in a few lines of code.
- *Online simulation*: Permits building simulations of online algorithms that code how the network *reacts* to different events generated by built-in or user-developed event generation modules. For instance, it can be used to evaluate network recovery schemes that react to failures and repairs or dynamic provisioning algorithms that allocate resources reacting to time-varying traffic demands. The distributed algorithms presented in Part II of the book are implemented as online algorithms, where nodes asynchronously iterate to adapt to network conditions.
- *Automatic report generation*: Net2Plan permits the generation of built-in or user-defined reports, from any network design.
- *Traffic matrix generation*: Net2Plan assists users in the process of generating and normalizing traffic matrices.

We remark that every algorithm and report in Net2Plan can be either built-in or user-made. For a full description of Net2Plan functionalities, and how to program algorithms in Net2Plan, please refer to Net2Plan documentation (including video tutorials) on the website.

D.2   On the Role of Net2Plan in this Book

Net2Plan enables the hands-on approach in network optimization targeted in this book:

- The book materials indexed repository in www.net2plan.com/ocn-book includes all the examples of models and algorithms in book Part I and Part II, suitable for Net2Plan v0.4.0 or later. No Java programming skills are needed to use these algorithms, repeating the tests found throughout the book or extending them to other network instances. The variety of built-in reports and performance metrics computed by Net2Plan can help the reader to further enrich the evaluation of the network designs.
- In addition, standard Java programming skills permit the reader to develop their own algorithms, apply the techniques described, and then test the produced methods in Net2Plan. This is actually the final goal of the book and how Net2Plan is being already used in several network optimization courses. For this, algorithms in the repository can be used as

templates or new algorithms can be created from scratch. Net2Plan includes various useful libraries for developing network algorithms, easing tasks like the manipulation of candidate path lists, conversion between different routing types (destination-based and flow-based routing), computation of incidence and adjacency matrices, multiple variants of shortest path algorithms, computation of performance metrics, generation and normalization of traffic matrices, or efficient algorithms for solving the simple projections or regularization examples in the book. Technology-related libraries are included to ease the development of algorithms for IP, wireless and optical networks.

The reader is encouraged to access

`www.net2plan.com/ocn-book`

and follow the instructions there to use the full Net2Plan resources, and get hands-on with optimization of computer networks!

Index

Optimization of Computer Networks – Modeling and Algorithms: A Hands-On Approach,
First Edition. Pablo Pavón Mariño.
© 2016 John Wiley & Sons, Ltd. Published 2016 by John Wiley & Sons, Ltd.
Companion Website: www.wiley.com/go/PavonMarinoSol16

Printed and bound by CPI Group (UK) Ltd, Croydon, CR0 4YY

27/10/2024

14580169-0002